Corien Bakermans (Ed.)
Microbial Evolution under Extreme Conditions
Life in Extreme Environments

Life in Extreme Environments

Edited by
Dirk Wagner

Volume 2

Microbial Evolution under Extreme Conditions

—

DE GRUYTER

Editor
Corien Bakermans
Altoona College
The Pennsylvania State University
3000 Ivyside Park
Altoona, PA 16601, USA
cub21@psu.edu

ISBN 978-3-11-033506-4
e-ISBN (PDF) 978-3-11-034071-6
e-ISBN (EPUB) 978-3-11-038964-7
ISSN 2197-9227

Library of Congress Cataloging-in-Publication Data
A CIP catalog record for this book has been applied for at the Library of Congress.

Bibliographic information published by the Deutsche Nationalbibliothek
The Deutsche Nationalbibliothek lists this publication in the Deutsche Nationalbibliografie;
detailed bibliographic data are available on the Internet at http://dnb.dnb.de.

© 2015 Walter de Gruyter GmbH, Berlin/Munich/Boston
Typesetting: le-tex publishing services GmbH, Leipzig
Printing and binding: Hubert & Co. GmbH & Co. KG, Göttingen
♾ Printed on acid-free paper
Printed in Germany

www.degruyter.com

Preface

"On any possible, reasonable or fair criterion, bacteria are – and always have been – the dominant forms of life on Earth."

– Stephen Jay Gould

Nearly 4 billion years of evolution have produced a captivating array of organisms that live in a variety of environments both exotic and mundane to us. The exotic environments have a special appeal for understanding how organisms, microorganisms in particular, push the physicochemical boundaries of biomolecules and are able to thrive. How evolution has resulted in and continues to shape the microbial genes, genomes, species, and communities in these extreme environments is certainly a drama of epic proportions that microbiologists are continuing to unravel and describe.

This volume explores the current state of knowledge about microbial evolution under extreme conditions and addresses the following questions: What is known about the processes of evolution that produce extremophiles and adaptations to extreme conditions? Can this knowledge be applied to other systems? What is the broader relevance? What remains unknown and requires future research? These questions are addressed from the perspectives of different extreme environments, organisms, and evolutionary processes. The information compiled in this volume reveals that there are disparate levels of knowledge about the different extreme environments and their inhabitants; yet, as noted in many of the chapters, genomics and metagenomics are having a significant impact on our understanding of microorganisms and microbial processes, including evolution, in extreme environments. It seems that microbial evolution and ecology are poised for a significant gain in comprehension through the synthesis and integration of data and hypotheses that will likely lead to new insights into evolution, as well as the redefinition of species and extreme environments. It is my hope that this volume will facilitate that synthesis and advance understanding of the evolution of microorganisms and evolution in general.

When I accepted the invitation to edit this volume, I didn't appreciate what an undertaking it would be. I was concerned that there was not enough material for an entire volume yet inspired to attempt a broader look at "extreme" environments, since extremophile is to some degree an anthropocentric term. What resulted has reminded me that there is so much to learn and that microorganisms are fascinatingly complex.

Contents

Preface —— v

Contributing authors —— xii

Corien Bakermans
1 Extreme environments as model systems for the study of microbial evolution —— 1
1.1 Introduction —— 1
1.2 Extreme environments as model systems —— 1
1.3 What is known about microbial evolution? —— 4
1.3.1 Community diversity as a measure of evolution —— 7
1.3.2 Adaptive traits as a measure of evolution —— 8
1.4 Themes from extreme environments —— 9
1.5 Conclusions and open questions —— 11

Francisco J. López de Saro, Héctor Díaz-Maldonado, and Ricardo Amils
2 Microbial evolution: the view from the acidophiles —— 19
2.1 Introduction —— 19
2.2 Horizontal gene transfer —— 20
2.3 The mobilome —— 21
2.4 Phages —— 22
2.5 Plasmids —— 23
2.6 Transposons —— 24
2.7 Evolution and ecology: long term studies of genetic variation —— 25
2.8 Future directions —— 26

R. Eric Collins
3 Microbial Evolution in the Cryosphere —— 31
3.1 Overview —— 31
3.1.1 Cryospheric evironments —— 31
3.1.2 Modes of evolution —— 34
3.1.3 Adaptations to living with ice —— 37
3.2 Focus on sea ice —— 38
3.2.1 Sea ice characteristics —— 38
3.2.2 Evolutionary modes in sea ice —— 42
3.3 Ongoing work and future directions —— 43
3.3.1 Field work and experimentation —— 43
3.3.2 '-omics' in the cryosphere —— 44
3.3.3 Linking phenotype and genotype —— 46

Maximiliano J. Amenabar, Matthew R. Urschel, and Eric S. Boyd
4 **Metabolic and taxonomic diversification in continental magmatic hydrothermal systems —— 57**
4.1 Introduction —— 57
4.2 Geological drivers of geochemical variation in continental hydrothermal systems —— 59
4.3 Taxonomic and functional diversity in continental hydrothermal ecosystems —— 64
4.4 Application of phylogenetic approaches to map taxonomic and functional diversity on spatial geochemical landscapes —— 68
4.5 Molecular adaptation to high temperature —— 72
4.5.1 Lipids —— 72
4.5.2 Protein stability —— 73
4.5.3 Cytoplasmic osmolytes —— 74
4.5.4 Motility —— 76
4.6 Mechanisms of evolution in high temperature environments —— 78
4.7 Concluding remarks —— 81

Aharon Oren
5 **Halophilic microorganisms and adaptation to life at high salt concentrations – evolutionary aspects —— 97**
5.1 Phylogenetic and physiological diversity of halophilic microorganisms —— 97
5.2 What adaptations are necessary to become a halophile? —— 99
5.3 Is an acidic (meta)proteome indeed indicative for halophily and high intracellular ionic concentrations? —— 100
5.4 Genetic variation and horizontal gene transfer in communities of halophilic Archaea —— 101
5.5 *Salinibacter*: convergent evolution and the 'salt-in' strategy of haloadaptation —— 103
5.6 High intracellular K^+ concentrations but no acidic proteome? The case of the *Halanaerobiales* —— 104
5.7 Different modes of haloadaptation in closely related *Halorhodospira* species —— 105
5.8 Final comments —— 105

John R. Battista
6 **The origin of extreme ionizing radiation resistance —— 111**
6.1 Introduction and background —— 111
6.1.1 Ionizing radiation —— 111
6.1.2 Biological damage caused by electromagnetic radiations —— 112

6.1.3	Exposure to ionizing radiation selects for ionizing radiation resistant bacteria —— 113
6.1.4	The occurrence of extreme ionizing radiation resistance within the Bacteria and Archaea —— 114
6.1.5	Natural sources of ionizing radiation —— 115
6.2	The existence of extreme ionizing radiation resistance is difficult to reconcile with the natural history of the Earth —— 115
6.3	Proposed explanations for the existence of ionizing radiation resistance —— 116
6.3.1	Panspermia: the exchange of bacteria between planets —— 116
6.3.2	Man-made sources of ionizing radiation are the source of extreme ionizing radiation resistant microorganisms —— 118
6.3.3	Exaptation —— 118
6.4	Conclusions —— 120

Jennifer B. Glass, Cecilia Batmalle Kretz, Melissa J. Warren, and Claire S. Ting

7	**Current perspectives on microbial strategies for survival under extreme nutrient starvation: evolution and ecophysiology** —— 127
7.1	Introduction —— 127
7.2	Carbon —— 128
7.3	Nitrogen —— 133
7.4	Phosphorus —— 136
7.5	Iron —— 137
7.6	Other micronutrients —— 138
7.7	Conclusions —— 139

Joseph Seckbach and Pabulo Henrique Rampelotto

8	**Polyextremophiles** —— 153
8.1	Introduction —— 153
8.2	Bacteria —— 154
8.2.1	*Deinococcus radiodurans*: Conan the bacterium —— 154
8.2.2	*Chroococcidiopsis* —— 156
8.3	Archaea —— 157
8.3.1	*Halobacterium salinarum* NRC-1: a model organism —— 157
8.4	Eukaryota —— 158
8.4.1	*Cyanidiophyceae* —— 158
8.4.2	Lichens —— 159
8.4.3	Tardigrades: nature's toughest animal —— 161
8.5	Conclusion —— 162

William F. Martin, Sinje Neukirchen, and Filipa L. Sousa
9 Early life —— 171

Cene Gostinčar, Nina Gunde-Cimerman, and Martin Grube
10 Polyextremotolerance as the fungal answer to changing environments —— 185
10.1 Introduction —— 185
10.2 Extremes in nature —— 185
10.3 Anthropogenic extremes: indoor habitats —— 190
10.4 Coincidental opportunities: opportunistic infections —— 192
10.5 Conclusions: polyextremotolerance —— 197

Alexander I. Culley, Migun Shakya, and Andrew S. Lang
11 Viral evolution at the limits —— 209
11.1 Introduction —— 209
11.2 Acidic hot springs and hypersaline environments —— 209
11.3 The deep sea —— 212
11.4 Polar environments —— 213
11.5 Viruses and their effects on host organisms and communities —— 214
11.6 Future perspectives —— 216

Eva C. M. Nowack and Arthur R. Grossman
12 Evolutionary pressures and the establishment of endosymbiotic associations —— 223
12.1 Introduction —— 223
12.2 Diversity, evolution, and stability of endosymbiotic relationships —— 227
12.2.1 Diversity of endosymbionts and their physiological functions —— 227
12.2.2 Evolutionary routes to establish and maintain endosymbiosis —— 228
12.2.3 Stability and the age of endosymbioses —— 230
12.3 Genome evolution in endosymbiotic bacteria —— 230
12.3.1 Reductive genome evolution in endosymbionts —— 230
12.3.2 Evolution toward an organelle and beyond —— 232
12.4 Evolution of the host genome as shaped by endosymbiosis —— 235
12.4.1 Complementarity of host and endosymbiont metabolic abilities —— 235
12.4.2 Acquisition of symbiotic potential —— 236
12.4.3 Redefinition of immune functions —— 238
12.5 Conclusions and future directions —— 239

Fabia U. Battistuzzi and Anais Brown
13 Rates of evolution under extreme and mesophilic conditions —— 247
13.1 Overview —— 247

13.2	How do we estimate rates of genetic change? —— 253
13.2.1	Relative rate estimation —— 254
13.2.2	Absolute rate estimation —— 255
13.3	How do we model evolutionary rates? —— 257
13.4	Environments and evolutionary rates —— 257
13.4.1	Evolutionary rates of pathogens —— 258
13.5	Large-scale genomic changes: duplications/loss and horizontal gene acquisition —— 259
13.5.1	Rates of gene duplication and loss —— 259
13.5.2	Highways of horizontal gene transfers —— 261
13.6	Conclusions —— 263

Index —— 269

Contributing authors

Maximiliano Amenabar
Montana State University
Department of Microbiology and Immunology
109 Lewis Hall,
Bozeman, MT 59717, USA
e-mail: maximiliano.amenabar@msu.montana.edu

Ricardo Amils
Centro de Astrobiología (INTA-CSIC)
Ctra. de Ajalvir, Km 4
28850 Torrejón de Ardoz, Spain
Centro de Biología Molecular Severo Ochoa
Universidad Autónoma de Madrid
28049 Madrid, Spain
e-mail: ramils@cab.inta-csic.es

Corien Bakermans
Altoona College
The Pennsylvania State University
3000 Ivyside Park
Altoona, PA 16601, USA
e-mail: cub21@psu.edu

John Battista
Louisiana State University
Biological Sciences
202 Life Sciences Bldg.
Baton Rouge, USA
e-mail: jbattis@lsu.edu

Fabia Ursula Battistuzzi
Oakland University
Department of Biological Sciences
Rochester, MI 48309, USA
e-mail: battistu@oakland.edu

Eric Boyd
University of Wisconsin Astrobiology Research Consortium
Montana State University
Department of Microbiology and Immunology
109 Lewis Hall,
Bozeman, MT 59717, USA
e-mail: eboyd@montana.edu

Anais Cay Brown
Oakland University
Department of Biological Sciences
Rochester, MI 48309, USA
e-mail: acbrown@oakland.edu

R. Eric Collins
Institute of Marine Science
University of Alaska Fairbanks
905 N. Koyukuk Dr.
Fairbanks, AK 99775-7220, USA
e-mail: recollins@alaska.edu

Alexander Culley
Université Laval
Département de biochimie, de microbiologie et de bio-informatique
Pavillon de Médecine dentaire
2420, rue de la Terrasse
QC, G1V 0A6, Québec, Canada
e-mail: alexander.culley@bcm.ulaval.ca

Héctor Díaz-Maldonado
Centro de Astrobiología (INTA-CSIC)
Ctra. de Ajalvir, Km 4
28850 Torrejón de Ardoz, Spain
e-mail: hector.diaz.maldonado@gmail.com

Jennifer B. Glass
Georgia Institute of Technology
311 Ferst Drive
Atlanta GA 30332, USA
e-mail: jennifer.glass@eas.gatech.edu

Cene Gostinčar
National Institute of Biology
Večna pot 111,
SI-1000 Ljubljana, Slovenia
e-mail: cene.gostincar@bf.uni-lj.si

Arthur R. Grossmann
Carnegie Institution for Science
Department of Plant Biology
260 Panama Street
94305 Stanford, CA, USA
e-mail: arthurg@stanford.edu

Martin Grube
Institut für Pflanzenwissenschaften
Karl-Franzens-Universität Graz
Holteigasse 6
8010 Graz, Austria
e-mail: martin.grube@uni-graz.at

Nina Gunde-Cimerman
Department of Biology, Biotechnical Faculty
University of Ljubljana
Centre of Excellence for Integrated Approaches in Chemistry and Biology of Proteins (CIPKeBiP)
Večna pot 111, SI-1000
Ljubljana, Slovenia
e-mail: nina.Gunde-Cimerman@bf.uni-lj.si

Cecilia Batmalle Kretz
Georgia Institute of Technology
311 Ferst Drive
Atlanta GA 30332, USA
e-mail: cecilia.kretz@eas.gatech.edu

Andrew Lang
Memorial University of Newfoundland
Department of Biology
232 Elizabeth Ave.
St. John's, NL, A1B 3X9, Canada
e-mail: aslang@mun.ca

Francisco López de Saro
Centro de Astrobiología (INTA-CSIC)
Ctra. de Ajalvir, Km 4
28850 Torrejón de Ardoz, Spain
e-mail: lopezsfj@cab.inta-csic.es

William F. Martin
Institut für Molekulare Evolution
Heinrich-Heine-University
Universitätsstr.1
Building 26.13.01, Room 34
40225 Düsseldorf, Germany
e-mail: bill@hhu.de

Sinje Neukirchen
Institut für Molekulare Evolution
Heinrich-Heine-University
Universitätsstr. 1
40225 Düsseldorf, Germany
e-mail: sinje.neukirchen@hhu.de

Eva C. M. Nowack
Emmy Noether Group Microbial Symbiosis and Organelle Evolution
Heinrich Heine University Düsseldorf
Universitätsstr. 1
Building 26.12, Room 01.70
40255 Düsseldorf, Germany
e-mail: e.nowack@hhu.de

Aharon Oren
Department of Plant and Envirommental Sciences
The Institute of Life Sciences
The Hebrew University of Jerusalem
Edmond J. Safra Campus, Givat Ram 91904
Jerusalem, Israel
e-mail: aharon.oren@mail.huji.ac.il

Pabulo Henrique Rampelotto
Interdisciplinary Center for Biotechnology Research
Federal University of Pampa
Antônio Trilha Avenue, PO Box 1847
97300-000, São Gabriel – RS, Brazil
e-mail: pabulo@lacesm.ufsm.br

Joseph Seckbach
(Retired from: The Hebrew University of Jerusalem)
Mevo Hadas 20, PO Box 1132
Efrat, 90435, Israel
e-mail: joseph.seckbach@mail.huji.ac.il

Migun Shakya
Dartmouth College
Department of Biological Sciences,
78 College St.
Hanover, NH, 03755, USA
e-mail: migun.shakya@dartmouth.edu

Filipa L. Sousa
Institut für Molekulare Evolution
Heinrich-Heine-University
Universitätsstr. 1
40225 Düsseldorf, Germany
e-mail: filipa.sousa@hhu.de

Claire S. Ting
Williams College
59 Lab Campus Drive
Williamstown MA 01267, USA
e-mail: claire.s.ting@williams.edu

Matthew Urschel
Montana State University
Department of Microbiology and Immunology
109 Lewis Hall
Bozeman, MT 59717, USA
e-mail: matthewur@gmail.com

Melissa J. Warren
Georgia Institute of Technology
311 Ferst Drive
Atlanta GA 30332, USA
e-mail: mwarren38@gatech.edu

Corien Bakermans
1 Extreme environments as model systems for the study of microbial evolution

1.1 Introduction

Today's microorganisms represent the vast majority of biodiversity on Earth and have survived nearly 4 billion years of evolutionary change. Microbial evolution occurred and continues to take place in a vast variety of environmental conditions that range, for example, from anoxic to oxic, from hot to cold, and from free-living to symbiotic. Some of these physicochemical conditions are considered "extreme", particularly when inhabitants are limited to microorganisms. It is easy to imagine that microbial life in extreme environments is somehow more constrained and perhaps subjected to different evolutionary pressures. But what do we actually know about microbial evolution under extreme conditions and how can the knowledge gained from extreme environments as model systems be applied to other conditions?

1.2 Extreme environments as model systems

Extreme environments generally have physicochemical conditions that are challenging to cells and their macromolecules (▶ Tab. 1.1). High temperatures cause the denaturation of proteins, membranes, and DNA, while also increasing rates of detrimental reactions (low temperatures have the opposite, equally detrimental, effect of "freezing" biomolecules and slowing reactions to a near halt) [1]. Macromolecules are also disrupted by high pH, low pH, and high salt concentrations, which also reduce water availability. Extreme conditions also affect the stability of free DNA, which degrades faster at higher temperatures, depurinates at low pH, and denatures at high pH [2, 3]. (DNA is most stable at a slightly alkaline pH of 8 and at low temperatures and high salt concentrations [2].) Environments with physicochemical extremes (i.e. pH or temperature) are considered in this volume, as well as environments with extremely low nutrient concentrations (▶ Chapter 7) and extraordinary cell-cell interactions (symbiosis, ▶ Chapter 12). Most of the extreme conditions examined occur in the "natural" (versus human created) world; however, anthropogenic environments such as acid mine drainage (AMD, ▶ Chapter 2) and extreme conditions that occur in our homes (such as extreme heat of dishwashers, extreme dryness, and variations therein) are included (▶ Chapter 10).

Table 1.1. Characteristics of extreme environments.

Environment	Defining characteristic	Examples	Example organisms from the three domains[a]
Acidic	< pH 5	sulfuric pools, hot springs, geysers, acid mine drainage	B: *Acidithiobacillus* A: *Ferroplasma, Picrophilus* E: *Cyanidium, Hortaea*
Alkaline	> pH 9	soda lakes, hot springs	B: *Bacillus alcalophilus, Natranaerobius* A: *Natronococcus occultus*
Hypersaline	> 35 ‰ salt	salterns, evaporite ponds	B: *Salinibacter ruber* A: *Halobacteriales* E: *Dunaliella, Hortaea*
Extremely hot	> 70 °C	geothermally heated springs, vents, sediments	B: *Aquifex, Thermotoga* A: *Sulfolobus, Methanothermus* E: none
Cold	< 5 °C	polar and alpine regions (glaciers, sea ice, permafrost), deep sea	B: *Colwellia, Psychroflexus* A: *Methanogenium frigidum, Methanococcoides burtonii* E: *Chlamydomonas nivalis*
High pressure	> 10 MPa	deep sea, sediments deep subsurface	B: *Shewanella benthica* A: *Pyrococcus abyssi*
Dry, Arid	a_w < 0.80	hot and cold deserts	B: *Microcoleus* E: *Xeromyces bisporus*
High radiation	UV, ionizing radiation	outer space	B: *Deinococcus radiodurans* A: *Thermococcus gammatolerans* E: *Histoplasma capsulatum*
Oligotrophic	low nutrient concentrations	open ocean, deserts	B: *Prochlorococcus, Pelagibacter* A: methanogens
Endosymbiosis	living within another cell/organism	various animal, plant, protozoan hosts	B: *Buchnera* A: *Methanosaeta sp.* E: *Symbiodinium*

[a] B = Bacteria, A = Archaea, E = Eukaryota

The organisms that inhabit extreme environments are called extremophiles because they thrive in the physicochemical conditions of extreme environments. These extremophiles can be found in all three domains (Archaea, Bacteria, and Eukaryota), although sometimes with a limited phylogenetic distribution. Some extremophiles are polyextremophiles (▶ Chapters 8 and 10) that thrive in multiple extreme conditions (e.g. high temperature, pH, and salt concentrations [4]); and many more organisms may be extremotolerant or even polyextremotolerant (i.e. tolerating, but not thriving in, the extreme conditions). While extremophiles succeed in extreme environments, the challenges presented by the physicochemical conditions of extreme environments commonly manifest as a lower abundance and diversity of inhabitants. For example, diversity decreases with increasing salinity in hypersaline environments ([5] and ▶ Chapter 5) and with increasing acidity and temperature in acidic hot springs ([6] and ▶ Chapter 4).

With their relatively limited numbers of inhabitants, extreme environments can serve as good model systems for the study of evolutionary processes [7]. Lab systems are often highly simplified, while many environments are exceedingly complex; extreme environments offer a middle ground that spans the habitat space between simple and complex. Important characteristics that make extreme environments good model systems include: fewer overall species (and/or fewer dominant species), fewer predators, fewer multicellular organisms, limited resources, fewer physicochemical variables, and often less heterogeneity. The diversity of communities in extreme environments varies considerably from practically a single species (symbionts [8]) to several dominant organisms (AMD [9, 10], salterns [11], and hot springs [12]) to hundreds of species (pelagic zones [13]). Communities in extreme environments also span a range of cell densities from 10^5 cfu/ml in soda lakes [14] to 10^7–10^8 cells/ml in salterns [11] to the very dense biofilms found in AMD and hot springs. Often the lack of multicellular organisms allows for the formation of extensive biofilms and mats in extreme environments [15]. The diversity and density of cells in an environment will impact the frequency and nature of interactions between cells (including the potential for gene transfer). Moreover, an overall low diversity means assembly of sequences from metagenomic studies is more possible as has been demonstrated in AMD [16], a hypersaline lake [17], and a hot spring [12], helping to reduce the need for cultivation. Extreme environments often have fewer physicochemical variables in the sense that one condition (such as pH or temperature) may dominate, thereby reducing variability in the system.

Extreme environments are also good models seeing as their geographic boundaries, and often their geological history, can be mapped. The history and distribution of extreme environments will affect the evolutionary history of extremophiles (and their adaptive traits). For example, currently high temperature (hydrothermal) systems are relatively isolated from one another geographically, yet plate tectonics and volcanism have insured that high temperature environments, which likely dominated the early earth (▶ Chapter 9), have existed throughout earth's history [18], thus we expect that hyperthermophiles (and hyperthermophilic traits) are ancient with the potential for divergence due to isolation. In contrast, while low temperatures (< 5 °C) are relatively widespread today (although seasonal) [19], frozen environments (▶ Chapter 3) have been less common throughout geologic history [20, 21]. It is likely that today's polar sea ice has existed only for the last 35–47 Ma, with multiyear ice persisting only in the last 2.5–3 Ma [22, 23], hence psychrophiles may be more recently evolved. While acidic and alkaline microniches are pervasive, suggesting that acidophiles and alkaliphiles may be ancient and could have lots of interactions with other acidophiles and alkaliphiles [24, 25].

All of these factors allow extreme environments to serve as model systems. Ecosystems with a dominant species or physicochemical condition can be very useful for disentangling the effects of different factors on the processes and mechanisms of microbial evolution. In particular, the variety of extreme environments, and especially comparisons between environments, will further facilitate the examination of microbial evolution and evolutionary processes across an entire spectrum of conditions from low to high diversity, low to high cell density, and low to high variability of physicochemical or biological conditions.

1.3 What is known about microbial evolution?

The study of evolution in microbes is complicated by their small size, rapid growth, asexual reproduction, horizontal gene transfer (HGT), living in close contact with diverse organisms, and the variability of their microhabitats. While an exhaustive review of microbial evolution is beyond the scope of this chapter, some key aspects of microbial evolution will be summarized here.

Our current understanding of microbial evolution suggests that it is complex and that there is no one evolutionary process that applies to and explains all microbes in all contexts (for current perspectives see [26–30]), but rather that there is a range of evolutionary lifestyles from nearly clonal (like *Bacillus anthracis* [31] and *Staphylococcus aureus* [32]) to highly sexual (like *Neisseria meningitides* [33] and *Helicobacter pylori* [34]). To some degree, the evolutionary history of microorganisms can be represented by phylogenetic trees. While every gene has its own phylogenetic tree and history, the core information-processing machinery of microbial cells generally adheres to linear (vertical) lines of descent as evidenced by the consensus topology of trees constructed from these genes [35, 36]. However, webs or networks may better represent the evolutionary history of modern prokaryotes given the pervasiveness of HGT [29, 37, 38]. Regardless of whether trees or webs best represent the evolutionary history of prokaryotes, vertical descent sustained by binary fission is an important component of microbial evolution.

Evolution cannot ensue without genetic variation, which can be introduced into microbial cells by a variety of mechanisms that include replication errors, stress-induced mutations [39, 40], recombination [41], HGT [37], *de novo* gene creation [42], and gene duplication [43]. Which mechanisms contribute the most to genetic diversity in any one organism is highly variable; for example, variation is predominately introduced by integrative conjugative and transposable elements in the intracellular parasitic *Orientia tsutsugamuchi* [44]; by integrons in the pathogenic marine *Vibrio cholera* [45]; and by nucleotide substitutions and small indels in the symbiont *Buchnera aphidicola* [46]. HGT is a major contributor to genetic variation in prokaryotes [47–49] with horizontally transferred genes accounting for 0.5 to 25% (average of 12%) of the genes in prokaryotic genomes [50]. Some categories of genes are apparently more likely to be transferred than others and include genes for cell surface proteins, DNA binding proteins, or pathogen-related functions [50]. Which mechanism of HGT (transformation [51], transduction, or conjugation) predominates likely varies by organism and conditions, with viruses potentially playing a large role ([49, 52] and ▶ Chapter 11). Prokaryotes themselves may be promoting HGT through the use of gene transfer agents (defective bacteriophage under the control of the host cell which randomly package host DNA [49, 53]) and type IV secretion systems (which secrete and take up DNA [54, 55]). While HGT can occur between highly divergent species, recombination rates drop exponentially as sequence divergence increases [56], limiting the frequency of recombination between distantly related species and contributing to the clustering of genotypes [26]. Genetic variation, however it is introduced, is fundamental to the ability of microorganisms to evolve.

Gene gain and loss are also important evolutionary processes in prokaryotes. Genome expansion can occur through the gain of genes via HGT (as discussed above), *de novo* gene creation, or gene duplication. Gene gain can be extensive as each of these processes may contribute up to 41% of the genes in prokaryotic genomes. *De novo* created genes (a.k.a. orphan genes or ORFans) have no homologs in other species and may constitute up to 14% of prokaryote genomes [29, 42]. While the function of most orphan genes is not known, orphan genes in *Escherichia coli* were shown to be short, AT-rich, functional proteins that evolve quickly [57]. It is likely that orphan genes provide species-specific traits [58] and environment-specific adaptations [59]. Genes can also be gained by duplication to produce paralogs that are most often genes for amino acid metabolism, transcription, and inorganic ion metabolism (often transport proteins) and may be used for adaptation to a constantly changing environment [60]. The extent of gene duplication within bacterial genomes ranges from 7% to 41% with many gene duplications (25%) located within tandem duplicates or block duplicated segments [60]. While gene gain is clearly an advantage for the acquisition of new functions (e.g. metabolism, pathogenicity, stress resistance), gene loss may be equally important for providing a fitness advantage [61]. Gene loss has been documented in a variety of organisms (e.g. *Lactobacillus* [62], *Prochlorococcus* [63], *B. aphidicola* [64], and various bacterial pathogens [65]) with the greatest losses evident in obligate intracellular pathogens and endosymbionts [66]. Indeed, mutational bias toward deletions (as documented in pseudogenes [67, 68]) and high levels of genetic drift [69] apparently contribute to the compact gene-rich genomes typical of prokaryotes.

Subsequently, natural selection can act on genetic variation to form clusters of genotypes that may represent species or ecotypes within communities. In positive selection traits that confer an advantage survive and increase within a population, while in negative (or purifying) selection traits that do not confer an advantage are removed from the population. Selective sweeps will promote retention of advantageous traits and removal of non-advantageous traits from the population, although selective sweeps likely affect microorganisms more at the level of genes than entire genomes [70–73]. Neutral processes such as genetic drift, population bottlenecks, and geographic isolation can also lead to population differentiation [8, 41, 74–76], although geographic isolation of prokaryotes may be rare. Moreover, environmental selection may be more important than history in affecting population and community structure on a small scale [77].

Measures of population and community structure are of course affected by the definition of microbial species (for current perspectives see [78–82]). Presently, microbial species are recognized to contain a core genome, which consists of the high-frequency, highly-conserved genes shared by all members of the species, and a flexible genome, which consists of the medium-frequency genes shared by a subset of the members of the species and the low-frequency genes present in only one member of the species (together, the core and flexible genomes make up the pan genome). The portion of a genome that contains the genes of the flexible genome ranges from 54% in *Pseudomonas fluorescens* [83] and 53% in *E. coli* [84] to 22% in *B. aphidicola* [85] and 20% in *Streptococcus agalactiae* [86]. It is likely that the flexible genome represents genes that are gained and lost via recombination and involved in the "partitioning of functional roles within the population" [27] and represents genes with high turnover that are often associated with mobile elements and involved in the response to changes in abiotic and biotic conditions (like nutrient abundance or bacteriophage infection) [27]. Thus prokaryotic populations are often clusters of highly similar genotypes. Ideally, genotypic clusters would represent cohesive ecologically relevant and functional populations and it is likely that the simultaneous examination of ecological and genetic clusters will be essential to defining microbial species [27, 82]. While microbiologists generally agree that there is a large amount of variability within microbial species, debate continues on how to delineate cutoffs in variability that correspond to species and how these cutoffs should relate to functional differences.

From the above summary, it is apparent that many variables, mechanisms, and processes are at work during microbial evolution contributing to its complexity. Reductionist studies in the laboratory contribute a great deal to our understanding of microbial evolution (e.g. [87]) as do the many studies examining the results of evolution in the environment (including extreme environments). With the advent of genomics and metagenomics [16, 88–90] along with a range of environments (extreme and otherwise) to sample, evolutionary microbiology is poised for significant advancement and synthesis in our understanding of how evolutionary and ecological processes (fine scale to large scale) determine the structure of microbial species and communities.

1.3.1 Community diversity as a measure of evolution

Diversity is one measure of the current state of microbial communities and resulted from evolution under complex conditions and over long time periods. Given the technological difficulties in culturing and studying microbes, the full diversity of microorganisms and microbial communities is still being uncovered and described (▶ Fig. 1.1) with the aid of metagenomic approaches [91, 92]. It is expected that the current structure of microbial populations and communities reflects their ecological context and evolutionary history [26, 77, 93]. A low diversity community can indicate stressful conditions; certainly, the diversity of extreme environments is generally lower than "non-extreme" environments presumably due to abiotic stress [94, 95]. For example, in a study of geothermal soils, sediments, and mats diversity correlated with temperature and, to a lesser extent, pH, peaking at 24°C and neutral pH [96]. Increased salinity has also been demonstrated to decrease species diversity and richness [5, 97]. Soil diversity is primarily affected by pH, peaking at pH 7 [98]. While extreme environments can be dominated by a relatively low number of species, new sequencing approaches are helping to identify the many less abundant species within these communities [99–101].

Many questions remain about how evolution versus ecology shapes microbial communities, particularly those in extreme environments. Ultimately, the relative rates of forces that increase variation (e.g. mutation, recombination [102, 103], spatial and temporal heterogeneity, physical isolation [104], and immigration) and of forces that decrease variation (e.g. selective sweeps, abiotic stress, predation/infection, bottlenecks, limited resources) will determine what process(es) maintain community structure in any given environment [105]. While the answers are likely to be complex

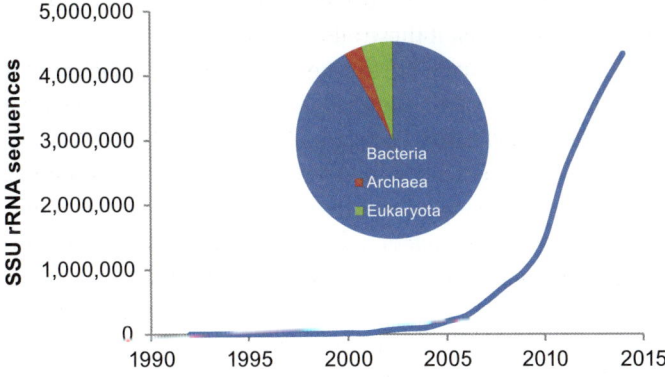

Fig. 1.1. Growth of SSU ribosomal RNA gene sequence data in the Ribosomal Database Project (1992–2006) and the Silva database (2006–present). Inset shows the distribution of sequences between the three domains for 2014. Data compiled from the Silva website http://www.arb-silva.de/documentation/release-119/ [108].

and specific to each community, themes will likely emerge. One would imagine that given the fast pace of evolution (even in extreme environments [106, 107]) and enough time, the diversity of extreme environments should approach that of "non-extreme" environments. Perhaps the low diversity of extreme environments is maintained by more frequent selective sweeps or bottlenecks possibly aided by relative temporal and spatial homogeneity of the environment and/or geographic isolation which limits immigration of extremophilic organisms and adaptive traits. Extreme environments, with their relatively low diversity, are good models for determining how ecology versus evolution shapes microbial communities.

1.3.2 Adaptive traits as a measure of evolution

The traits that enable current microbes to thrive in extreme environments are another measure of evolution that is perhaps more directly related to the physicochemical stresses of extreme environments. An excellent compilation and review of adaptive traits of extremophiles can be found in the *Extremophiles Handbook* [109]. Adaptive traits are very useful for discerning evolutionary processes in extreme environments; for example, adaptive traits that consist of one gene or a few genes, such as metal resistance [110, 111], are useful for tracking HGT and recombination events. Those adaptive traits that are found in many genes (and sometimes genome-wide), such as the pI of proteins [112], are useful for examining evolutionary processes such as mutation rates and genetic drift. All adaptive traits can inform ideas about convergent versus divergent evolution as well as thinking about the likelihood of evolving new functions and the evolutionary history of organisms.

The phylogenetic distribution of traits can assist the identification of convergent or divergent evolution. Similar traits found in distantly-related lineages can be indicative of convergent evolution; although if these adaptive traits are very similar on a molecular level then HGT is likely responsible for the phylogenetic distribution. If adaptive traits are found on most, but not all, members of a group of organisms then divergent evolution with gene loss is indicated. For example, convergent evolution was shown in piezophiles from different genera of *Gammaproteobacteria* that had different adaptions in their 16S ribosomal RNA gene sequence that enhanced ribosome function at high pressure [113]. In contrast, the paraphyletic and highly conserved methanogenesis genes and pathways of modern methanogens suggests a singular ancient origin of methanogenesis (▶ Chapter 9) within the *Euryarchaeota* with subsequent divergence into Class I and Class II methanogens and other closely-related groups of Archaea (*Halobacteria, Thermoplasmata,* and *Archaeoglobi*) which lost the genes for methanogenesis [114, 115]. Of course, different adaptive traits in the same organism can have different evolutionary histories, for example: the halophilic bacterium *Salinibacter ruber* (phylum *Bacteroidetes*) has evolved salt tolerance both through independent, convergent evolution of a proteome with a low pI (also characteristic of haloarchaea) and through the acquisition of several genes for K^+ uptake/efflux via HGT from haloarchaea [116]. The convergence or divergence of adaptive traits can sometimes reflect ecological history or likelihood of evolution of the trait.

The prevalence and conservation of adaptive traits can also indicate the likelihood of evolving these traits. The broad phylogenetic distribution and convergent evolution of adaptive traits required for psychrophiles, piezophiles, halophiles, and endosymbionts suggests that these extremophiles are highly likely to evolve (see examples above). In contrast, the highly-conserved nature and sometimes narrow distribution of methanogenesis and nitrogen fixation suggest that these traits are less likely to evolve. The highly conserved nitrogen fixation (*nif*) genes likely originated in hydrogenotrophic methanogens and spread through HGT to various groups of Bacteria with subsequent duplication and divergence (with a relatively recent acquisition of *nif* genes apparent in some *Aquificales* species) [117]. Not surprisingly, adaptive traits are very useful for examining the processes and history of microbial evolution.

1.4 Themes from extreme environments

Many of the evolutionary processes and mechanisms discussed above were elucidated through the examination of organisms from extreme environments. In many ways evolution in extreme environments occurs similarly to evolution in "non-extreme" environments. While stress-induced mutations might be expected to be more prevalent in extreme environments given that some extreme conditions cause DNA damage that can lead to mutation, this has not been demonstrated to date. Overall mutation and substitution rates are primarily affected by the stability of conditions (more stable conditions result in lower mutation and substitution rates, ▶ Chapter 13). Moreover, rates of homologous recombination tend to be higher for prokaryotes in marine and aquatic systems versus terrestrial systems, regardless of whether or not organisms are extremophiles [106]. Undoubtedly, studies of microbial communities from extreme environments have highlighted the prevalence of HGT and underscored the importance of exaptation, while endosymbionts have provided an important reference point.

HGT has been reported at high levels in extreme environments and is likely to be high in most environments. In one example, extensive HGT was documented between slow-growing genera in the cold, hypersaline Deep Lake in Antarctica [118]. Some adaptive traits important for survival in extreme environments are known to be conferred by HGT and include metal resistance genes on plasmids in acidophiles [110, 111], transport proteins in hyperthermophiles [119], the hypersalinity gene island in *S. ruber* [116], nutrient acquisition genes in *Prochlorococcus* [120, 121], and symbiosis island genes found in *Rhizobia* [122]. The predominant mechanisms of HGT are not well known for many habitats, but mechanisms that minimize DNA exposure to the environment may be more important in extreme environments where the stability of free DNA is diminished (see above). For example, plasmids may be important in acidic environments [123], while highly efficient uptake systems likely facilitate HGT in hot environments [3]. Viruses likely play a major role as well (▶ Chapter 11). A higher number of viral defense systems in archaea than bacteria and in hyperthermophiles than mesophiles suggests that there are more viruses and thus a higher likelihood for transduction in extreme environments where archaea dominate and temperatures are high [124]. Extreme conditions can also affect the stability of the virus particle; however, stability can be maintained via structural modification as seen in phages from hypersaline environments [125] and acidic hot springs [126, 127]. Alternately, phages can circumvent stressful conditions by avoiding the lytic lifestyle as seen in the higher abundance of prophages in communities from the Arctic ocean [128], hydrothermal vents [129], and the deep ocean [130]. The depth of our understanding of HGT in extreme environments varies from extensive (halophiles and thermophiles) to limited (psychrophiles).

Exaptation, in which a trait has a function for which it was not originally selected or evolved, is also very important in extreme environments. For example, compatible solutes can protect against salt stress and cold stress [131, 132]. In many cases, it may not be clear which function was originally selected. However, it is likely that extreme ionizing radiation resistance is an exaptation given that there are no terrestrial sources of high doses of ionizing radiation (▶ Chapter 6). Extreme radiation resistance may have originated from resistance to heat and/or aridity [133] given that high temperatures and extremely dry conditions induce double stranded breaks in DNA [134] that are similar to those induced by high doses of radiation. Similarly, cold tolerance can be a foundation for piezotolerance as the increased flexibility of proteins required to tolerant low temperatures is also useful at high pressures [135]. Often adaptation to one extreme can offer tolerance to other extremes through similar mechanisms (e.g. DNA repair, chaperones, or organic solutes). Compatible solutes can offer protection against osmotic stress (high salt), desiccation, freezing, and heat shock [136, 137] through their interaction with water (and proteins). As a result, many extremophiles are polyextremophiles like the halophilic alkalithermophile *Natranaerobius thermophilus* [4], or at least polyextremotolerant such as the psychrotolerant haloalkiliphilic *Nesterenkonia* sp. AN1 [138]. Not surprisingly, polyextremophiles inhabit extreme environments that are extreme in more than one condition, such as the haloalkiliphilic *Natronomonas pharaonis* isolated from a highly saline soda lake [139].

Endosymbionts have been excellent models for the study of evolution in relatively stable environments with interactions between a highly limited number of species (▶ Chapters 12 and 13). Given the stability of the host cell environment (in terms of pH, salinity, nutrient levels, and sometimes temperature), the loss of genes that are redundant with host genes is common among endosymbionts and results in genome reduction [64]. Endosymbionts have also been good models for the study of mutation rates and gene loss given that evolving in the absence of other bacteria limits the ability of endosymbionts to acquire genes from outside sources (in order to maintain genes against mutation and loss) [46]. The study of endosymbionts illustrates how the comparison of evolutionary processes in extreme and "non-extreme" environments may lead to the redefinition of what constitutes an extreme environment for microorganisms. Extreme environments should perhaps be expanded to include environments in which unusual (outliers) rates or processes of evolution can be identified.

1.5 Conclusions and open questions

A look at the microbial inhabitants of today's extreme environments provides a snapshot in time of evolution and adaptation to extreme conditions. These adaptations manifest at different levels from established communities and species to genome content and changes in specific genes that result in altered function or gene expression. But as a recent (2011) report titled *Microbial Evolution* from the American Academy of Microbiology observes: "A complex issue in the study of microbial evolution is unraveling the process of evolution from that of adaptation. In many cases, microbes have the capacity to adapt to various environmental changes by changing gene expression or community composition as opposed to having to evolve entirely new capabilities."

Extreme environments often have a dominant species or physicochemical condition that enables the separate examination of factors that influence microbial evolution and therefore can serve as good model systems. Clear advances have been made in understanding microbial adaptations to extreme conditions and how these adaptations have evolved. Certainly, there are disparate levels of knowledge about the different extreme environments. For example, thermophiles, acidophiles, and halophiles are particularly well characterized, having benefitted from many man-hours devoted to their study. On the other hand, piezophiles, oligotrophs, and organisms of the deep subsurface are less well-characterized in part due to difficulty in cultivating these organisms. Relatively less is known about the evolutionary mechanisms that led to these adaptations and how commonly those mechanisms are employed. That is, are the same mechanisms used everywhere, when do difference arise and why? How did the different processes of evolution such as mutation, immigration, HGT, recombination, genetic drift, fixation, positive and negative selection, selective sweeps, and bottlenecks contribute to the evolution of these genes, genomes, microbial species, communities, and functions? More longitudinal surveys that examine how populations and communities change over time are needed, as is the continued exploration of how to define microbial species. Further advances will be gained by determining the evolutionary histories of a larger variety of organisms and traits, establishing the rates and relative contributions of evolutionary mechanisms under different conditions, and synthesizing these data to ascertain if there are common themes in evolutionary lifestyles or an infinite variety dependent on local conditions. Conversely, current understanding of microbial evolution can be used to enhance the study of microorganisms in extreme environments. For example, knowing that HGT is widespread, previously unrecognized adaptations to extreme conditions might be identified by examining horizontally transferred genes. Furthermore, the comparison of evolutionary processes in extreme and "non-extreme" environments may lead to the redefinition of what constitutes an extreme environment for microorganisms.

This book explores the current state of knowledge about microbial evolution under extreme conditions and endeavors to answer the following questions: What is known about the processes of microbial evolution (mechanisms, rates, etc.) under extreme conditions? Can this knowledge be applied to other systems and what is the broader relevance? What remains unknown and requires future research? These questions will be addressed from several perspectives including different extreme environments, specific organisms, and specific evolutionary processes. From these and other studies, it is becoming apparent that, evolutionarily, extreme environments and their microbial communities are perhaps not so different from microbial communities in other "non-extreme" environments, but that the conditions of extreme environments render them more tractable systems for study. With the support of genomic and metagenomic examination of extreme environments, evolutionary microbiology is poised for significant advancement and synthesis in understanding how evolutionary and ecological processes (fine scale to large scale) determine the structure of microbial traits, species, and communities.

References

[1] G. Feller. Protein stability and enzyme activity at extreme biological temperatures. J Phys-Condens Mat 2010,22,323101.
[2] T. Lindahl. Instability and decay of the primary structure of DNA. Nature 1993,362,709–715.
[3] M. van Wolferen, M. Ajon, A. J. Driessen, S. V. Albers. How hyperthermophiles adapt to change their lives: DNA exchange in extreme conditions. Extremophiles 2013,17,545–563.
[4] B. Zhao, N. M. Mesbah, E. Dalin, et al. Complete genome sequence of the anaerobic, halophilic alkalithermophile *Natranaerobius thermophilus* JW/NM-WN-LF. J Bacteriol 2011,193,4023–4024.
[5] A. Oren. The bioenergetic basis for the decrease in metabolic diversity at increasing salt concentrations: implications for the functioning of salt lake ecosystems. Hydrobiologia 2001,466,61–72.
[6] E. B. Alsop, E. S. Boyd, J. Raymond. Merging metagenomics and geochemistry reveals environmental controls on biological diversity and evolution. BMC Ecol 2014,14,16.
[7] V. J. Denef, R. S. Mueller, J. F. Banfield. AMD biofilms: using model communities to study microbial evolution and ecological complexity in nature. ISME J 2010,4,599–610.
[8] D. J. Funk, J. J. Wernegreen, N. A. Moran. Intraspecific variation in symbiont genomes: bottlenecks and the aphid-buchnera association. Genetics 2001,157,477–489.
[9] B. J. Baker, J. F. Banfield. Microbial communities in acid mine drainage. FEMS Microbiol Ecol 2003,44,139–152.
[10] B. J. Baker, G. W. Tyson, L. Goosherst, J. F. Banfield. Insights into the diversity of eukaryotes in acid mine drainage biofilm communities. Appl Environ Microbiol 2009,75,2192–2199.
[11] A. Oren. Diversity of halophilic microorganisms: environments, phylogeny, physiology, and applications. J Ind Microbiol Biotechnol 2002,28,56–63.

[12] C. G. Klatt, J. M. Wood, D. B. Rusch, et al. Community ecology of hot spring cyanobacterial mats: predominant populations and their functional potential. ISME J 2011,5,1262–1278.
[13] J. C. Venter, K. Remington, J. F. Heidelberg, et al. Environmental genome shotgun sequencing of the Sargasso Sea. Science 2004,304,66–74.
[14] B. E. Jones, W. D. Grant, A. W. Duckworth, G. G. Owenson. Microbial diversity of soda lakes. Extremophiles 1998,2,191–200.
[15] A. Orell, S. Fröls, S.-V. Albers. Archaeal biofilms: the great unexplored. Annu Rev Microbiol 2013,67,337–354.
[16] G. W. Tyson, J. Chapman, P. Hugenholtz, et al. Community structure and metabolism through reconstruction of microbial genomes from the environment. Nature 2004,428,37–43.
[17] P. Narasingarao, S. Podell, J. A. Ugalde, et al. De novo metagenomic assembly reveals abundant novel major lineage of Archaea in hypersaline microbial communities. ISME J 2012,6,81–93.
[18] E. G. Nisbet, N. H. Sleep. The habitat and nature of early life. Nature 2001,409,1083–1091.
[19] N. Russell, T. Hamamoto. Psychrophiles. In: K. Horikoshi, W. D. Grant, eds. Extremophiles: Microbial life in extreme environments. New York: Wiley-Liss, Inc., 1998,25–45.
[20] F. Robert, M. Chaussidon. A palaeotemperature curve for the Precambrian oceans based on silicon isotopes in cherts. Nature 2006,443,969–972.
[21] A. J. Kaufman, A. H. Knoll, G. M. Narbonne. Isotopes, ice ages, and terminal Proterozoic earth history. Proc Natl Acad Sci U S A 1997,94,6600–6605.
[22] L. Polyak, R. B. Alley, J. T. Andrews, et al. History of sea ice in the Arctic. Quaternary Science Reviews 2010,29,1757–1778.
[23] R. M. DeConto, D. Pollard. Rapid Cenozoic glaciation of Antarctica induced by declining atmospheric CO_2. Nature 2003,421,245–249.
[24] E. V. Pikuta, R. B. Hoover, J. Tang. Microbial extremophiles at the limits of life. Crit Rev Microbiol 2007,33,183–209.
[25] I. Yumoto, K. Hirota, K. Yoshimune. Environmental Distribution and Taxonomic Diversity of Alkaliphiles. In: K. Horikoshi, ed. Extremophiles Handbook: Springer Japan, 2011,55–79.
[26] M. F. Polz, E. J. Alm, W. P. Hanage. Horizontal gene transfer and the evolution of bacterial and archaeal population structure. Trends Genet 2013,29,170–175.
[27] O. X. Cordero, M. F. Polz. Explaining microbial genomic diversity in light of evolutionary ecology. Nat Rev Microbiol 2014,12,263–273.
[28] E. V. Koonin, Y. I. Wolf. Evolution of microbes and viruses: a paradigm shift in evolutionary biology? Front Cell Infect Microbiol 2012,2,119.
[29] V. Merhej, D. Raoult. Rhizome of life, catastrophes, sequence exchanges, gene creations, and giant viruses: how microbial genomics challenges Darwin. Front Cell Infect Microbiol 2012,2,113.
[30] J. Wiedenbeck, F. M. Cohan. Origins of bacterial diversity through horizontal genetic transfer and adaptation to new ecological niches. FEMS Microbiol Rev 2011,35,957–976.
[31] M. E. Zwick, F. McAfee, D. J. Cutler, et al. Microarray-based resequencing of multiple *Bacillus anthracis* isolates. Genome Biol 2005,6,R10.
[32] E. J. Feil, J. E. Cooper, H. Grundmann, et al. How clonal is *Staphylococcus aureus*? J Bacteriol 2003,185,3307–3316.
[33] C. E. Corless, E. Kaczmarski, R. Borrow, M. Guiver. Molecular characterization of *Neisseria meningitidis* isolates using a resequencing DNA microarray. J Mol Diagn 2008,10,265–271.
[34] S. Suerbaum, M. Achtman. *Helicobacter pylori*: recombination, population structure and human migrations. Int J Med Microbiol 2004,294,133–139.
[35] P. Puigbo, Y. I. Wolf, E. V. Koonin. Genome-wide comparative analysis of phylogenetic trees: the prokaryotic forest of life. Methods Mol Biol 2012,856,53–79.

[36] K. Schliep, P. Lopez, F. J. Lapointe, E. Bapteste. Harvesting evolutionary signals in a forest of prokaryotic gene trees. Mol Biol Evol 2011,28,1393–1405.
[37] J. P. Gogarten, W. F. Doolittle, J. G. Lawrence. Prokaryotic evolution in light of gene transfer. Mol Biol Evol 2002,19,2226–2238.
[38] O. Popa, E. Hazkani-Covo, G. Landan, W. Martin, T. Dagan. Directed networks reveal genomic barriers and DNA repair bypasses to lateral gene transfer among prokaryotes. Genome Res 2011,21,599–609.
[39] R. G. Ponder, N. C. Fonville, S. M. Rosenberg. A switch from high-fidelity to error-prone DNA double-strand break repair underlies stress-induced mutation. Mol Cell 2005,19,791–804.
[40] R. S. Galhardo, P. J. Hastings, S. M. Rosenberg. Mutation as a stress response and the regulation of evolvability. Crit Rev Biochem Mol Biol 2007,42,399–435.
[41] C. Fraser, W. P. Hanage, B. G. Spratt. Recombination and the nature of bacterial speciation. Science 2007,315,476–480.
[42] D. Cortez, P. Forterre, S. Gribaldo. A hidden reservoir of integrative elements is the major source of recently acquired foreign genes and ORFans in archaeal and bacterial genomes. Genome Biol 2009,10,R65.
[43] H. Innan, F. Kondrashov. The evolution of gene duplications: classifying and distinguishing between models. Nat Rev Genet 2010,11,97–108.
[44] K. Nakayama, A. Yamashita, K. Kurokawa, et al. The whole-genome sequencing of the obligate intracellular bacterium *Orientia tsutsugamushi* revealed massive gene amplification during reductive genome evolution. DNA Res 2008,15,185–199.
[45] J. Chun, C. J. Grim, N. A. Hasan, et al. Comparative genomics reveals mechanism for short-term and long-term clonal transitions in pandemic *Vibrio cholerae*. Proc Natl Acad Sci U S A 2009,106,15442–15447.
[46] N. A. Moran, H. J. McLaughlin, R. Sorek. The dynamics and time scale of ongoing genomic erosion in symbiotic bacteria. Science 2009,323,379–382.
[47] J. G. Lawrence, A. C. Retchless. The interplay of homologous recombination and horizontal gene transfer in bacterial speciation. Methods Mol Biol 2009,532,29–53.
[48] H. Ochman, J. G. Lawrence, E. A. Groisman. Lateral gene transfer and the nature of bacterial innovation. Nature 2000,405,299–304.
[49] L. D. McDaniel, E. Young, J. Delaney, F. Ruhnau, K. B. Ritchie, J. H. Paul. High frequency of horizontal gene transfer in the oceans. Science 2010,330,50.
[50] Y. Nakamura, T. Itoh, H. Matsuda, T. Gojobori. Biased biological functions of horizontally transferred genes in prokaryotic genomes. Nat Genet 2004,36,760–766.
[51] J. C. Mell, R. J. Redfield. Natural competence and the evolution of DNA uptake specificity. J Bacteriol 2014,196,1471–1483.
[52] C. Canchaya, G. Fournous, S. Chibani-Chennoufi, M.-L. Dillmann, H. Brüssow. Phage as agents of lateral gene transfer. Curr Opin Microbiol 2003,6,417–424.
[53] A. S. Lang, J. T. Beatty. The gene transfer agent of *Rhodobacter capsulatus* and "constitutive transduction" in prokaryotes. Arch Microbiol 2001,175,241–249.
[54] H. L. Hamilton, N. M. Dominguez, K. J. Schwartz, K. T. Hackett, J. P. Dillard. *Neisseria gonorrhoeae* secretes chromosomal DNA via a novel type IV secretion system. Mol Microbiol 2005,55,1704–1721.
[55] C. E. Alvarez-Martinez, P. J. Christie. Biological diversity of prokaryotic type IV secretion systems. Microbiol Mol Biol Rev 2009,73,775–808.
[56] J. Majewski. Sexual isolation in bacteria. FEMS Microbiol Lett 2001,199,161–169.
[57] V. Daubin, H. Ochman. Bacterial genomes as new gene homes: the genealogy of ORFans in *E. coli*. Genome Res 2004,14,1036–1042.

[58] A.-R. Carvunis, T. Rolland, I. Wapinski, et al. Proto-genes and de novo gene birth. Nature 2012,487,370–374.
[59] D. Tautz, T. Domazet-Lošo. The evolutionary origin of orphan genes. Nat Rev Genet 2011,12,692–702.
[60] D. Gevers, K. Vandepoele, C. Simillion, Y. Van de Peer. Gene duplication and biased functional retention of paralogs in bacterial genomes. Trends Microbiol 2004,12,148–154.
[61] G. D'Souza, S. Waschina, S. Pande, K. Bohl, C. Kaleta, C. Kost. Less is more: Selective advantages can explain the prevalent loss of biosynthetic genes in bacteria. Evolution 2014,68,2559–2570.
[62] K. Makarova, A. Slesarev, Y. Wolf, et al. Comparative genomics of the lactic acid bacteria. Proc Natl Acad Sci U S A 2006,103,15611–15616.
[63] Z. Sun, J. L. Blanchard. Strong genome-wide selection early in the evolution of *Prochlorococcus* resulted in a reduced genome through the loss of a large number of small effect genes. PLoS One 2014,9,e88837.
[64] N. A. Moran, A. Mira. The process of genome shrinkage in the obligate symbiont *Buchnera aphidicola*. Genome Biol 2001,2,Research0054.
[65] N. A. Moran. Microbial minimalism: genome reduction in bacterial pathogens. Cell 2002,108,583–586.
[66] H. Ochman. Genomes on the shrink. Proc Natl Acad Sci U S A 2005,102,11959–11960.
[67] A. Mira, H. Ochman, N. A. Moran. Deletional bias and the evolution of bacterial genomes. Trends Genet 2001,17,589–596.
[68] J. O. Andersson, S. G. Andersson. Pseudogenes, junk DNA, and the dynamics of *Rickettsia* genomes. Mol Biol Evol 2001,18,829–839.
[69] C.-H. Kuo, N. A. Moran, H. Ochman. The consequences of genetic drift for bacterial genome complexity. Genome Res 2009,19,1450–1454.
[70] B. J. Shapiro, J. Friedman, O. X. Cordero, et al. Population genomics of early events in the ecological differentiation of bacteria. Science 2012,336,48–51.
[71] H. Cadillo-Quiroz, X. Didelot, N. L. Held, et al. Patterns of gene flow define species of thermophilic Archaea. PLoS Biol 2012,10,e1001265.
[72] V. J. Denef, L. H. Kalnejais, R. S. Mueller, et al. Proteogenomic basis for ecological divergence of closely related bacteria in natural acidophilic microbial communities. Proc Natl Acad Sci U S A 2010,107,2383–2390.
[73] D. S. Guttman, D. E. Dykhuizen. Detecting selective sweeps in naturally occurring *Escherichia coli*. Genetics 1994,138,993–1003.
[74] R. J. Whitaker. Allopatric origins of microbial species. Philos Trans R Soc Lond B Biol Sci 2006,361,1975–1984.
[75] F. M. Cohan. Towards a conceptual and operational union of bacterial systematics, ecology, and evolution. Philos Trans R Soc Lond B Biol Sci 2006,361,1985–1996.
[76] P. Escobar-Páramo, S. Ghosh, J. DiRuggiero. Evidence for genetic drift in the diversification of a geographically isolated population of the hyperthermophilic archaeon *Pyrococcus*. Mol Biol Evol 2005,22,2297–2303.
[77] J. L. Macalady, T. L. Hamilton, C. L. Grettenberger, D. S. Jones, L. E. Tsao, W. D. Burgos. Energy, ecology and the distribution of microbial life. Philos Trans R Soc Lond B Biol Sci 2013,368,20120383.
[78] W. F. Doolittle, O. Zhaxybayeva. On the origin of prokaryotic species. Genome Res 2009,19,744–756.
[79] M. Vos. A species concept for bacteria based on adaptive divergence. Trends Microbiol 2011,19,1–7.

[80] W. F. Doolittle, R. T. Papke. Genomics and the bacterial species problem. Genome Biol 2006,7,116.
[81] D. M. Ward, F. M. Cohan, D. Bhaya, J. F. Heidelberg, M. Kuhl, A. Grossman. Genomics, environmental genomics and the issue of microbial species. Heredity (Edinb) 2008,100,207–219.
[82] B. J. Shapiro, M. F. Polz. Ordering microbial diversity into ecologically and genetically cohesive units. Trends Microbiol 2014,22,235–247.
[83] J. E. Loper, K. A. Hassan, D. V. Mavrodi, et al. Comparative genomics of plant-associated *Pseudomonas* spp.: insights into diversity and inheritance of traits involved in multitrophic interactions. PLoS Genet 2012,8,e1002784.
[84] D. A. Rasko, M. J. Rosovitz, GS. A. Myers, et al. The pangenome structure of *Escherichia coli*: Comparative genomic analysis of *E-coli* commensal and pathogenic isolates. J Bacteriol 2008,190,6881–6893.
[85] R. C. H. J. van Ham, J. Kamerbeek, C. Palacios, et al. Reductive genome evolution in *Buchnera aphidicola*. Proceedings of the National Academy of Sciences 2003,100,581–586.
[86] H. Tettelin, V. Masignani, M. J. Cieslewicz, et al. Genome analysis of multiple pathogenic isolates of *Streptococcus agalactiae*: Implications for the microbial "pan-genome". Proc Natl Acad Sci U S A 2005,102,13950–13955.
[87] A. F. Bennett, R. E. Lenski. Phenotypic and evolutionary adaptation of a model bacterial system to stressful thermal environments. EXS 1997,83,135–154.
[88] O. Beja, L. Aravind, E. V. Koonin, et al. Bacterial rhodopsin: evidence for a new type of phototrophy in the sea. Science 2000,289,1902–1906.
[89] Frias-J. Lopez, Y. Shi, G. W. Tyson, et al. Microbial community gene expression in ocean surface waters. Proc Natl Acad Sci U S A 2008,105,3805–3810.
[90] N. C. VerBerkmoes, V. J. Denef, R. L. Hettich, J. F. Banfield. Systems biology: Functional analysis of natural microbial consortia using community proteomics. Nat Rev Microbiol 2009,7,196–205.
[91] P. Hugenholtz, B. Goebel, N. Pace. Impact of culture-independent studies on the emerging phylogenetic view of bacterial diversity. J Bacteriol 1998,180,4765–4774.
[92] B. P. Hedlund, J. A. Dodsworth, S. K. Murugapiran, C. Rinke, T. Woyke. Impact of single-cell genomics and metagenomics on the emerging view of extremophile "microbial dark matter". Extremophiles 2014.
[93] S. O'Brien, D. J. Hodgson, A. Buckling. The interplay between microevolution and community structure in microbial populations. Curr Opin Biotechnol 2013,24,821–825.
[94] K. J. Edwards, T. M. Gihring, J. F. Banfield. Seasonal variations in microbial populations and environmental conditions in an extreme acid mine drainage environment. Appl Environ Microbiol 1999,65,3627–3632.
[95] Q. Huang, H. Jiang, B. R. Briggs, et al. Archaeal and bacterial diversity in acidic to circumneutral hot springs in the Philippines. FEMS Microbiol Ecol 2013,85,452–464.
[96] C. E. Sharp, A. L. Brady, G. H. Sharp, S. E. Grasby, M. B. Stott, P. F. Dunfield. Humboldt's spa: microbial diversity is controlled by temperature in geothermal environments. ISME J 2014,8,1166–1174.
[97] S. Benlloch, A. Lopez-Lopez, E. O. Casamayor, et al. Prokaryotic genetic diversity throughout the salinity gradient of a coastal solar saltern. Environ Microbiol 2002,4,349–360.
[98] N. Fierer, R. B. Jackson. The diversity and biogeography of soil bacterial communities. Proc Natl Acad Sci U S A 2006,103,626–631.
[99] A. Lanzen, A. Simachew, A. Gessesse, D. Chmolowska, I. Jonassen, L. Ovreas. Surprising prokaryotic and eukaryotic diversity, community structure and biogeography of Ethiopian soda lakes. PLoS One 2013,8,e72577.

[100] B. J. Baker, G. W. Tyson, R. I. Webb, et al. Lineages of acidophilic Archaea revealed by community genomic analysis. Science 2006,314,1933–1935.
[101] M. L. Sogin, H. G. Morrison, J. A. Huber, et al. Microbial diversity in the deep sea and the underexplored "rare biosphere". Proc Natl Acad Sci U S A 2006,103,12115–12120.
[102] R. J. Whitaker, J. F. Banfield. Population genomics in natural microbial communities. Trends Ecol Evol 2006,21,508–516.
[103] R. T. Papke, J. E. Koenig, F. Rodriguez-Valera, W. F. Doolittle. Frequent recombination in a saltern population of *Halorubrum*. Science 2004,306,1928–1929.
[104] R. T. Papke, D. M. Ward. The importance of physical isolation to microbial diversification. FEMS Microbiol Ecol 2004,48,293–303.
[105] R. J. Whitaker, D. W. Grogan, J. W. Taylor. Recombination shapes the natural population structure of the hyperthermophilic archaeon *Sulfolobus islandicus*. Mol Biol Evol 2005,22,2354–2361.
[106] M. Vos, X. Didelot. A comparison of homologous recombination rates in bacteria and archaea. ISME J 2009,3,199–208.
[107] V. J. Denef, J. F. Banfield. *In situ* evolutionary rate measurements show ecological success of recently emerged bacterial hybrids. Science 2012,336,462–466.
[108] C. Quast, E. Pruesse, P. Yilmaz, et al. The SILVA ribosomal RNA gene database project: improved data processing and web-based tools. Nucleic Acids Res 2013,41,D590-D6.
[109] K. Horikoshi, G. Antranikian, A. T. Bull, F. T. Robb, K. O. Stetter, eds. Extremophiles Handbook. New York: Springer, 2011.
[110] F. Arsene-Ploetze, S. Koechler, M. Marchal, et al. Structure, function, and evolution of the *Thiomonas* spp. genome. PLoS Genet 2010,6,e1000859.
[111] I. M. Tuffin, P. de Groot, S. M. Deane, D. E. Rawlings. An unusual Tn21-like transposon containing an ars operon is present in highly arsenic-resistant strains of the biomining bacterium *Acidithiobacillus caldus*. Microbiology 2005,151,3027–3039.
[112] R. Deole, J. Challacombe, D. W. Raiford, W. D. Hoff. An extremely halophilic proteobacterium combines a highly acidic proteome with a low cytoplasmic potassium content. J Biol Chem 2013,288,581–588.
[113] F. M. Lauro, R. A. Chastain, L. E. Blankenship, A. A. Yayanos, D. H. Bartlett. The unique 16S rRNA genes of piezophiles reflect both phylogeny and adaptation. Appl Environ Microbiol 2007,73,838–845.
[114] E. Bapteste, C. Brochier, Y. Boucher. Higher-level classification of the Archaea: evolution of methanogenesis and methanogens. Archaea 2005,1,353–363.
[115] G. Borrel, P. W. O'Toole, H. M. Harris, P. Peyret, J. F. Brugere, S. Gribaldo. Phylogenomic data support a seventh order of methylotrophic methanogens and provide insights into the evolution of methanogenesis. Genome Biol Evol 2013,5,1769–1780.
[116] E. F. Mongodin, K. E. Nelson, S. Daugherty, et al. The genome of *Salinibacter ruber*: Convergence and gene exchange among hyperhalophilic bacteria and archaea. Proc Natl Acad Sci U S A 2005,102,18147–18152.
[117] E. Boyd, J. W. Peters. New insights into the evolutionary history of biological nitrogen fixation. Frontiers in Microbiology 2013,4.
[118] M. Z. DeMaere, T. J. Williams, M. A. Allen, et al. High level of intergenera gene exchange shapes the evolution of haloarchaea in an isolated Antarctic lake. Proc Natl Acad Sci U S A 2013,110,16939–16944.
[119] K. E. Nelson, R. A. Clayton, S. R. Gill, et al. Evidence for lateral gene transfer between Archaea and bacteria from genome sequence of *Thermotoga maritima*. Nature 1999,399,323–329.

[120] A. C. Martiny, M. L. Coleman, S. W. Chisholm. Phosphate acquisition genes in *Prochlorococcus* ecotypes: evidence for genome-wide adaptation. Proc Natl Acad Sci U S A 2006,103,12552–12557.
[121] M. L. Coleman, M. B. Sullivan, A. C. Martiny, et al. Genomic islands and the ecology and evolution of *Prochlorococcus*. Science 2006,311,1768–1770.
[122] J. T. Sullivan, C. W. Ronson. Evolution of rhizobia by acquisition of a 500-kb symbiosis island that integrates into a phe-tRNA gene. Proc Natl Acad Sci U S A 1998,95,5145–5149.
[123] L. G. Acuña, J. P. Cárdenas, P. C. Covarrubias, et al. Architecture and gene repertoire of the flexible genome of the extreme acidophile *Acidithiobacillus caldus*. PLoS ONE 2013,8,e78237.
[124] K. S. Makarova, Y. I. Wolf, S. Snir, E. V. Koonin. Defense islands in bacterial and archaeal genomes and prediction of novel defense systems. J Bacteriol 2011,193,6039–6056.
[125] H. Schnabel, W. Zillig, M. Pfaffle, R. Schnabel, H. Michel, H. Delius. *Halobacterium halobium* phage oH. EMBO J 1982,1,87–92.
[126] M. Haring, G. Vestergaard, R. Rachel, L. Chen, R. A. Garrett, D. Prangishvili. Virology: independent virus development outside a host. Nature 2005,436,1101–1102.
[127] A. C. Ortmann, B. Wiedenheft, T. Douglas, M. Young. Hot crenarchaeal viruses reveal deep evolutionary connections. Nat Rev Microbiol 2006,4,520–528.
[128] F. E. Angly, B. Felts, M. Breitbart, et al. The marine viromes of four oceanic regions. PLoS Biol 2006,4,e368.
[129] S. J. Williamson, S. C. Cary, K. E. Williamson, et al. Lysogenic virus-host interactions predominate at deep-sea diffuse-flow hydrothermal vents. ISME J 2008,2,1112–1121.
[130] M. G. Weinbauer, I. Brettar, M. G. Höfle. Lysogeny and virus-induced mortality of bacterioplankton in surface, deep, and anoxic marine waters. Limnol Oceanogr 2003,48,1457–1465.
[131] A. U. Kuhlmann, T. Hoffmann, J. Bursy, M. Jebbar, E. Bremer. Ectoine and hydroxyectoine as protectants against osmotic and cold stress: uptake through the SigB-controlled betaine-choline- carnitine transporter-type carrier EctT from *Virgibacillus pantothenticus*. J Bacteriol 2011,193,4699–4708.
[132] T. Hoffmann, E. Bremer. Protection of *Bacillus subtilis* against cold stress via compatible-solute acquisition. J Bacteriol 2011,193,1552–1562.
[133] V. Mattimore, J. R. Battista. Radioresistance of *Deinococcus radiodurans*: functions necessary to survive ionizing radiation are also necessary to survive prolonged desiccation. J Bacteriol 1996,178,633–637.
[134] K. Dose, A. Bieger-Dose, M. Labusch, M. Gill. Survival in extreme dryness and DNA-single-strand breaks. Adv Space Res 1992,12,221–229.
[135] E. F. Delong, D. G. Franks, A. A. Yayanos. Evolutionary relationships of cultivated psychrophilic and barophilic deep-sea bacteria. Appl Environ Microbiol 1997,63,2105–2108.
[136] M. S. da Costa, H. Santos, E. A. Galinski. An overview of the role and diversity of compatible solutes in Bacteria and Archaea. Adv Biochem Eng Biotechnol 1998,61,117–153.
[137] D. T. Welsh. Ecological significance of compatible solute accumulation by micro-organisms: from single cells to global climate. FEMS Microbiol Rev 2000,24,263–290.
[138] H. Aliyu, P. De Maayer, J. Rees, M. Tuffin, D. A. Cowan. Draft genome sequence of the Antarctic polyextremophile *Nesterenkonia* sp. strain AN1. Genome Announc 2014,2.
[139] M. Falb, F. Pfeiffer, P. Palm, et al. Living with two extremes: conclusions from the genome sequence of *Natronomonas pharaonis*. Genome Res 2005,15,1336–1343.

Francisco J. López de Saro, Héctor Díaz-Maldonado, and Ricardo Amils
2 Microbial evolution: the view from the acidophiles

2.1 Introduction

Acidophilic organisms have provided a highly fertile ground for research into microbial evolution. Their low-biodiversity communities have allowed for extensive metagenomic, metatranscriptomic, and metaproteomic analysis [1]. A wealth of data from comparative genomics of closely related strains is beginning to reveal the evolutionary processes that allow for genotypic change, and how they relate to selective pressures. In the last decade it has become evident that the genetic diversity available in bacterial communities is vast and in constant flow. DNA is constantly mobilized by plasmids and phage, and recombination occurs at high rates. Recent studies in acidophiles have described not only the type of events that are taking place, but also to begin to make a quantitative assessment of their predominance and rates.

The acidophiles have been mainly studied in two scenarios. The first one, acid mine drainage (AMD) environments, are areas in which organisms rely on chemoautotrophic production mainly based on iron and sulphur oxidation. In addition to very low pH, there are often high concentrations of heavy metals such as iron, zinc or arsenic [1–3]. Well-characterized examples of these environments include the Río Tinto in Southern Spain [4] and Iron Mountain in California, USA [5]. The main actors of these studies have been *Leptospirillum* (*Nitrospira*), *Acidithiobacillus* (*Gammaproteobacteria*) and *Ferroplasma* (Archaea, *Thermoplasmata*). The second scenario is the volcanic springs or "mud pots" generated by geothermal activity, in which, in addition to extreme acidity, organisms must contend with temperatures that can reach 80 °C. These environments are dominated by thermoacidophilic Archaea and the main actor of evolutionary studies has been the genus *Sulfolobus* (*Crenarchaeota*).

Although it had been assumed that extremely acidic environments could prove hostile or limit DNA exchange, there is no evidence that mechanisms of gene transfer or genomic change are different from those operating in other less-extreme habitats. Numerous phage, plasmids, and mobile elements have been described in association with acidophilic communities or as part of the genomes of acidophiles, as well as mechanisms for DNA uptake, DNA secretion, or CRISPR (clusters of regularly interspaced short palindromic repeats) defense systems [see reviews in 6–9]. Indeed, phage are abundant and diverse in all environments where acidophilic prokaryotes have been found. For example, the optimal growth conditions of the *Sulfolobus* turreted icosahedral virus (STIV) are pH 3.3 and 80 °C [10]. There is some evidence, however, that DNA exchange among acidophilic organisms, even when not closely related phylogenetically, is more frequent than DNA exchange from organisms from other habitats, pointing to the relative isolation of these extreme ecosystems. Further, the conditions endured by acidophiles were thought initially to impose special requirements for DNA repair or damage tolerance, but no significant differences have been found if compared with organisms living in less-extreme habitats. For example, the mutation rate in the hyperthermophilic archaeon *Sulfolobus acidocaldarius* was shown to be equal to mesophilic organisms [11].

Interest in the evolutionary mechanisms present in acidophiles is twofold. First, acidophiles living in AMD environments are relevant in biomining (bioleaching and bio-oxidation) for the extraction of copper and gold [12]. However, the manipulation of cultures of acidophilic organisms for use in biomining is still in its infancy. Since microbial consortia are involved, often consisting of one or few dominant species accompanied by low-abundant but diverse bacteria, the problem to be addressed is the understanding of how communities change given a set of environmental parameters. Second, *Sulfolobus* and other thermoacidophilic Archaea have been studied for their potential in production of thermostable enzymes and processes of interest in biotechnology [13]. However, despite recent advances, the genetic engineering of *Sulfolobus* is also limited [14, 15].

In this review we describe the major advances of recent years in the study of the evolutionary mechanisms that shape the genomes of acidophilic organisms, as well as the ecological scenarios in which these changes take place. Finally, we suggest future avenues of research into this fascinating group of microorganisms.

2.2 Horizontal gene transfer

Horizontal gene transfer (HGT), the transference of genetic material between organisms not directly related genealogically, is pervasive in bacterial communities. It has been widely documented among the acidophiles, a phenomenon that could point to the fact that acidic environments are probably relatively closed habitats. For example, the sequencing of the *Picrophilus* genome has shown that it contains almost as many genes in common with the phylogenetically close organism *Thermoplasma* (Euryarchaeota) as with the phylogenetically distant *Sulfolobus solfataricus* (Crenarchaeota) [16]. Also, *Sulfolobus islandicus* strains often share genes with other *Sulfolobus* species [17], possibly highlighting the fact that these environments are quite refractory to foreign genetic input.

An extraordinary example of HGT has been demonstrated recently with the sequencing of the genome of acidophilic red algae *Galdieria* [18]. Comparative genomics have shown that adaptation to the acidic environment, heavy-metal resistance, and metabolic versatility can be directly attributed to at least 75 separate gene acquisitions from Archaea and Bacteria. For example, the bacterial arsenic membrane protein pump ArsB likely was acquired from thermoacidophilic Bacteria, and the acetate permeases and polyamine transporters present in *G. sulphuraria*'s genome originate from Bacteria and Archaea, respectively [18]. Interestingly, genes recruited by HGT into the algae's genome are specially enriched in those from extremophilic Bacteria, again suggesting the closed ecosystem idea. Typically eukaryotes evolve via gene duplications and neofunctionalizations, but this alga has adapted the prokaryotic way, by appropriation of genes by HGT.

HGT seems the key to explain fast adaptation and evolution of genomes in short evolutionary time-scales. Genes which may provide an ecological advantage or are estimated to be physiologically relevant have often been observed in recently transferred blocks of DNA in the genomes of acidophiles. These include, for example, quorum sensing genes of the LuxIR system in *Leptospirillum* [19], metabolic genes (toluene monooxygenase and nitrate reductase) in *S. islandicus* [20], or genes allowing adaptation to metal tolerance and acidity (arsenic-specific operons *ars2* and *aox*, biofilm formation, and motility) in *Thiomonas* [21] and *Acidithiobacillus* [22]. Many of these genes are likely niche-specific which could contribute to adaptation in restrictive ecosystems.

Of considerable interest is to ascertain the vehicles for genetic flow in natural populations, and to quantify their relative importance. Of the three classical mechanisms, natural transformation, conjugation (plasmid-mediated), and transduction (phage-mediated), transformation by free DNA is probably limited among the acidophiles due to the hydrolysis of DNA in acidic conditions with high metal content [7]. However, the analysis of the half-life of free DNA in acidic environments clearly requires further quantitative studies. Wide host-range phage could be responsible for some limited HGT but a recent study indicated that they could be poor vectors in prokaryotes due to the tight packing in their genomes [23]. However, the extraordinary abundance and variety of phage in natural ecosystems make any conclusions in this respect highly speculative. Conjugation, mediated by plasmids or ICEs (integrative conjugative elements), which require physical contact between donor and recipient cells, seems to be the main mechanism of HGT [24]. The host range is wider for conjugation than for transduction. In this regard, the genes required for DNA transfer and processing have been found in the genomes of most acidophiles.

2.3 The mobilome

The sequencing of full genomes of diverse strains of the same species has revealed the existence of a core genome that is common to all of them, and a variable or 'flexible' genome that is strain-specific and the product of genetic exchange and recombination. For example, a study of 7 strains of *S. islandicus* isolated from 3 different locations showed that their variable genome accounts for 20–30% of the genes [17]. A recent study comparing two strains of *Acidithiobacillus caldus* has shown that about 20% of the genes present in their genomes were strain-specific, and that a large number of mobile elements (including plasmids, transposons, and integrative elements) suggests a high degree of genetic flux [25].

The sequencing of various genomes from acidophiles have revealed the presence of genomic islands, segments of the chromosome of up to 200 kb in length and which are distinct from the core genome [26]. Genomic islands identified in most sequenced genomes in acidophiles are often associated with prophages, transposable elements, or plasmid mobilization elements. Many genomic islands contain a 'recombination module' containing at least an integrase, which can be used to track their evolutionary history, and its attachment sites [27]. Genomic islands have often been transferred via HGT in the recent evolutionary past, as revealed by their anomalous nucleotide composition (G+C content or codon usage). They can also be identified by comparative genomics of phylogenetically close relatives. Interestingly, genomic islands often encode genes that are directly selectable and environmentally relevant, such as those for heavy-metal resistance, DNA repair, biofilm formation, or motility. It is for this reason that they are considered critical during adaptation to changing environments, ecological differentiation, and, in general, genome evolution. For example, a comparative analysis of genomic islands found in two strains of *Acidithiobacillus ferrooxidans* showed major differences in gene content, with predominance of metal resistance (e.g., mercury detoxification, copper transport) and metabolic genes of ecological relevance [28].

2.4 Phages

Phages are essential players in microbial ecosystems and genome evolution, yet little is known about their role, dynamics, and impact in acidic environments. The study of phages is now being revolutionized by the use of high-throughput sequencing techniques and bioinformatics. Phages have been found in all environments where bacteria can be found and there are clearly no physical limitations to the maintenance of large and highly diverse phage communities in extreme acidic conditions. Phage predation is a major selective force for the evolution of bacterial populations, and some authors believe that they serve to preserve metabolic diversity by allowing the coexistence of multiple bacterial strains rather than just one [29]. On the other hand, it has been suggested that the huge phage 'metavirome' could serve as a genetic reservoir and allow the quick retrieval of advantageous genes under shifting environmental conditions. Another possible function of phage is as vehicles and facilitators for intra- and inter-species HGT. Often phage have broad host ranges, as has been shown for prophages found in the genomes of *Ferroplasma* and G-plasma, and in the genomes of *Leptospirillum* Groups II and III [30, 31]. Also, it has been suggested that phage transduction could promote the rise of phenotypic differences and fast and adaptive evolution in coexisting populations of *Ferroplasma* [32].

The study of several phages that infect *Sulfolobus* species has revealed two major forms that are genus-specific, the spindle-shaped viruses (SSVs) and the rod-shaped viruses (SIRVs), which could be specific to *S. islandicus* [33]. Local studies of these phages at specific hot spring locations throughout the world revealed high sequence diversity, suggesting that the combination of high isolation and mutation drove phage population evolution. Surprisingly, a recent study of SSV and SIRV population dynamics over a period of two years in the hot and acidic springs of Yellowstone National Park has demonstrated that phage migration from distant [global] locations contributes critically to maintaining local diversity [34]. Indeed, the rate of virus immigration and colonization, followed by extensive recombination, was significantly higher than mutation, and the reason for the high local genetic diversity. This study highlighted the fact that the fast-evolving phage populations must be studied from a temporal point of view, with repeated sampling. Further, this study demonstrated that phages, which could most likely travel long distances by air currents, could be critical in gene shuffling for bacterial populations at a global scale [34].

The highly complex relationship between phages and bacteria has been enhanced by the discovery and study of the CRISPR loci in bacterial genomes. These chromosomal regions consist of hypervariable arrays of short segments of phage, plasmid, or transposon DNA, and are thought to reflect the history of infections of that specific genome [35]. CRISPR loci serve as templates for the synthesis of antisense RNA that is used to target and destroy incoming foreign DNA, and have been used to analyze the dynamics of virus and hosts in AMD populations [36, 37]. CRISPR loci, whose study was pioneered in acidophiles, are currently being successfully developed into the newest-generation of tools for gene manipulation in eukaryotes [38].

2.5 Plasmids

Plasmids are the most important vehicles for HGT and critically contribute to host-cell adaptability and fitness. Many plasmids have been described for most acidophilic Bacteria [6] and thermoacidophilic Archaea [7]. They often carry niche-specific genes that point to their contribution to ecological adaptation as, for example, the arsenic-resistant genes contained in plasmids isolated from *A. caldus* [39], or genes mediating conjugation. The comparative analysis of two strains of *A. caldus* has shown the presence of a megaplasmid (> 150 Kb) in both strains, and a large number of metabolic genes also present in the chromosome, suggesting that the exchange of genes from chromosomes to plasmids and vice versa is fluid [25]. In *Leptospirillum*, however, a large plasmid was described containing a large number of proteins of unknown function, making the contribution of these elements to the metabolic potential difficult to deduce [40]. Plasmids isolated from *Acidiphilium symbioticum* have been reported to provide resistance to cadmium and zinc, and to carry a multi-drug efflux system [41]. Although the potential of plasmids as biotechnology tools is obvious, their use as genetic tools for manipulation of organisms with bioleaching potential is still in development.

2.6 Transposons

Although phages and plasmids can ferry genes between cells in the bacterial population, another class of mobile elements, transposons, facilitate the flow of genes between replicons. When present in high copy number, transposons create instability in the chromosome by increasing recombination, deletions, and chromosomal rearrangements. The insertion sequences (ISs) are the smallest of transposons, often carrying only one gene encoding the transposase required for their movement. Since insertion sequences are ubiquitous and very abundant, they facilitate DNA exchange within species in bacterial communities [42, 43]. Transposon activity could be an indication of accelerated change in genomes, or of stress under fluctuating environmental conditions. However, studies of these small mobile elements have been limited by their high diversity and the difficulties in tracing their genealogies and behaviour at the population-level. On the other hand, since ISs are extremely variable, they are excellent bacterial strain-level genetic markers. For example, substantial differences in IS content patterns can be found in strains of *S. islandicus* [17] or in *A. caldus* [8, 25].

Transposon expression was detected by proteomics in AMD biofilms dominated by *Leptospirillum* in early-stage biofilms (see below, [44]) and by microarray hybridization in *Leptospirillum ferrooxidans* [45], highlighting the fast dynamics of these elements. Changes in IS composition have also been observed in laboratory settings [46] or during cultivation of *Ferroplasma* [32]. Further studies of changes in IS copy number and location in chromosomes will be required to examine the impact of ISs in adaptation.

In addition to their effects as catalysts of recombination and genetic change, ISs can be carriers of genes of ecological relevance. For example, an IS21-derived transposon was found to contain a nine-gene operon containing arsenic resistance genes [22] in a strain of *A. caldus* isolated in an arsenopyrite bio-oxidation tank, and later was found in a strain of *Leptospirillum ferriphilum* from the same tank [47]. Recent research has shown that ISs interact with their hosts via an essential and universal replication factor, thus facilitating their movement between cohabiting but phylogenetically distant microorganisms [48]. The importance of IS elements is highlighted by the fact that they are often abundant in genomic islands and plasmids, possibly facilitating integration and recombination events.

2.7 Evolution and ecology: long term studies of genetic variation

The study of the genomic mechanisms of change shows a picture of great plasticity and adaptive potential. Due to their relatively rare habitats and limited biodiversity, acidic environments have been excellent settings for the initial dissection of evolving microbial communities, as they can be readily analyzed using high-throughput techniques. These techniques have revolutionized ecological analysis because they allow for the relatively unbiased coverage of DNA, RNA, and proteins by metagenomics, metatranscriptomics, and metaproteomics, respectively.

Mueller and colleagues studied a large set of biofilms from an acid mine, with the objective of identifying physiological changes and ecological interactions among organisms [44]. These biofilms were analyzed at different stages of development, generating a detailed view of how metabolism and biodiversity change with time. Bacteria from the genus *Leptospirillum* were always the founders and dominant species within the biofilms but as they aged, the biodiversity increased greatly. Interestingly, the physiology of *Leptospirillum* also changed with the maturity of the biofilms, likely as a result of interactions with other organisms [44]. Mobile DNA elements were especially overrepresented in low-diversity biofilms, probably reflecting a lower selective pressure. Importantly, it also showed that mobile elements, facilitators of genetic diversity and evolutionary change, could be modulated environmentally. It is tempting to speculate that transposons and other mobile elements become active to provide genomic plasticity in a regulated fashion, perhaps as a result of stress.

In another landmark study, Denef and Banfield analyzed underground acidic biofilms in a nine-year period by metagenomics [49]. These biofilms were mostly dominated by six genotypes of *Leptospirillum*. The authors could assemble different genotypes and reconstruct their evolutionary history and relative abundance over time. The conclusion is that successive prevalence of one genotype over the others could happen relatively quickly and be determined by a major recombination event. Indeed, evolution of *Leptospirillum* consisted of a periodic succession of events of HGT, recombination, and selective sweeps. The authors suggest that the evolutionary advantage that determines dominance of a genotype over others could often be determined by just a few genetic changes. In addition, Denef and Banfield could derive for the fist time the single-nucleotide substitution rate of a free-living organism, in a cultivation-independent manner. This rate, 1.4×10^{-9} ($\pm 0.2 \times 10^{-9}$) substitutions per nucleotide per generation, is consistent with other estimates of mutation rates in bacterial chromosomes [49].

Detailed metagenomic studies combined with comparative genomics have also been carried out for populations of *S. islandicus*. Whittaker and collaborators have studied two strains of *S. islandicus* that grow in isolated mud pots in the Mutnovsky Volcano in Russia [20]. The patterns of homologous gene flow among genomes of 12 strains show strong signs of sympatric speciation into two groups. These groups show a declining exchange of DNA among them, suggesting that divergence is increasing with time. Multilocus sequence analysis of many strains collected in 2000 and 2010 show that the two groups are coexisting with no signs of competitive exclusion resulting in extinction of one of them, at least in this time-scale [20]. The nature of the barriers to genetic exchange are obscure, but ecological specialization due to large genomic islands (genomic continents) is most likely responsible. In this case the comparison between aligned genomes and geological records produced an average rate of single nucleotide substitution per site per year of 4.66×10^{-9} ($\pm 6.76 \times 10^{-10}$) [17].

While studies of *S. islandicus* have shown that isolation and geographical distance between the volcanic springs correlates with genetic divergence, and that allopatric diversification is possible, the opposite pattern has been observed in studies of *S. acidocaldarius* growing in acidic springs separated by thousands of kilometres [50]. The analysis of these strains has shown near identical genotypes, suggesting rapid, global gene flow among them. These results have led to suggest that, somewhat surprisingly and contrary to what had been often argued, distance does not restrict gene flow among Bacteria or Archaea [17, 50].

2.8 Future directions

The study of acidophiles will continue to provide strong insights into evolutionary processes through the integration of high-throughput techniques with ecology studies. This will in turn allow the production of predictive models for community composition and change. Long-term studies aimed at the observation of change in natural settings are essential to address the impact of phage predation on bacterial populations, or the effect of fluctuating environmental conditions on community composition.

Detailed field studies can be complemented with results obtained in controlled laboratory conditions. For example, recombination has been extensively studied in the laboratory for *S. acidocaldarius* [51], and the study of recombination mechanisms of *S. islandicus* using genetic markers is starting to yield insights into their pathways of allopatric speciation [52]. Much remains to be learned about the mechanisms that promote or restrict gene flow on a global scale, and acidophiles could provide an excellent tool for these types of studies. While AMD or volcanic acidic springs could seem isolated from an anthropocentric point of view, genetic exchange between them, perhaps promoted by air-borne bacteria or phages, could be rapid and frequent. In this regard, sampling must be considered carefully when addressing questions of bacterial population dynamics.

The relatively low-biodiversity natural populations of acidophilic organisms could also be a fertile ground for the analysis of the dynamics of transposable elements. Most major questions remain unanswered regarding these critical actors of genomic evolution, such as their major routes of propagation, their capacity to expand explosively and cause lineage extinction, the correlation between their abundance and genomic evolvability, or their role in the growth and reduction of chromosomes.

Transposable elements thrive and evolve on host chromosomes in a fashion that could resemble that of viruses, but very little is known about their dynamics in natural environments, or about their impact on bacterial populations.

The described analysis of the genome of *G. sulphuraria* suggests that transfer of DNA between prokaryotic and eukaryotic organisms could be more frequent than previously expected. Very little is known about the mechanisms of transfer of this genetic material, the selective pressures that favour this exchange, or how the incorporated genes adapt to the host genome. Detailed knowledge of these processes could open new ways to genetic engineering of eukaryotic organisms to perform functions otherwise restricted to Bacteria and Archaea. A largely unexplored eukaryotic diversity is present in acidic environments [53, 53] and proteomic analysis in laboratory settings has shown possible adaptations of these organisms to fluctuating stress conditions [55].

Finally, an area of great interest is the study of acidic underground habitats. Little is known about the structure and biodiversity of the bacterial communities that thrive deep in the earth's crust [3]. Certainly the slow metabolism and growth, coupled to the restrictions in the movement of individual organisms and genetic material, impose a completely distinct evolutionary dynamics compared to organisms living in water-rich, open environments. Although there are major technological challenges in these types of studies, they will ultimately provide essential insights into the role of acidophiles in geochemical processes on Earth and, perhaps, in evolutionary potential beyond our planet [56].

Acknowledgments

This work was supported by grants ERC-250350/IPBSL from the European Research Council and CGL2010-17384 from the Spanish Government.

References

[1] V. J. Denef, R. S. Mueller, J. F. Banfield. AMD biofilms, using model communities to study microbial evolution and ecological complexity in nature. ISME J 2010,4,599–610.
[2] E. Gonzalez-Toril, E. Llobet-Brossa, E. O. Casamayor, R. Amann, R. Amils. Microbial ecology of an extreme acidic environment, the Tinto River. Appl Environ Microbiol 2003,69,4853–4865.
[3] D. B. Johnson. Geomicrobiology of extremely acidic subsurface environments. FEMS Microbiol Ecol 2012,81,2–12.
[4] L. A. Amaral-Zettler, E. R. Zettler, S. M. Theroux, C. Palacios, A. Aguilera, R. Amils. Microbial community structure across the tree of life in the extreme Rio Tinto. ISME J 2011,5,42–50.
[5] P. Wilmes, S. L. Simmons, V. J. Denef, J. F. Banfield. The dynamic genetic repertoire of microbial communities. FEMS Microbiol Rev 2009,33,109–132.

[6] J. P. Cárdenas, J. Valdes, R. Quatrini, F. Duarte, D. S. Holmes. Lessons from the genomes of extremely acidophilic bacteria and archaea with special emphasis on bioleaching microorganisms. Appl Microbiol Biotechnol 2010,88,605–620.
[7] M. van Wolferen, M. Ajon, A. J. M. Driessen, S. V. Albers. How hyperthermophiles adapt to change their lives, DNA exchange in extreme conditions. Extremophiles 2013,17,545–563.
[8] F. J. López de Saro, M. Gómez, E. González-Tortuero, V. Parro. The dynamic genomes of acidophiles. In, J. Seckbach, A. Oren, H. Stan-Lotter, eds. Polyextremophiles, Life under multiple forms of stress. 1st ed. Dordrecht, Springer, 2013,83–97.
[9] R. A. Garrett, S. A. Shah, G. Vestergaard, L. Deng, S. Gudbergsdottir, C. S. Kenchappa, et al. CRISPR-based immune systems of the Sulfolobales, complexity and diversity. Biochem Soc Trans 2011,39,51–57.
[10] G. Rice, L. Tang, K. Stedman, F. Roberto, J. Spuhler, E. Gillitzer, et al. The structure of a thermophilic archaeal virus shows a double-stranded DNA viral capsid type that spans all domains of life. Proc Natl Acad Sci U S A 2004,101,7716–7720.
[11] D. W. Grogan, G. T. Carver, J. W. Drake. Genetic fidelity under harsh conditions, analysis of spontaneous mutation in the thermoacidophilic archaeon *Sulfolobus acidocaldarius*. Proc Natl Acad Sci U S A 2001,98,7928–7933.
[12] D. E. Rawlings, D. B. Johnson. The microbiology of biomining, development and optimization of mineral-oxidizing microbial consortia. Microbiology 2007,153,315–324.
[13] A. Sharma, Y. Kawarabayasi, T. Satyanarayana. Acidophilic bacteria and archaea, acid stable biocatalysts and their potential applications. Extremophiles 2012,16,1–19.
[14] Y. Maezato, K. Dana, P. Blum. Engineering thermoacidophilic archaea using linear DNA recombination. Methods in molecular biology 2011,765,435–445.
[15] J. A. Leigh, S. V. Albers, H. Atomi, T. Allers. Model organisms for genetics in the domain Archaea, methanogens, halophiles, Thermococcales and Sulfolobales. FEMS Microbiol Rev 2011,35,577–608.
[16] O. Fütterer, A. Angelov, H. Liesegang, G. Gottschalk, C. Schleper, B. Schepers, et al. Genome sequence of *Picrophilus torridus* and its implications for life around pH 0. Proc Natl Acad Sci U S A 2004,101,9091–9096.
[17] M. L. Reno, N. L. Held, C. J. Fields, P. V. Burke, R. J. Whitaker. Biogeography of the *Sulfolobus islandicus* pan-genome. Proc Natl Acad Sci U S A 2009,106,8605–8610.
[18] G. Schönknecht, W.-H. Chen, C. M. Ternes, G. G. Barbier, R. P. Shrestha, M. Stanke, et al. Gene transfer from bacteria and archaea facilitated evolution of an extremophilic Eukaryote. Science 2013,339,1207–1210.
[19] S. L. Simmons, G. Dibartolo, V. J. Denef, D. S. Goltsman, M. P. Thelen, J. F. Banfield. Population genomic analysis of strain variation in *Leptospirillum* group II bacteria involved in acid mine drainage formation. PLoS Biol 2008,6,e177.
[20] H. Cadillo-Quiroz, X. Didelot, N. L. Held, A. Herrera, A. Darling, M. L. Reno, et al. Patterns of gene flow define species of thermophilic Archaea. PLoS Biol 2012,10,e1001265.
[21] F. Arsene-Ploetze, S. Koechler, M. Marchal, J. Y. Coppee, M. Chandler, V. Bonnefoy, et al. Structure, function, and evolution of the *Thiomonas* spp. genome. PLoS Genet 2010,6,e1000859.
[22] I. M. Tuffin, P. de Groot, S. M. Deane, D. E. Rawlings. An unusual Tn21-like transposon containing an ars operon is present in highly arsenic-resistant strains of the biomining bacterium *Acidithiobacillus caldus*. Microbiology 2005,151,3027–3039.
[23] S. Leclercq, R. Cordaux. Do phages efficiently shuttle transposable elements among prokaryotes? Evolution 2011,65,3327–3331.
[24] R. A. Wozniak, M. K. Waldor. Integrative and conjugative elements, mosaic mobile genetic elements enabling dynamic lateral gene flow. Nat Rev Microbiol 2010,8,552–563.

[25] L. G. Acuña, J. P. Cardenas, P. C. Covarrubias, et al. Architecture and gene repertoire of the flexible genome of the extreme acidophile *Acidithiobacillus caldus*. PLoS One 2013,8,e78237.

[26] M. Juhas, J. R. van der Meer, M. Gaillard, R. M. Harding, D. W. Hood, D. W. Crook. Genomic islands, tools of bacterial horizontal gene transfer and evolution. FEMS Microbiol Rev 2009,33,376–393.

[27] E. F. Boyd, S. Almagro-Moreno, M. A. Parent. Genomic islands are dynamic, ancient integrative elements in bacterial evolution. Trends Microbiol 2009,17,47–53.

[28] J. Valdés, J. P. Cárdenas, R. Quatrini, M. Esparza, H. Osorio, F. Duarte, et al. Comparative genomics begins to unravel the ecophysiology of bioleaching. Hydrometallurgy 2010,104,471–476.

[29] F. Rodriguez-Valera, A. B. Martin-Cuadrado, B. Rodriguez-Brito, L. Pasic, T. F. Thingstad, F. Rohwer, et al. Explaining microbial population genomics through phage predation. Nat Rev Microbiol 2009,7,828–836.

[30] G. W. Tyson, J. Chapman, P. Hugenholtz, E. E. Allen, R. J. Ram, P. M. Richardson, et al. Community structure and metabolism through reconstruction of microbial genomes from the environment. Nature 2004,428,37–43.

[31] R. J. Ram, N. C. Verberkmoes, M. P. Thelen, G. W. Tyson, B. J. Baker, R. C. Blake, et al. Community proteomics of a natural microbial biofilm. Science 2005,308,1915–1920.

[32] E. E. Allen, G. W. Tyson, R. J. Whitaker, J. C. Detter, P. M. Richardson, J. F. Banfield. Genome dynamics in a natural archaeal population. Proc Natl Acad Sci U S A 2007,104,1883–1888.

[33] J. C. Snyder, M. J. Young. Advances in understanding archaea-virus interactions in controlled and natural environments. Curr Opin Microbiol 2011,14,497–503.

[34] J. C. Snyder, B. Wiedenheft, M. Lavin, F. F. Roberto, J. Spuhler, A. C. Ortmann, et al. Virus movement maintains local virus population diversity. Proc Natl Acad Sci U S A 2007,104,19102–19107.

[35] R. Barrangou, C. Fremaux, H. Deveau, M. Richards, P. Boyaval, S. Moineau, et al. CRISPR provides acquired resistance against viruses in prokaryotes. Science 2007,315,1709–1712.

[36] G. W. Tyson, J. F. Banfield. Rapidly evolving CRISPRs implicated in acquired resistance of microorganisms to viruses. Environ Microbiol 2008,10,200–207.

[37] A. F. Andersson, J. F. Banfield. Virus population dynamics and acquired virus resistance in natural microbial communities. Science 2008,320,1047–1050.

[38] T. R. Sampson, D. S. Weiss. Exploiting CRISPR/Cas systems for biotechnology. Bioessays 2014,36,34–38.

[39] L. J. van Zyl, S. M. Deane, L. A. Louw, D. E. Rawlings. Presence of a family of plasmids (29 to 65 kilobases) with a 26-kilobase common region in different strains of the sulfur-oxidizing bacterium Acidithiobacillus caldus. Appl Environ Microbiol 2008,74,4300–4308.

[40] D. S. Goltsman, V. J. Denef, S. W. Singer, et al. Community genomic and proteomic analyses of chemoautotrophic iron-oxidizing "*Leptospirillum rubarum*" (Group II) and "*Leptospirillum ferrodiazotrophum*" (Group III) bacteria in acid mine drainage biofilms. Appl Environ Microbiol 2009,75,4599–4615.

[41] S. K. Singh, A. Singh, P. C. Banerjee. Plasmid encoded AcrAB-TolC tripartite multidrug-efflux system in *Acidiphilium symbioticum* H8. Curr Microbiol 2010,61,163–168.

[42] A. Toussaint, M. Chandler. Prokaryote genome fluidity, toward a system approach of the mobilome. Methods in molecular biology 2012,804,57–80.

[43] R. K. Aziz, M. Breitbart, R. A. Edwards. Transposases are the most abundant, most ubiquitous genes in nature. Nuc Acids Res 2010,38,4207–4217.

[44] R. S. Mueller, V. J. Denef, L. H. Kalnejais, et al. Ecological distribution and population physiology defined by proteomics in a natural microbial community. Mol Syst Biol 2010,6,374.

[45] V. Parro, M. Moreno-Paz, E. Gonzalez-Toril. Analysis of environmental transcriptomes by DNA microarrays. Environ Microbiol 2007,9,453–464.
[46] T. F. Kondrat'eva, V. N. Danilevich, S. N. Ageeva, G. I. Karavaiko. Identification of IS elements in *Acidithiobacillus ferrooxidans* strains grown in a medium with ferrous iron or adapted to elemental sulfur. Arch Microbiol 2005,183,401–410.
[47] I. M. Tuffin, S. B. Hector, S. M. Deane, D. E. Rawlings. Resistance determinants of a highly arsenic-resistant strain of *Leptospirillum ferriphilum* isolated from a commercial biooxidation tank. Appl Environ Microbiol 2006,72,2247–2253.
[48] M. Gómez, H. Díaz-Maldonado, E. González-Tortuero, F. J. López de Saro. Chromosomal replication dynamics and interaction with the ß sliding clamp determine orientation of bacterial transposable elements. Genome Biol Evol 2014,6,727–740.
[49] V. J. Denef, J. F. Banfield. In situ evolutionary rate measurements show ecological success of recently emerged bacterial hybrids. Science 2012,336,462–466.
[50] D. Mao, D. Grogan. Genomic evidence of rapid, global-scale gene flow in a *Sulfolobus* species. ISME J 2012,6,1613–1616.
[51] D. Mao, D. W. Grogan. Heteroduplex formation, mismatch resolution, and genetic sectoring during homologous recombination in the hyperthermophilic archaeon *Sulfolobus acidocaldarius*. Front Microbiol 2012,3,192.
[52] C. Zhang, D. J. Krause, R. J. Whitaker. *Sulfolobus islandicus*, a model system for evolutionary genomics. Biochemical Society Transactions 2013,41,458–462.
[53] A. Aguilera, V. Souza-Egipsy, F. Gomez, R. Amils. Development and structure of eukaryotic biofilms in an extreme acidic environment, Rio Tinto (SW, Spain). Microb Ecol 2007,53,294–305.
[54] I. Nancucheo, D. B. Johnson. Acidophilic algae isolated from mine-impacted environments and their roles in sustaining heterotrophic acidophiles. Front Microbiol 2012,3,325.
[55] C. Cid, L. Garcia-Descalzo, V. Casado-Lafuente, R. Amils, A. Aguilera. Proteomic analysis of the response of an acidophilic strain of *Chlamydomonas* sp. (Chlorophyta) to natural metal-rich water. Proteomics 2010,10,2026–2036.
[56] R. Amils, E. Gonzalez-Toril, A. Aguilera, et al. From Rio Tinto to Mars, the terrestrial and extraterrestrial ecology of acidophiles. Adv Appl Microbiol 2011,77,41–70.

R. Eric Collins
3 Microbial Evolution in the Cryosphere

3.1 Overview

Icy environments have appeared repeatedly through Earth's history and exist today primarily at high latitudes and high altitudes. Microbial communities within Earth's cryosphere, including Bacteria, Archaea, Eukarya, and viruses, play important roles in both natural and built environments by mediating biogeochemical cycles in the polar oceans, sea ice, permafrost, snow, alpine glaciers, and ice sheets, in addition to affecting food safety and as sources of novel biotechnology. These microorganisms and their biomolecules are often concentrated into the physical matrix of ice, where environmental conditions are challenging. Extremes of temperature, osmotic pressure, water activity, oxidative stress, radiation, and pH are common. We now know that other planetary bodies in our solar system, prominently Mars, Europa, Enceladus, and Titan, also harbor extensive frozen habitats where microorganisms might conceivably survive and evolve. To understand the mechanisms by which the microbial inhabitants of cold environments, on Earth or possibly elsewhere, adapt to thrive in those environments, we must understand how these extreme environmental conditions affect the evolutionary landscape and how microorganisms have overcome those challenges.

This chapter is organized into three parts: first, a broad overview of current knowledge, as well as open questions, regarding microbial evolution in frozen environments; second, a case study using microbial evolution in sea ice as an example; and third, a look into modern methods of research on microbial evolution in the cold, including suggestions for areas of future research. Each of the first two parts feature narratives on cryospheric environments as they pertain to microbial communities, an evaluation of evolutionary processes influenced by the physical environment, and a survey of adaptations that enable the survival of microbes in these environments. The taxonomic focus is on prokaryotic microorganisms, but eukaryotic protists can also be important components of icy communities.

3.1.1 Cryospheric evironments

Frozen habitats come in many different forms, but in common they share low temperatures and, except in the driest deserts, the presence of water ice. From the perspective of microbial evolution, these habitats can be grouped or separated based on many abiotic factors, including seasonality, age, temperature, salinity, water ice fraction, liquid water fraction, mineral fraction, and organic matter content. While each of these perspectives is valuable, this review will explore differences between seasonally and perennially frozen environments, emphasizing the importance of seasonality as a determinant of microbial evolution in the cryosphere.

Seasonally frozen environments, like sea ice, snow, and the active layer overlying permafrost, persist continuously on Earth for millions of years but are also spatially and geographically ephemeral. These elements of the cryosphere appear to be highly suitable for rapid microbial evolution, with communities exhibiting larger population densities, faster metabolic rates, and more widespread dispersal than those in comparable perennially frozen environments like ice sheets and permafrost. In contrast, perennially frozen environments act as long-lived refugia for ancient microbial communities, providing a biological repository extending millions of years into the past through which we can observe evolution.

3.1.1.1 Seasonally frozen environments: sea ice, snow, and the active layer

Sea ice has been present intermittently for at least 40 million years and perennially for over 10 million years, but recent warming has caused dramatic changes to the sea ice cover in the Arctic [1–3]. In 1981, over 90% of the Arctic pack ice was at least 10 years old, but by 2011 less than 10% was even 5 years old [4, 5]. The average lifespan of Arctic sea ice is now less than 2 years, and of Antarctic sea ice less than 1 year [5], so microbes entrained into the ice during freeze-up derive primarily from seawater, to which they ultimately return. Sea ice is characterized by strong gradients and highly variable conditions of temperature and salinity, which can range from 0 °C to below −50 °C and from 0 to 300 ‰, respectively [6].

Snow forms seasonally at high altitudes and high latitudes, but doesn't persist long as snow per se, as it either melts or is compressed and recrystallized into ice in the form of a glacier. Many bacteria nucleate ice formation and appear to produce specialized ice-nucleating proteins for this purpose; these nuclei are easily dispersed by wind [7–9]. The microbial ecology of snow is still not well known, but there is evidence that the metapopulation of microbes in snow derive from various source populations, including marine, terrestrial, and atmospheric environments [10, 11]. The concentration of microorganisms in snow is generally between 10^2 and 10^3 cells/ml, but can range from 10^4 to 10^5 cells/ml in saline snow and during snow melt [12–14].

Soil is a highly complex mixture of minerals, water, and biological matter, with a multitude of differences occurring due to stratigraphic, spatial, and temporal variations. When these diverse soils become frozen, they harbor similarly diverse microbial communities [15, 16]. Seasonally frozen ground, including the active layer overlying permafrost (generally comprising the uppermost 30–100 cm), is the region of ground that freezes in winter and thaws in the summer. In the polar regions, the active layer may be a relatively thin slice of the total frozen ground (which may reach to hundreds of meters deep). Overall, seasonally frozen ground covers a huge area, amounting to 50% of the non-glaciated land cover of the Northern hemisphere [17]. Seasonally frozen ground is subject to summer melt and thus the possibility exists for microbial dispersal with groundwater movement.

Seasonally frozen environments like sea ice, snow, and the active layer are likely to be prime sites for rapid microbial evolution in the cryosphere due to strong selective pressures combined with greater dispersal potential and faster generation times compared to perennially frozen environments. These environments might also act as 'training grounds' for the adaptation of microorganisms to perennially frozen environments.

3.1.1.2 Perennially frozen environments: glaciers, ice sheets, and permafrost

Glaciers are complex flowing masses of snow and ice that encompass a variety of microbial habitats, including surficial, englacial, and subglacial (basal) zones [18]. Seasonal communities of cryoconites, snow algae, and ice worms form on the surface of glaciers (and can be detected with remote sensing) [19, 20]. Depending on the size of the glacier, residence times for microbes entrained into the glacier may be decades to centuries. Microbial communities in subglacial lakes or rivers can be isolated from the surface for thousands of years [21, 22], and microbes accreted with sediment into basal ice can remain active at levels well above those of englacial communities [23, 24].

Ice sheets are large continental glaciers like those that currently cover Greenland and Antarctica. Consider the Greenland ice sheet, which grows by accumulation and compaction of snowfall over thousands of years, rising to a height of over 3,000 m. Of the snow-inhabiting microbes that fall onto and are frozen into the ice sheet, some cells may remain metabolically active (at very slow rates), while others form spores or resting stages that persist in the ice for many thousands of years [25, 26]. The relatively small amount of liquid water present within glacial ice ($\sim 0.0001\%$ compared to $\sim 1\%$ in sea ice) is found in micrometer-scale veins at ice crystal interfaces [27, 28]. Fluid flow is negligible through these veins in glacial ice [29], while temperature and metabolic limitations inhibit the growth and motility of microorganisms there [26–30], leading to limited internal dispersal once entrained into the glacier.

Depending on where it falls, snow deposited on an ice sheet may survive only seasonally (in the ablation zone) or be compressed into ice that survives for many thousands of years (in the accumulation zone). In Greenland, the oldest ice is over 100,000 years old and the average age of the ice is tens of thousands of years [31]. Antarctic glacier ice up to 8 million years old has been discovered [32], but on average the ice is much younger, around 100,000 years old [33]. Most of the individual cells that survive their glacial residence will end up in the ocean, either via glacial melt, runoff or iceberg calving. Of those cells that survive the journey to the sea, only a minute fraction can be expected to be re-entrained into snow. At a glacial abundance of only 10^2–10^3 per ml to begin with, far less than 1 cell per ml might be re-deposited onto a glacier to continue its evolutionary pathway.

Permafrost is another long-lived frozen environment contributing to the 'deep cold biosphere' [34, 35]. Defined as soil that has been frozen continuously for two or more years, permafrost is different from ice sheets in many ways, but similar life history characteristics and evolutionary processes might occur in the microbial communities of each. Permafrost, like an ice sheet, is stable over the course of hundreds of thousands or millions of years, with subsequently little or no dispersal of microorganisms during that time [16, 36]. Permafrost generally has orders of magnitude higher concentrations of cells than ice sheets, but both are host to viable organisms that survive their entombment [37].

The natural histories of microbes residing within perennially frozen environments vary considerably from those in seasonally frozen environments, and these differences may impact their evolution [38]. Perennially frozen environments act as long-lived repositories for ancient genes and genomes that no longer exist in the 'active' biosphere, which may have consequences on ecosystems as these environments continue melting [39]. Surprising findings are being made in the health sciences, as ancient permafrost bacteria provide insight into the evolution of antibiotic resistance genes since the rise of widespread antibiotic usage [40]. This repository also has astrobiological implications: since perennially frozen environments are favorable to the preservation of ancient microbial communities on Earth, they may also be critical environments in which to search for extraterrestrial life in Martian permafrost or the Europan ice shell, for example [37, 41, 42]. Icy environments like permafrost and subglacial lakes are thus used as 'analog environments' to prepare for eventual exploration of extraplanetary environments on Mars, Europa, Enceladus, and Titan [43–45].

3.1.2 Modes of evolution

The processes by which microorganisms evolve and colonize new environments are not well understood, and many fundamental questions remain: What (if anything) defines a microbial species, or a microbial population? Is there a 'tree' of life, a 'ring' of life, or a 'forest'? What are the most important abiotic and biotic selective factors? How fast are rates of mutation in the environment? What are the rates of horizontal gene transfer (HGT)? What mechanisms of HGT are most important? How do these rates and mechanisms vary with environmental conditions or with microbial community structure? What are the roles of viruses and predators? What is the role of the rare biosphere? These questions and others lead us forward in our attempts to quantitatively understand microbial evolution.

Like all known life on earth, microorganisms adapt by gradual change through mutation and natural selection; however, another mode of evolution called HGT is now known to strongly influence the evolution of microorganisms – in short, making genetic leaps where none were previously predicted [46, 47]. While mechanisms for HGT have been known for decades from laboratory investigations, the relative importance of genetic recombination via HGT in icy environments is essentially unknown.

3.1.2.1 Darwinian Processes

From the individual perspective, only actively reproducing cells can adapt to a new environment via positive selection (i.e. mutation followed by selection for an advantageous trait). However, from a broader perspective, a population of cells can evolve by negative selection, in which environmental conditions act as a filter to remove the unfit individuals. As an example, the latter mode might be expected to dominate within populations of cells deposited onto an ice sheet, a very harsh environment in which metabolic activity is confined to maintenance and repair.

If we consider the fixation of an advantageous allele into a microbial population as a signature of the adaptation of that population (e.g. positive selection during a selective sweep), then frozen environments may have lower rates of adaptation than warmer environments. Even isolates of psychrophiles, which by definition grow optimally below 15 °C and exclusively below 20 °C, tend to grow slowly at subzero temperatures with doubling times in the range of days to weeks [48–51]. Unlikely mutations, e.g. those involving multiple substitutions, may thus require many generations to occur, such that decreased growth rates at low temperatures act as impediments to the adaptation of microbes in the cryosphere.

Of course, temperature is not the only limiting factor in microbial growth, and may not even be the most important – nutrient and organic matter availability are also critical for determining growth rates [52, 53]. In this regard frozen environments vary widely, from oligotrophic environments with few nutrients, like snow, glacial ice, ice sheets, desert permafrost, and winter sea ice, to copiotrophic environments with many nutrients, like frozen foods, some permafrost, and highly productive summer sea ice.

In addition to slow rates of growth, dispersal can be highly limited in frozen environments. In particular, perennially frozen environments like permafrost and ice sheets, which persist over geological time scales, are poor environments for the dispersal of microbial lineages that have evolved an improved fitness under extreme conditions. The limited opportunities for microbes to disperse suggests that populations will be highly fragmented due to genetic drift, a hypothesis that is now feasible to test with next-generation sequencing techniques. Over time, the members of these communities might arrive at similar solutions to survival in an exhibition of convergent evolution. While this hypothesis also remains to be tested, a model for this type of evolution in the cryosphere is found in antifreeze proteins, which are distributed widely through the tree of life and share similar functions, but have profoundly different evolutionary histories [54].

Contributing to genetic drift, microbial communities may undergo partitioning or experience population bottlenecks when transitioning from non-frozen to frozen environments, or from seasonally to perennially frozen environments. For example, polar surface seawater generally contains ~ 10^5 cells/ml, which is reduced to ~ 10^4 cells/ml during entrainment into sea ice due to brine expulsion [55]. Likewise, seasonally frozen soils host bacterial abundances of 10^7–10^9 cells/g, while abundances in perennially frozen ground are reduced to 10^3–10^7 cells/g [36]. In permafrost-affected soils of northeast Greenland, bacterial direct counts dropped by 60–99% within the upper half meter of the surface [56]. These reductions may additionally be biased by selective mortality or physical enrichment, as in the case of EPS-producing bacteria which may physically associate with ice crystals [57], or gas vacuolate bacteria, which have been hypothesized to use positive buoyancy to selectively entrain into sea ice [58].

3.1.2.2 Horizontal gene transfer

The direct movement of genetic material from one organism to another in the absence of reproduction is called HGT. Complete genome sequencing of thousands of microorganisms confirms the pivotal role that HGT has played in the evolution of extant microbial genomes, including its important role in recombination [59, 60]. This genomic evidence leads to the conclusion that HGT is widespread among all three domains of life and intensive, with gene insertion and deletion events taking place roughly as often as nucleotide substitutions in bacterial genomes [61]. Although genome sequencing data indicate that HGT must occur in the natural environment, to date relatively few studies have investigated even the potential for HGT in a natural ecological setting [62–64].

Unknown to date is the relative importance of each of the known mechanisms of HGT on the evolution of microbial genomes [65], much less in frozen ecosystems. Three universal mechanisms of HGT, found in each of the domains of life, have been discovered: transformation is the direct uptake and integration of extracellular donor DNA by a 'competent' host cell [66, 67]; conjugation requires direct contact between donor and host cells [68] and is mediated by mobile genetic elements like plasmids; and transduction is based on phage transfer of genes from donor to host [69].

Each of these common mechanisms of HGT might play important roles in the evolution of microbes in the cryosphere, but as yet there is no data on the frequency of HGT in frozen environments. In laboratory experiments, the frequency of transformation is proportional to the abundance of naturally competent microorganisms and extracellular DNA [70], either or both of which are potentially concentrated within brine inclusions in sea ice or permafrost, for example. Experiments also demonstrate that transformation and transduction occur more frequently between closely related microorganisms, though the transferred DNA can theoretically originate from any organism in the environment, including other viruses [71]. While some bacteriophage may be generalists capable of infecting a wide range of hosts, most are probably fairly specific in their host requirements [72].

The proteins required for transformation and conjugation are encoded by known genes and can be studied with genomics, proteomics, and transcriptomics [73]. Although the diversity of viruses is at present too large to comprehensively investigate, genes and gene fragments related to insertion sequences, transposons, integrases, and other mobile genetic elements leave tell-tale traces of this process in most microbial genomes [74, 75], including genomes of ice-associated bacteria and archaea [76, 77]. Lysogenic phage or CRISPR elements incorporated into a host genome may tell a more complete story of viral involvement in genomic evolution [78, 79].

3.1.3 Adaptations to living with ice

While much focus has been placed on the evolution of cold tolerance in psychrophilic and psychrotolerant microbes, low temperature is only one of the challenges faced by organisms in icy environments. Low water activity, oxidative stress, low pH, and threats of physical disruption by ice crystals can also pose challenges in icy settings, defining inhabitants of these environments as polyextromphiles [80]. An open question is the extent to which ice-associated microbes might have adaptations that differ from microbes that are active at low temperature but which do not interact with ice, e.g. deep sea psychrophiles or pathogenic marine psychrophiles.

Common phenotypes observed in microbes adapted to life in frozen environments include the release of extracellular polymeric substances (EPS), accumulation of compatible solutes and efflux of salts, changes in the saturation state of membrane lipids, and the production of antifreeze proteins, ice-binding proteins, and cold shock proteins [81–84], many of which have biotechnological potential [85–87]. Because EPS and ice binding proteins are extracellular, and compatible solutes may escape from cells through lysis or leakage, these adaptations may afford protection to an unadapted fraction of a natural ice community as well [88–90].

Mutations that may act in aggregate to define a complex adaptive phenotype like low growth temperature include changes in amino acid composition, protein flexibility, and enzyme efficiency [91–93]. This chapter briefly describes some adaptations based on the evolution of genetic pathways, e.g. by insertion or deletion of genes, especially by HGT. For more detailed information on mutational processes that enable cold adaptation, the reader is referred to a number of useful literature reviews [94–97].

Compatible solutes are generally small neutral or zwitterionic molecules that are highly concentrated intracellularly to maintain turgor pressure in the presence of low temperatures or high salt concentrations [88, 98]. They can be synthesized de novo or imported from the medium. Some stabilize proteins and DNA, others protect DNA from cleavage by endonucleases [99].

EPS is a common, chemically complex material with a composition that varies among species, strains, and environmental conditions, though it is found in all three domains of life. The mechanism of action of EPS is not fully understood but it may depress the freezing point of a liquid and alter the microstructure of ice to prevent impingement of ice crystals onto cell surfaces [81, 100, 101]. The production of EPS is critical to biofilm formation in many environments; in the psychrophile *Colwellia psychrerythraea* 34H its production can be triggered by ice formation [102, 103].

Ice-binding proteins and antifreeze proteins are the subject of intense research by both ecologists and industrialists because of their interesting roles in depressing the freezing point of water and preventing the recrystallization of ice [104–106]. These proteins are diverse in origin, found in all three domains of life, and may have played a role in allowing ice algae to colonize ice [82, 107, 108].

3.2 Focus on sea ice

Sea ice is a particularly interesting environment in which to study microbial evolution because it is geographically extensive, has a large liquid water fraction, and hosts seasonally dense microbial populations compared to other parts of the cryosphere. In addition, sea ice is seasonally and geologically dynamic, and is relatively accessible for field work. Many of the challenges presented to microbes in sea ice are common to other elements of the cryosphere. Although the modes of evolution may vary from habitat to habitat, there are likely recurrent themes that are shared by all.

3.2.1 Sea ice characteristics

3.2.1.1 Physical properties shape the biological communities

Arctic and Antarctic sea ice cover about 15 million and 18 million square kilometers of ocean, respectively, at their maximal extent during the winter, providing an extensive (but shrinking) habitat for microorganisms [109]. Sea ice is a dynamic habitat, changing dramatically over the course of days, weeks, and seasons (▶ Fig. 3.1). The ice pack is also geographically variable, as Arctic and Antarctic sea ice behave differently at regional scales due to their opposing oceanographic contexts: Arctic sea ice is encircled by land, while Antarctic pack ice surrounds a continent and is bordered by the Southern Ocean to the north. At smaller scales, local climatological conditions affect the ice, e.g. heavy precipitation on Antarctic sea ice can lead to slushy ice and microbial assemblages on the surface of the ice that are rarely seen in Arctic sea ice [110]. It is not yet established whether the Arctic and Antarctic are geographically isolated enough to limit microbial dispersal, but there is some evidence of endemism within bacterial communities at the species level [76, 111].

First-year sea ice forms in the autumn as the air temperature begins to drop below the freezing point of seawater (−1 to −2 °C), cooling the surface layer of the ocean. Minute ice crystals form at the surface and can be mixed by wind and waves into the upper water column. As the crystals rise to the surface they scavenge and concentrate particles, including microbes. Phytoplankton and bacteria attached to the particles can be actively entrained into the growing ice, while other constituents like salts are passively incorporated from the seawater [112–114]. As the ice grows into a contiguous sheet, turbulent mixing is depressed in the water column and the ice begins to grow in

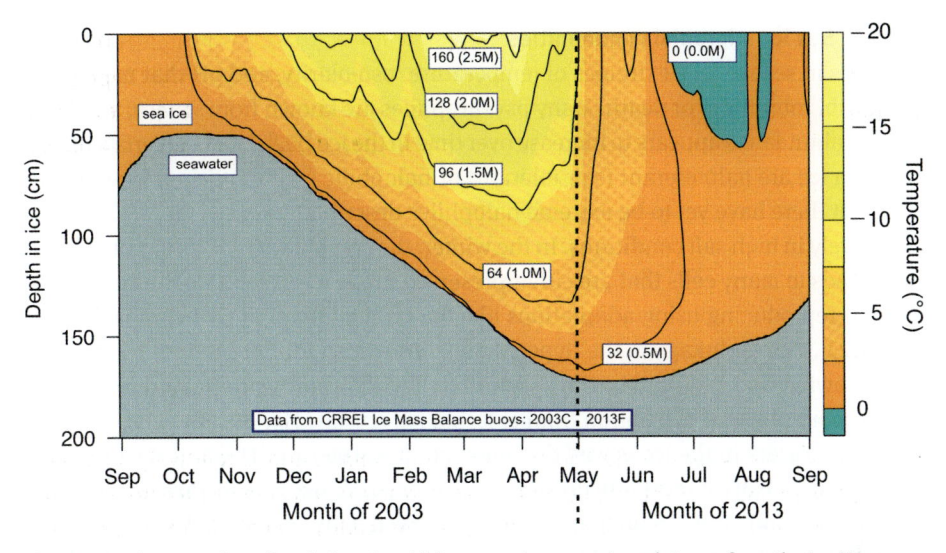

Fig. 3.1. Sea ice growth and evolution. Ice thickness and temperature data are from the Ice Mass Balance group at the Cold Regions Research and Engineering Laboratory, US Army [121]. Data from two buoys was combined to cover a yearly cycle of sea ice growth and decay. Color gradients depict changes in temperature; solid lines show depths of constant brine salinity (calculated from the temperature) [6], which are labeled with the salinity in parts per thousand (‰) and molarity (M).

a slower, more orderly fashion, as seawater freezes to the base of the ice sheet. Ice crystal growth excludes most non-water components so that salts, cells, viruses, inorganic particles, organic matter, and dissolved nutrients all become enriched in pore spaces between ice crystals called brine inclusions [115, 116]. The temperature of the ice determines both the brine volume fraction (porosity) and the salinity of the inclusions: colder ice has smaller, more saline inclusions than warmer ice. Salts and unattached cells (and presumably viruses) in the brine are drained from the growing ice as a result of brine expulsion. At colder temperatures, constriction of brine channels causes the cessation of brine movement near a bulk salinity of 5‰ and a temperature of −5 °C, at which point the porosity is about 5% [117]. At temperatures below this threshold, microbes inside the ice become 'trapped' and subject to the volatility of the harsh winter environment [55, 118]. The temperature in the uppermost ice can drop below −25 °C for days or weeks, with corresponding brine salinities above 225‰, nearly 7× the salinity of the underlying seawater. The base of the ice is nearly constantly at the freezing point of the seawater, creating a strong gradient of temperature and salinity within the ice, which may grow up to 2 meters in a season. Ice breakup proceeds when snow melt leads to the formation of melt ponds, inducing an overpressure that drives brine out of the ice. The melt ponds expand to cover the surface, and eventually melt through to seawater, leaving rotten ice that is disrupted and dispersed by wind and tide [119–121].

3.2.1.2 Sea ice microbial communities

Microbes in sea ice face extremes of temperature, osmolarity, and pH that may result in growth, inhibition, or death. Many psychrophiles are known from sea ice, and their relative abundance appears to increase over time in the ice [122]. Most bacteria isolated from sea ice are halotolerant (salt-tolerant) or halophilic (salt-requiring) to some degree, but there have yet to be extreme halophiles described from sea ice, which grow exclusively in high salt conditions. In the winter, the combination of cold, salty conditions leads to many cells that are either preserved in the ice or tolerate the conditions without reproducing using adaptations like the production of EPS or compatible solutes [123]. Cell shrinkage in the hyperosmotic brine may inhibit internal ice crystals from growing and perforating the cell wall, while EPS production could be a strategy to avoid physical disruption from external ice crystals [14, 89, 124, 125]. Lysis by protistan grazers in the ice is well documented in warmer ice [113] and is probably a viable mechanism for mortality down to −10 °C or so. Viruses are always found in sea ice [126–128] and viral production has been observed in incubations of natural sea ice brine communities at −7 to −11 °C [115, 126, 129]. Since it is the active fraction of the microbial population that is susceptible to phage infection, viral production also implies microbial activity under these extreme conditions. Evolutionary consequences of cell lysis in sea ice include the release of phage (potentially including transducing phage) and the release of nucleic acids that could be utilized as an energy source or for HGT via transformation [73].

Of the 10^4–10^6 bacteria and archaea per ml in polar seawater, the majority are likely to be adapted for pelagic growth, but some may be adapted for growth in sea ice and simply persist once they are released during ice melt. This hypothesis has not been tested quantitatively, but advances in sequencing technology are enabling deeper investigations of the 'rare biosphere' that should eventually answer this question. Multiyear sea ice may act as a refugium for microbes, particularly in the Arctic, where historically up to 50% of sea ice survives the summer melt, while less than 20% of Antarctic sea ice survives one summer [130–132]. However, large recent decreases in Arctic multiyear ice extent may foretell the demise of these refugia, which are acutely threatened by global warming [109, 133].

Microbial communities persist in sea ice throughout the harsh winter, with the main taxonomic groups varying by season. During winter the communities consist of clades like *SAR11* (*Pelagibacter*), *Thaumarchaeota* (formerly Marine Group I *Crenarchaeota*), Marine Group II *Euryarchaeota*, and *Polaribacter* [123, 134–136]. Many of these clades are oligotrophic and found ubiquitously in seawater around the world. Markedly different are spring and summer sea ice communities, which are typically dominated by sea ice specific psychrophiles among the *Alteromonadales* (e.g. *Colwellia*, *Glaciecola*), *Flavobacteriales* (e.g. *Polaribacter*, *Psychromonas*), and marine *Roseobacter* (e.g. *Octadecabacter*) [50, 137]. These copiotrophic 'r-strategists' are highly culturable and are commonly epiphytes (attached to algal cells) feeding off exudates from the abundant primary productivity in the lower layers of the ice [138, 139].

The presence of common seawater microbes in winter sea ice indicates that at least some of the cells are passively entrained during sea ice formation, becoming stuck in the ice along with salts and other impurities. However, potential mechanisms exist for active entrainment of ice-associated bacteria during the initial stages of sea ice growth, including gas vacuole formation, EPS attachment, and ice nucleation. While no bias towards these phenotypes has yet been proven, sensitive next-generation sequencing techniques will soon allow researchers to definitively demonstrate if and when these entrainment mechanisms are employed by marine bacteria [57, 58, 140].

If the extreme environmental conditions during winter act as a selective pressure, then common seawater microorganisms should preferentially die off over time, enriching the community for those microorganisms capable of surviving and multiplying in the ice, i.e. those with adaptations to the sea ice environment. Alternately, in the case that the winter conditions fail to impose a strong selective pressure on the community, the expectation is that no change in diversity should be detected, and the microorganisms would perhaps be best described as 'preserved.' In the only seasonal study of microbial abundance and diversity in Arctic sea ice using community fingerprinting approaches, the majority of seawater microorganisms frozen in from seawater in the autumn persisted through the winter [55]. These findings lead to the conclusion that rather than the extreme winter environment 'weeding out' the seawater-adapted fraction of the microbial community, the ice-adapted microbes may inadvertently aid the survival of others by the production of EPS, which was observed to take place throughout the winter and into the spring [135]. Under this hypothesis, only later in the season, with the return of the light, photosynthesis, and algal exudates, would the *SAR11*, *Thaumarchaeota*, and other oligotrophs be outcompeted.

Contrasting with the apparent preservative qualities of winter sea ice, frost flowers comprise a strongly selective habitat that forms on thin, cold sea ice [141, 142]. These 1–5 cm crystalline structures are formed by brine freezing during expulsion to the ice surface. Additional brine is wicked up the fine crystals, resulting in an extremely cold, salty habitat that may persist for several days before being blown away or covered by snow. The microbial communities associated with frost flowers are highly divergent from seawater communities, implying a significant selective effect due to the extremely low temperatures and high brine salinities associated with these ice structures. Frost flower communities are likely highly dispersible due to the flowers' frangibility and prominence above the ice, allowing intensive winter storms to carry entrained contaminants and microbes long distances in short periods of time.

At the approach of spring, highly enriched sea ice microbial communities form, commonly near the base of the ice, built upon a foundation of primary production by sea ice algae [143, 144]. Ice algal communities can be extraordinarily productive – within the ice sheet they are able to maintain themselves in a stable light environment relative to phytoplankton in the underlying seawater, and receive replenishing nutrients from seawater flushing the ice. While sea ice-specific diatoms typically dominate these communities, cyanobacterial sequences have recently been reported at both poles [130, 145]. Unique and highly culturable sea ice bacteria are often isolated from spring and summer sea ice [146, 147], which may persist until the breakup of the ice sheet in late summer. Archaea have been detected in summer sea ice in the Antarctic but not the Arctic [148].

3.2.2 Evolutionary modes in sea ice

Sea ice is a habitat that is well known for hosting populations of psychrophilic microorganisms, which grow profusely in spring and summer sea ice. Each of the adaptations to ice that were previously detailed have been identified in sea ice microorganisms, but the evolutionary processes that led to those adaptations have in most cases not been identified. Among other evidence, phylogenetic investigations into two of these adaptations – ice-binding proteins and compatible solute metabolism – have demonstrated that HGT is a dominant force in the evolution of sea ice microorganisms. These studies are described in more detail in the following section on genomics; here the focus is on the environmental conditions that might allow sea ice to be a hotspot for HGT.

3.2.2.1 Horizontal gene transfer

The potential for HGT in an environment depends on the presence of the necessary raw materials, and sea ice is replete with these materials. Transformation requires free DNA, either dissolved or bound, and transformation frequency is generally proportional to DNA concentration [149, 150]. Measurements of dissolved DNA concentration in sea ice range from 1 to 135 µg/l brine, on average much greater than in Arctic seawater (1–15 µg/l) or other cold waters, with mean values < 16 µg/l [116, 151]. Free DNA might accumulate in sea ice brines by concentration from the source seawater, but ice samples were frequently more enriched than expected assuming only passive entrainment, suggesting additional inputs due to cell lysis by phage or grazers, or the release of DNA-containing EPS [116]. Naturally competent cells (genetically and phenotypically able to take up DNA) are also required for transformation. A pair of studies from the early 1990s found that about 15% of a marine microbial community was naturally competent to take up plasmid DNA from a marine *Vibrio* strain, but comparable experiments have yet to be conducted in the cryosphere [152, 153].

Transduction requires the presence of phage, which have often been observed and enumerated in melted sea ice samples [115, 126, 154–156]. However, transducing phage (the agents of transduction) have yet to be identified from sea ice, though they are well known from the marine environment [72, 157]. Wells and Deming [129] first calculated contact rates between cells and viruses in sea ice, finding that they were 13–600 × higher at temperatures below −24 °C than in the underlying seawater, due to the concentration of brine by solute exclusion during freezing. Similar findings were reported by Collins and Deming [135], with the additional discovery of virus to bacteria ratios exceeding 1 000 in first year sea ice in the Beaufort Sea.

Conjugation requires a donor cell containing a mobile genetic element (i.e. a conjugative plasmid) and an adjacent recipient cell (lacking the mobile genetic element). In the only study of its kind to date, numerous plasmids were identified from cultured isolates of bacteria from Antarctic seawater, sediment, and sea ice, but conjugative capacity was not determined [158]. Reports of HGT in the environmental literature focus primarily on the terrestrial environment, in the contexts of genetically modified agricultural products and bioremediation. Studies of HGT in the marine environment are sparse, however, several microcosm studies have demonstrated that marine microbial communities can and do undergo gene exchange under permissible conditions, although the mechanisms are not always known [63, 71, 151, 157, 159, 160].

Surfaces associated with sea ice, including ice crystals, biota, or other entrained particles, are an important aspect of the microbial ecology in this frozen environment. Microbes attached to surfaces are more likely to be active in sea ice, and in very cold sea ice are the only active members [139]. Additionally, surface attachment often increases the frequency of gene transfer in microcosm and laboratory studies by stabilizing extracellular DNA, attracting high cell densities, and inducing natural competence [62, 161].

The rapid redistribution of microorganisms from sea ice also has potential evolutionary significance. Via aerial dispersal by frost flowers or direct transport by ocean currents, Arctic sea ice can distribute material throughout the Arctic Ocean, into the Northern Atlantic, and from there via deep thermohaline circulation throughout the world oceans [162]. This rapid and widespread dispersal contrasts with perennially frozen environments, and evolutionary processes enabled by high dispersal, like selective sweeps, may be detectable in these populations with next generation sequencing techniques [163, 164].

3.3 Ongoing work and future directions

3.3.1 Field work and experimentation

To experimentally investigate evolutionary processes on natural ice communities requires the isolation and characterization of new bacteria and archaea, new phage, and new mobile genetic elements like plasmids. Exciting advancements have been made in the development of new low-temperature model systems, including a high-frequency-of-transformation strain of *Pseudoalteromonas* [165], and psychrophilic phage-host systems, including the model gammaproteobacterium *C. psychrerythraea* 34H and its bacteriophage 9A [127, 166]. The genome sequence for bacterophage 9A has been reported but it has not been demonstrated to be transducing [167]. Significant advances in the understanding of HGT processes in frozen environments could be made if future cultivation efforts were focused on isolating transducing phage, psychrophilic conjugative plasmids, and naturally competent bacteria and archaea from the cryosphere. With these key pieces in hand, experimental methods like green fluorescent protein reporting could be applied *in situ* to detect HGT as it happens [160].

3.3.2 '-omics' in the cryosphere

The use of bioinformatics techniques to detect microbial evolution is quickly becoming the standard, but the application of these techniques to the cryosphere is still in the early stages. Approaches commonly used in molecular microbial ecology include barcoding, genomics, transcriptomics, and proteomics (and their 'meta-' instantiations), but due to the low biomass present in many frozen environments, microbiological approaches that utilize cultivation or nucleotide amplification (e.g. polymerase chain reaction or multiple displacement amplification) have been preferred to date. There are few published studies using metagenomic, metatranscriptomic, or metaproteomic approaches from icy environments, though relatively greater numbers of studies have utilized metabarcoding (generally using the 16S/18S ribosomal RNA gene) and genome sequencing of isolates from culture libraries, of which many isolates remain to be sequenced.

The genomes of many cold-adapted bacteria are now publicly available, but considering the extensive size and diversity of the cryosphere, the number of genomes lags behind other environments. Of more than 20,000 complete and draft microbial genomes available in public databases, only about 50 (0.25%) are derived from icy environments (▶ Fig. 3.2). While the coming tidal wave of genome sequences will enable more extensive analyses of evolution in frozen environments, a concerted effort should be made by the community to increase the presence of cryosphere-derived genomes in the public databases.

One application of genome sequencing is for the detection of HGT – if HGT occurs in the cryosphere, it should be evident in the genomes of microorganisms that live there. Genomic methods to identify HGT include analyzing %G+C nucleotide composition, codon usage patterns, and most convincingly, phylogenetic analysis to identify

▶ **Fig. 3.2.** The evolutionary relationships among 48 ice-associated microorganisms in the *hima* database (http://cryomics.org/hima). The region and habitat of isolation are shown, as are the minimum, optimum, and maximum growth temperatures reported in the literature, and the total genome size in megabasepairs. Single copy genes from each genome were extracted and translated into amino acid sequences, which were aligned, trimmed, and concatenated prior to maximum likelihood phylogenetic estimation. Thirty-one genes were used in the analysis: *dnaG, frr, infC, nusA, pgk, pyrG, rpmA, rpoB, rpsM, rpsS, smpB, tsf*, large subunit ribosomal proteins *rplA, B, C, D, E, F, K, L, M, N, P, S, T*, and small subunit ribosomal proteins *rpsB, C, E, I, J, K*.

3.3 Ongoing work and future directions — 45

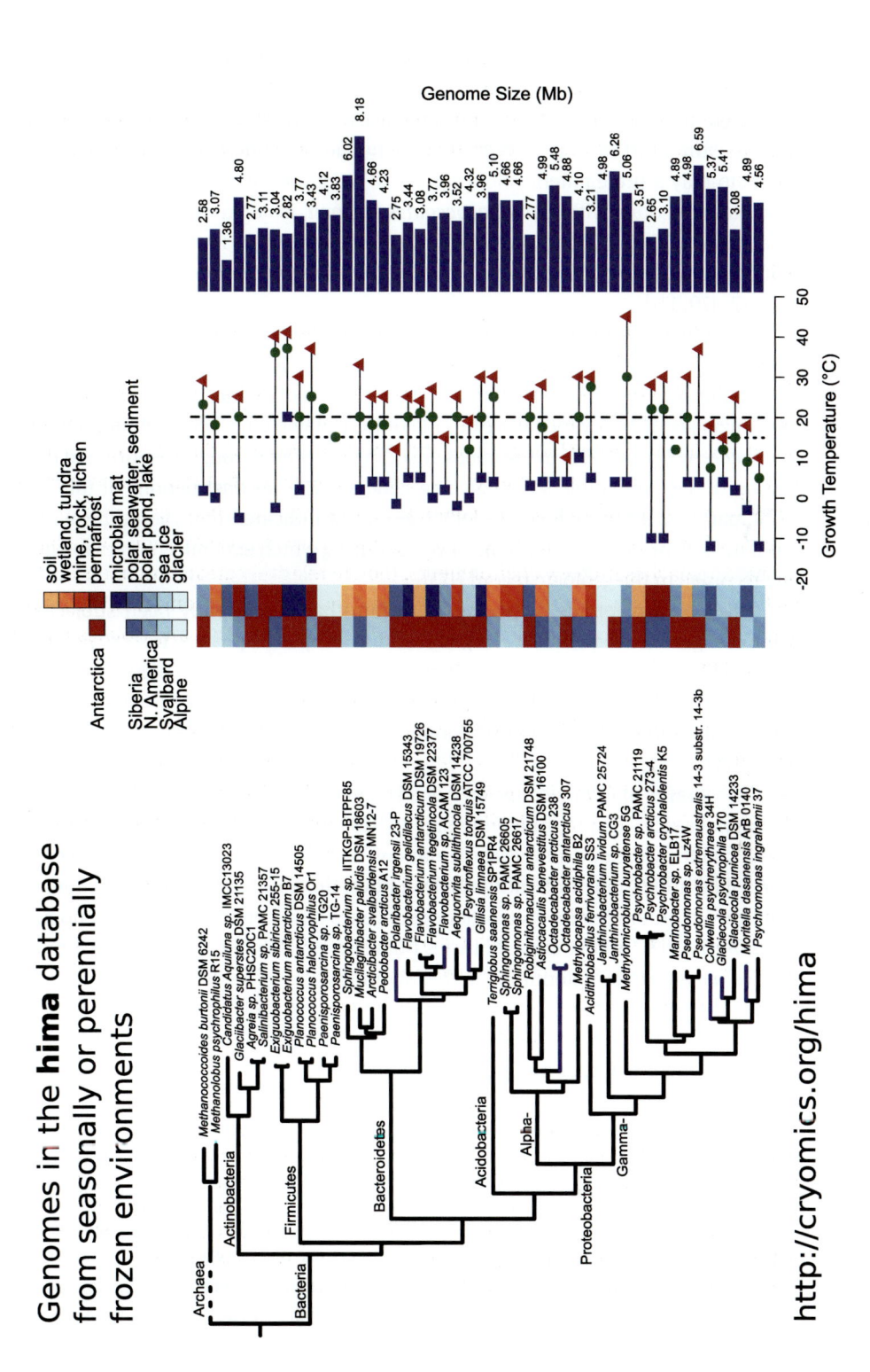

Genomes in the hima database from seasonally or perennially frozen environments

http://cryomics.org/hima

genes having inconsistent phylogeny with a conserved marker like the 16S ribosomal RNA gene [65, 168, 169].

There are now enough genome sequences of ice-associated microorganisms to start teasing apart the adaptations that define the cryospheric lifestyle, and HGT appears to play a pre-eminent role. Impressive work on the genomics of cold adaptation has been done with *Methanococcoides burtonii*, isolated from Ace Lake, Antarctica, showing that this archaeon has utilized extensive HGT in its adaptations to low temperatures, as well as a host of changes to gene expression and amino acid usage [77, 170, 171].

The genome of the model psychrophile *C. psychrerythraea* strain 34H also shows plentiful evidence of adaptation to cold environments, including genes involved in the maintenance of membrane fluidity and the synthesis of extracellular polysaccharides and enzymes [172]. A very clear example of HGT playing a role in cold-adaptation in *C. psychrerythraea* 34H is in the genes encoding the production, transport, and degradation of the compatible solute glycine betaine [173]. Reconstructions of this metabolic pathway using molecular clock techniques indicated that the pathways were built up over hundreds of millions of years, during which extreme changes in climate are recorded in the geologic record [173].

Psychroflexus torquis, a sea ice bacterium that has recently been subjected to complete genome sequencing [174], is found in a separate phylum from *C. psychrerythraea* 34H but uses similar adaptations, including the production of EPS, polyunsaturated fatty acids, and ice-binding proteins. Additionally, many of the genes encoding these adaptations are located on genomic islands, indicating xenologous origins via HGT [174].

Two strains of *Octadecabacter* – one isolated from Arctic sea ice, the other from Antarctic sea ice – display numerous signs of HGT in their genomic adaptation towards a sea ice lifestyle. Some of these adaptations include gas vesicle formation, compatible solute utilization, and, intriguingly, a novel proteorhodopsin that functions as a light-driven proton pump. Proteorhodopsin genes have been found in the genomes of *Polaribacter* [175] and in clone libraries from sea ice [147], but their specific utility in sea ice remains to be determined in the laboratory.

3.3.3 Linking phenotype and genotype

Placing phenotypic and ecological data into a phylogenomic context is a beginning towards a holistic view of the evolutionary forces at work in the adaptation of microbes in extreme environments. As a community we are a long way from predicting complex, quantitative phenotypes like optimal growth temperature from genomic information alone, thus there is still a strong need for phenotypic information in this field. One hundred years after Beijerinck ("everything is everywhere"), the modern flood of sequencing data has revolutionized the classically data-poor field of environmental microbiology. And yet, connecting this information with traditional phenotypic observations like those made by Baas Becking ("but the environment selects") has lagged behind the digital deluge. Traditionally this kind of data is compiled in *Bergey's Manual of Systematic Bacteriology*, but there is as yet no electronic database associated with that venerable publication. For example, no online database allows one to list all bacterial isolates known to grow at subzero temperature, much less to take the next step and link that trait to available phylogenomic data.

As a step towards a future where phenotypic, genomic, and phylogenetic information can be readily accessed for use in comparative genomics and predictive biology, I have undertaken to data mine the literature to compile a curated electronic database called *hima* linking these disparate datasets, using cold-active microbes as a testbed. Available parameters include sampling environment and location, type strain identification, optimal temperature and growth range, electron acceptors used, growth substrates used, genome sequences, gene clusters, mapped metagenomic reads, and various other traits. Currently, the database includes more than a dozen cold-active genera, encompassing 200 complete genome sequences and several hundred metagenomic and metabarcoding studies, including a number of transcriptomes and viromes. The *hima* database is navigable on the web (http://cryomics.org/hima) via a phylogenetic tree interface and will be updated into the future, as it is further expanded to include more analytical resources for use by researchers interested in the ecology and evolution of cold-loving microbes.

Acknowledgments

J. W. Deming and H. Eicken provided valuable input on a previous version of this manuscript. REC is supported by NSF PLR #1203267.

References

[1] C. E. Stickley, K. St John, N. Koç, et al. Evidence for middle Eocene Arctic sea ice from diatoms and ice-rafted debris. Nature 2009;460:376–379.
[2] L. Polyak, R. B. Alley, J. T. Andrews, et al. History of sea ice in the Arctic. Quaternary Sci Rev 2010;29:1757–1778.
[3] C. E. Stickley. Palaeoclimate: The sea ice thickens. Nature Geosci 2014;7:165–166.
[4] I. G. Rigor, J. M. Wallace. Variations in the age of Arctic sea-ice and summer sea-Ice extent. Geophys Res Lett 2004;31.
[5] J. Maslanik, J. Stroeve, C. Fowler, W. Emery. Distribution and trends in Arctic sea ice age through spring 2011. Geophys Res Lett 2011;38.
[6] G. F. N. Cox, W. F. Weeks. Changes in the salinity and porosity of sea-ice samples during shipping and storage. J Glaciol 1986;32:371–375.
[7] K. Junge, B. D. Swanson. High-resolution ice nucleation spectra of sea-ice bacteria: implications for cloud formation and life in frozen environments. Biogeosciences 2008;5:865–873.

[8] O. Möhler, D. G. Georgakopoulos, C. E. Morris, et al. Heterogeneous ice nucleation activity of bacteria: new laboratory experiments at simulated cloud conditions. Biogeosciences 2008;5:1425–1435.

[9] R. M. Bowers, S. McLetchie, R. Knight, N. Fierer. Spatial variability in airborne bacterial communities across land-use types and their relationship to the bacterial communities of potential source environments. ISME J 2011;5:601–612.

[10] K. A. Cameron, B. Hagedorn, M. Dieser, et al. Diversity and potential sources of microbiota associated with snow on western portions of the Greenland Ice Sheet. Environ Microbiol 2014:10.1111/462–2920.12446.

[11] C. Larose, S. Berger, C. Ferrari, et al. Microbial sequences retrieved from environmental samples from seasonal Arctic snow and meltwater from Svalbard, Norway. Extremophiles 2010;14:205–212.

[12] P. Amato, Hennebelle Rel, O. Magand, et al. Bacterial characterization of the snow cover at Spitzberg, Svalbard. FEMS Microbiol Ecol 2007;59:255–264.

[13] T. Harding, A. D. Jungblut, C. Lovejoy, W. F. Vincent. Microbes in High Arctic Snow and Implications for the Cold Biosphere. Appl Environ Microbiol 2011;77:3234–3243.

[14] M. Ewert, S. D. Carpenter, J. Colangelo-Lillis, J. W. Deming. Bacterial and extracellular polysaccharide content of brine-wetted snow over Arctic winter first-year sea ice. J Geophys Res-Oceans 2013;118:726–735.

[15] D. A. Gilichinsky, E. A. Vorobyova, L. G. Erokhina, D. G. Fyordorov-Dayvdov, N. R. Chaikovskaya. Long-term preservation of microbial ecosystems in permafrost. Adv Space Res 1992;12:255–263.

[16] J. K. Jansson, N. Taş. The microbial ecology of permafrost. Nat Rev Microbiol 2014.

[17] T. Zhang, R. G. Barry, K. Knowles, F. Ling, R. L. Armstrong. Distribution of seasonally and perennially frozen ground in the Northern Hemisphere. Proceedings of the 8th International Conference on Permafrost; 2003. p. 1289–1294.

[18] A. Hodson, A. M. Anesio, M. Tranter, et al. Glacial ecosystems. Ecol Monogr 2008;78:41–67.

[19] N. Takeuchi, R. Dial, S. Kohshima, T. Segawa, J. Uetake. Spatial distribution and abundance of red snow algae on the Harding Icefield, Alaska derived from a satellite image. Geophys Res Lett 2006;33.

[20] C. M. Foreman, B. Sattler, J. A. Mikucki, D. L. Porazinska, J. C. Priscu. Metabolic activity and diversity of cryoconites in the Taylor Valley, Antarctica. J Geophys Res-Biogeosciences 2007;112.

[21] B. C. Christner, M. L. Skidmore, J. C. Priscu, M. Tranter, C. M. Foreman. Bacteria in subglacial environments. In: R. Margesin, F. Schinner, J.-C. Marx, C. Gerday, eds. Psychrophiles: from biodiversity to biotechnology: Springer; 2008:51–71.

[22] B. C. Christner, J. C. Priscu, A. M. Achberger, et al. A microbial ecosystem beneath the West Antarctic ice sheet. Nature 2014;512:310–313.

[23] M. Skidmore, S. P. Anderson, M. Sharp, J. Foght, B. D. Lanoil. Comparison of microbial community compositions of two subglacial environments reveals a possible role for microbes in chemical weathering processes. Appl Environ Microbiol 2005;71:6986–6997.

[24] M. Tranter, M. Skidmore, J. Wadham. Hydrological controls on microbial communities in subglacial environments. Hydrol Process 2005;19:995–998.

[25] B. Lanoil, M. Skidmore, J. C. Priscu, et al. Bacteria beneath the West Antarctic ice sheet. Environ Microbiol 2009;11:609–615.

[26] P. B. Price, T. Sowers. Temperature dependence of metabolic rates for microbial growth, maintenance, and survival. Proc Natl Acad Sci USA 2004;101:4631–4636.

[27] P. B. Price. A habitat for psychrophiles in deep Antarctic ice. Proc Natl Acad Sci USA 2000;97:1247–1251.

[28] P. B. Price. Microbial life in glacial ice and implications for a cold origin of life. FEMS Microbiol Ecol 2007;59:217–231.
[29] A. G. Fountain, J. S. Walder. Water flow through temperate glaciers. Rev Geophys 1998;36:299–328.
[30] R. A. Rohde, P. B. Price. Diffusion-controlled metabolism for long-term survival of single isolated microorganisms trapped within ice crystals. Proc Natl Acad Sci USA 2007;104:16592–16597.
[31] Neem community members. Eemian interglacial reconstructed from a Greenland folded ice core. Nature 2013;493:489–494.
[32] D. E. Sugden, D. R. Marchant, N. Potter Jr, et al. Preservation of Miocene glacier ice in East Antarctica. Nature 1995;376:412–414.
[33] J.-R. Petit, J. Jouzel, D. Raynaud, et al. Climate and atmospheric history of the past 420,000 years from the Vostok ice core, Antarctica. Nature 1999;399:429–436.
[34] E. Vorobyova, V. Soina, M. Gorlenko, et al. The deep cold biosphere: facts and hypothesis. FEMS Microbiol Rev 1997;20:277–290.
[35] J. A. Mikucki, S. K. Han, B. D. Lanoil. Ecology of Psychrophiles: Subglacial and Permafrost Environments. Extremophiles Handbook 2011:755–775.
[36] D. Gilichinsky, T. Vishnivetskaya, M. Petrova, E. Spirina, V. Mamykin, E. Rivkina. Bacteria in permafrost. In: R. Margesin, F. Schinner, J.-C. Marx, C. Gerday, eds. Psychrophiles: from biodiversity to biotechnology: Springer; 2008:83–102.
[37] D. A. Gilichinsky, G. S. Wilson, E. I. Friedmann, et al. Microbial populations in Antarctic permafrost: biodiversity, state, age, and implication for astrobiology. Astrobiology 2007;7:275–311.
[38] A. M. Anesio, J. Laybourn-Parry. Glaciers and ice sheets as a biome. Trends Ecol Evol 2012;27:219–225.
[39] D. E. Graham, M. D. Wallenstein, T. A. Vishnivetskaya, et al. Microbes in thawing permafrost: the unknown variable in the climate change equation. ISME J 2011;6:709–712.
[40] G. Perron, L. Whyte, P. Turnbaugh, W. P. Hanage, G. Dantas, M. M. Desai. Functional characterization of Bacteria isolated from ancient Arctic soil exposes diverse resistance mechanisms to modern antibiotics. PLoS ONE 2013.
[41] T. M. Fisher, D. Schulze-Makuch. Nutrient and population dynamics in a subglacial reservoir: a simulation case study of the Blood Falls ecosystem with implications for astrobiology. Int J Astrobiology 2013;12:304–311.
[42] E. Gaidos, B. Lanoil, T. Thorsteinsson, et al. A viable microbial community in a subglacial volcanic crater lake, Iceland. Astrobiology 2004;4:327–344.
[43] J. C. Priscu, K. P. Hand. Microbial habitability of icy worlds. Microbe 2010;7:167–172.
[44] R. D. Lorenz, D. Gleeson, O. Prieto-Ballesteros, F. Gomez, K. Hand, S. Bulat. Analog environments for a Europa lander mission. Adv Space Res 2011;48:689–696.
[45] J. L. Heldmann, W. Pollard, C. P. McKay, et al. The high elevation Dry Valleys in Antarctica as analog sites for subsurface ice on Mars. Planet Space Sci 2013;85:53–58.
[46] E. V. Koonin, Y. I. Wolf. Is evolution Darwinian or/and Lamarckian? Biol Direct 2009;4:42.
[47] E. V. Koonin. Darwinian evolution in the light of genomics. Nucleic Acids Res 2009;37:1011–1034.
[48] J. M. Larkin, J. L. Stokes. Growth of psychrophilic microorganisms at subzero temperatures. Can J Microbiol 1968;14:97–101.
[49] R. Y. Morita. Psychrophilic bacteria. Bacteriol Rev 1975;39:144–167.
[50] E. Helmke, H. Weyland. Psychrophilic versus psychrotolerant bacteria–occurrence and significance in polar and temperate marine habitats. Cell Mol Biol 2004;50:553–561.

[51] S. J. Tuorto, P. Darias, L. R. McGuinness, et al. Bacterial genome replication at subzero temperatures in permafrost. ISME J 2013;8:139–149.

[52] W. J. Wiebe, W. M. Sheldon, L. R. Pomeroy. Evidence for an enhanced substrate requirement by marine mesophilic bacterial isolates at minimal growth temperatures. Microb Ecol 1993;25:151–159.

[53] L. R. Pomeroy, W. J. Wiebe. Temperature and substrates as interactive limiting factors for marine heterotrophic bacteria. Aquat Microb Ecol 2001;23:187–204.

[54] C.-H. C. Cheng. Evolution of the diverse antifreeze proteins. Curr Opin Genet Dev 1998;8:715–720.

[55] R. E. Collins, S. D. Carpenter, J. W. Deming. Spatial heterogeneity and temporal dynamics of particles, bacteria, and pEPS in Arctic winter sea ice. J Marine Syst 2008;74:902–917.

[56] L. Ganzert, F. Bajerski, D. Wagner. Bacterial community composition and diversity of five different permafrost-affected soils of Northeast Greenland. FEMS Microbiol Ecol 2014;89:426–441.

[57] M. Ewert, J. W. Deming. Selective retention in saline ice of extracellular polysaccharides produced by the cold-adapted marine bacterium Colwellia psychrerythraea strain 34H. Annal Glaciol 2011;52:111–117.

[58] J. J. Gosink, J. T. Staley. Biodiversity of gas vacuolate bacteria from Antarctic sea ice and water. Appl Environ Microbiol 1995;61:3486–3489.

[59] S. Garcia-Vallvé, A. Romeu, J. Palau. Horizontal gene transfer in bacterial and archaeal complete genomes. Genome Res 2000;10:1719–1725.

[60] K. T. Konstantinidis, E. F. DeLong. Genomic patterns of recombination, clonal divergence and environment in marine microbial populations. ISME J 2008;2:1052–1065.

[61] W. Hao, G. B. Golding. The fate of laterally transferred genes: life in the fast lane to adaptation or death. Genome Res 2006;16:636–643.

[62] M. Hermansson, C. Linberg. Gene transfer in the marine environment. FEMS Microbiol Ecol 1994;15:47–54.

[63] C. Dahlberg, M. Bergström, M. Andreasen, B. B. Christensen, S. Molin, M. Hermansson. Interspecies bacterial conjugation by plasmids from marine environments visualized by gfp expression. Mol Biol Evol 1998;15:385–390.

[64] A. Rizzi, A. Pontiroli, L. Brusetti, et al. Strategy for in situ detection of natural transformation-based horizontal gene transfer events. Appl Environ Microbiol 2008;74:1250–1254.

[65] J. R. Zaneveld, D. R. Nemergut, R. Knight. Are all horizontal gene transfers created equal? Prospects for mechanism-based studies of HGT patterns. Microbiology 2008;154:1–15.

[66] F. Griffith. Significance of pneumococcal types. J Hyg Cambridge 1928;27:113.

[67] O. T. Avery, C. M. MacLeod, M. McCarty. Studies on the chemical nature of the substance inducing transformation of pneumococcal types. I. Inductions of transformation by a desoxyribonucleic acid fraction isolated from Pneumococcus type III. J Exp Med 1944;79:137–157.

[68] J. Lederberg, E. L. Tatum. Gene Recombination in Escherichia coli. Nature 1946;158:558.

[69] N. D. Zinder, J. Lederberg. Genetic exchange in Salmonella. J Bacteriol 1952;64:679–699.

[70] M. E. Frischer, J. M. Thurmond, J. H. Paul. Factors affecting competence in a high-frequency of transformation marine Vibrio. J Gen Microbiol 1993;139:753–761.

[71] S. C. Jiang, J. H. Paul. Gene transfer by transduction in the marine environment. Appl Environ Microbiol 1998;64:2780–2787.

[72] K. E. Wommack, R. R. Colwell. Virioplankton: viruses in aquatic ecosystems. Microbiol Mol Biol Rev 2000;64:69–114.

[73] R. J. Redfield. Do bacteria have sex? Nature Rev Genet 2001;2:634–639.

[74] M. G. Weinbauer, F. Rassoulzadegan. Are viruses driving microbial diversification and diversity? Environ Microbiol 2004;6:1–11.

[75] J. L. Siefert. Defining the mobilome. Horizontal Gene Transfer: Springer; 2009:13–27.
[76] J. Vollmers, S. Voget, S. Dietrich, et al. Poles Apart: Arctic and Antarctic Octadecabacter strains share high genome plasticity and a new type of xanthorhodopsin. PLoS ONE 2013;8:e63422.
[77] M. A. Allen, F. M. Lauro, T. J. Williams, et al. The genome sequence of the psychrophilic archaeon, Methanococcoides burtonii: the role of genome evolution in cold adaptation. ISME J 2009;3:1012–1035.
[78] R. E. Anderson, W. J. Brazelton, J. A. Baross. Using CRISPRs as a metagenomic tool to identify microbial hosts of a diffuse flow hydrothermal vent viral assemblage. FEMS Microbiol Ecol 2011;77:120–133.
[79] M. T. Cottrell, D. L. Kirchman. Virus genes in Arctic marine bacteria identified by metagenomic analysis. Aquat Microb Ecol 2012;66:107–116.
[80] J. Goordial, G. Lamarche-Gagnon, C.-Y. Lay, L. Whyte. Left Out in the Cold: Life in Cryoenvironments. Polyextremophiles: Springer; 2013:335–363.
[81] C. Krembs, J. W. Deming. The role of exopolymers in microbial adaptation to sea ice. In: R. Margesin, F. Schinner, J.-C. Marx, C. Gerday, eds. Psychrophiles: from Biodiversity to Biotechnology: Springer-Verlag; 2008:247–264.
[82] M. Bayer-Giraldi, I. Weikusat, Besir Hs, G. Dieckmann. Characterization of an antifreeze protein from the polar diatom Fragilariopsis cylindrus and its relevance in sea ice. Cryobiol 2011;63:210–219.
[83] H. H. Wemekamp-Kamphuis, R. D. Sleator, J. A. Wouters, C. Hill, T. Abee. Molecular and physiological analysis of the role of osmolyte transporters BetL, Gbu, and OpuC in growth of Listeria monocytogenes at low temperatures. Appl Environ Microbiol 2004;70:2912–2918.
[84] H. A. Thieringer, P. G. Jones, M. Inouye. Cold shock and adaptation. BioEssays 1998;20:49–57.
[85] B. C. Christner. Bioprospecting for microbial products that affect ice crystal formation and growth. Appl Microbiol Biotechnol 2010;85:481–489.
[86] R. Cavicchioli, T. Charlton, H. Ertan, S. M. Omar, K. S. Siddiqui, T. J. Williams. Biotechnological uses of enzymes from psychrophiles. Micro Biotech 2011;4:449,60.
[87] D. de Pascale, C. D. Santi, J. Fu, B. Landfald. The microbial diversity of Polar environments is a fertile ground for bioprospecting. Mar Genom 2012;8:15–22.
[88] D. T. Welsh. Ecological significance of compatible solute accumulation by micro-organisms: from single cells to global climate. FEMS Microbiol Rev 2000;24:263–290.
[89] C. Krembs, H. Eicken, J. W. Deming. Exopolymer alteration of physical properties of sea ice and implications for ice habitability and biogeochemistry in a warmer Arctic. Proc Natl Acad Sci USA 2011;108:3653–3658.
[90] M. Ewert, J. W. Deming. Sea ice microorganisms: Environmental constraints and extracellular responses. Biol 2013;2:603–628.
[91] A. L. Huston, B. B. Krieger-Brockett, J. W. Deming. Remarkably low temperature optima for extracellular enzyme activity from Arctic bacteria and sea ice. Environ Microbiol 2000;2:383–388.
[92] C. Bakermans, P. W. Bergholz, Ayala-del-Río oH, J. Tiedje. Genomic insights into cold adaptation of permafrost bacteria. Permafrost soils: Springer; 2009:159–168.
[93] A. Casanueva, M. Tuffin, C. Cary, D. A. Cowan. Molecular adaptations to psychrophily: the impact of 'omic' technologies. Trends Microbiol 2010;18:374–381.
[94] S. D'Amico, P. Claverie, T. Collins, et al. Molecular basis of cold adaptation. Philos Trans R Soc London 2002;357:917–925.
[95] G. Feller, C. Gerday. Psychrophilic enzymes: hot topics in cold adaptation. Nat Rev Microbiol 2003;1:200–208.

[96] J. C. Marx, T. Collins, S. D'Amico, G. Feller, C. Gerday. Cold-adapted enzymes from marine Antarctic microorganisms. Mar Biotechnol 2007;9:293–304.
[97] F. M. Lauro, M. A. Allen, D. Wilkins, T. J. Williams, R. Cavicchioli. Psychrophiles: genetics, genomics, evolution. Extremophiles Handbook 2011:865–890.
[98] E. A. Galinski. Compatible solutes of halophilic eubacteria: molecular principles, water-solute interaction, stress protection. Experientia 1993;49:487–496.
[99] M. F. Roberts. Organic compatible solutes of halotolerant and halophilic microorganisms. Saline Systems 2005;1:5.
[100] K. Meiners, R. Gradinger, J. Fehling, G. Civitarese, M. Spindler. Vertical distribution of exopolymer particles in sea ice of the Fram Strait (Arctic) during autumn. Mar Ecol Prog Ser 2003;248:1–13.
[101] D. N. Thomas, R. J. Lara, H. Eicken, G. Kattner, A. Skoog. Dissolved organic matter in Arctic multiyear sea ice during winter: Major components and relationship to ice characteristics. Polar Biol 1995;15:477–483.
[102] M. Ewert, J. W. Deming. Bacterial responses to fluctuations and extremes in temperature and brine salinity at the surface of Arctic winter sea ice. FEMS Microbiol Ecol 2014.
[103] J. G. Marx, S. D. Carpenter, J. W. Deming. Production of cryoprotectant extracellular polysaccharide substances (EPS) by the marine psychrophilic bacterium Colwellia psychrerythraea strain 34H under extreme conditions. Can J Microbiol 2009;55:63–72.
[104] M. G. Janech, A. Krell, T. Mock, J. S. Kang, J. A. Raymond. Ice-binding proteins from sea ice diatoms (Bacillariophyceae). J Phycol 2006;42:410–416.
[105] J. A. Raymond, C. A. Knight. Ice binding, recrystallization inhibition, and cryoprotective properties of ice-active substances associated with Antarctic sea ice diatoms. Cryobiol 2003;46:174–181.
[106] J. A. Raymond, C. Fritsen, K. Shen. An ice-binding protein from an Antarctic sea ice bacterium. FEMS Microbiol Ecol 2007;61:214–221.
[107] J. A. Raymond, H. J. Kim. Possible Role of Horizontal Gene Transfer in the Colonization of Sea Ice by Algae. PLoS ONE 2012;7:e35968.
[108] M. Bayer-Giraldi, C. Uhlig, U. John, T. Mock, K. Valentin. Antifreeze proteins in polar sea ice diatoms: diversity and gene expression in the genus Fragilariopsis. Environ Microbiol 2010;12:1041–1052.
[109] J. C. Stroeve, M. C. Serreze, M. M. Holland, J. E. Kay, J. Malanik, A. P. Barrett. The Arctic's rapidly shrinking sea ice cover: a research synthesis. Climatic Change 2012;110:1005–1027.
[110] R. Horner, S. F. Ackley, G. S. Dieckmann, et al. Ecology of sea ice biota– 1. Habitat, terminology, and methodology. Polar Biol 1992;12:417–427.
[111] J. T. Staley, J. J. Gosink. Poles Apart: biodiversity and biogeography of sea ice bacteria. Annu Rev Microbiol 1999;53:189–215.
[112] C. Petrich, H. Eicken. Growth, Structure and Properties of Sea Ice. In: D. N. Thomas, G. S. Dieckmann, eds. Sea Ice: An Introduction to its Physics, Chemistry and Biology, 2nd Ed: Wiley-Blackwell Oxford; 2010:23–77.
[113] R. Gradinger, J. Ikavalko. Organism incorporation into newly forming Arctic sea ice in the Greenland Sea. J Plankton Res 1998;20:871–886.
[114] A. Riedel, C. Michel, M. Gosselin, B. LeBlanc. Enrichment of nutrients, exopolymeric substances and microorganisms in newly formed sea ice on the Mackenzie shelf. Mar Ecol Prog Ser 2007;342:55–67.
[115] R. E. Collins, J. W. Deming. Abundant dissolved genetic material in Arctic sea ice Part I: Extracellular DNA. Polar Biol 2011;34:1819–1830.
[116] R. E. Collins, J. W. Deming. Abundant dissolved genetic material in Arctic sea ice Part II: Viral dynamics during autumn freeze-up. Polar Biol 2011;34:1831–1841.

[117] K. M. Golden, S. F. Ackley, V. I. Lytle. The percolation phase transition in sea ice. Science 1998;282:2238–2241.
[118] D. Delille, C. Rosiers. Seasonal changes in microbial biomass in the first-year ice of the Terre Adelie area (Antarctica). Aquat Microb Ecol 2002;28:257–265.
[119] R. Brinkmeyer, F.-O. Glöckner, E. Helmke, R. Amann. Predominance of beta-proteobacteria in summer melt pools on Arctic pack. Limnol Oceanogr 2004;49:1013–1021.
[120] M. Kramer, R. Kiko. Brackish meltponds on Arctic sea ice: a new habitat for marine metazoans. Polar Biol 2011;34:603–608.
[121] D. Perovich, J. Richter-Menge, B. Elder, T. Arbetter, K. Claffey, C. Polashenski. Observing and understanding climate change: Monitoring the mass balance, motion, and thickness of Arctic sea ice. 2009, http://imb.erdc.dren.mil/pubs.htm.
[122] D. Delille, C. Rosiers. Seasonal changes of Antarctic marine bacterioplankton and sea ice bacterial assemblages. Polar Biol 1996;16:27–34.
[123] K. Junge, H. Eicken, J. W. Deming. Bacterial activity at −2 to −20 degrees C in Arctic wintertime sea ice. Appl Environ Microbiol 2004;70:550–557.
[124] K. Meiners, R. Brinkmeyer, M. A. Granskog, A. Lindfors. Abundance, size distribution and bacterial colonization of exopolymer particles in Antarctic sea ice (Bellingshausen Sea). Aquat Microb Ecol 2004;35:283–296.
[125] S. N. Aslam, G. J. C. Underwood, H. Kaartokallio, et al. Dissolved extracellular polymeric substances (dEPS) dynamics and bacterial growth during sea ice formation in an ice tank study. Polar Biol 2012;35:661–676.
[126] R. Maranger, D. F. Bird, S. K. Juniper. Viral and bacterial dynamics in Arctic sea-ice during the spring algal bloom near Resolute, NWT, Canada. Mar Ecol Prog Ser 1994;111:121–127.
[127] L. E. Wells, J. W. Deming. Characterization of a cold-active bacteriophage on two psychrophilic marine hosts. Aquat Microb Ecol 2006;45:15–29.
[128] A. M. Anesio, C. M. Bellas. Are low temperature habitats hot spots of microbial evolution driven by viruses? Trends Microbiol 2011;19:52–57.
[129] L. E. Wells, J. W. Deming. Modelled and measured dynamics of viruses in Arctic winter sea-ice brines. Environ Microbiol 2006;8:1115–1121.
[130] J. S. Bowman, S. Rasmussen, N. Blom, J. W. Deming, Rysgaard Sor, T. Sicheritz-Ponten. Microbial community structure of Arctic multiyear sea ice and surface seawater by 454 sequencing of the 16S rRNA gene. ISME J 2011;6:11–20.
[131] R. Gradinger, Q. Zhang. Vertical distribution of bacteria in Arctic sea ice from the Barents and Laptev Seas. Polar Biol 1997;17:448–454.
[132] I. Hatam, R. Charchuk, B. Lange, J. Beckers, C. Haas, B. Lanoil. Distinct bacterial assemblages reside at different depths in Arctic multiyear sea ice. FEMS Microbiol Ecol 2014.
[133] A. M. Comeau, W. K. W. Li, J.-E. Tremblay, E. C. Carmack, C. Lovejoy. Arctic Ocean microbial community structure before and after the 2007 record sea ice minimum. PLoS ONE 2011;6:e27492.
[134] E. Helmke, H. Weyland. Bacteria in sea ice and underlying water of the Eastern Weddell Sea in midwinter. Mar Ecol Prog Ser 1995;117:269–287.
[135] R. E. Collins, G. Rocap, J. W. Deming. Persistence of bacterial and archaeal communities in sea ice through an Arctic winter. Environ Microbiol 2010;12:1828–1841.
[136] J. S. Bowman, C. T. Berthiaume, E. Armbrust, J. W. Deming. The genetic potential for key biogeochemical processes in Arctic frost flowers and young sea ice revealed by metagenomic analysis. FEMS Microbiol Ecol 2014.
[137] K. Junge, B. Christner, J. T. Staley. Diversity of Psychrophilic Bacteria from Sea Ice-and Glacial Ice Communities. Extremophiles Handbook: Springer; 2011:793–815.

[138] R. E. H. Smith, P. Clement, G. F. Cota. Population dynamics of bacteria in Arctic sea ice. Microbial Ecol 1989;17:63–76.
[139] K. Junge, F. Imhoff, T. Staley, J. W. Deming. Phylogenetic diversity of numerically important Arctic sea-ice bacteria cultured at subzero temperature. Microb Ecol 2002;43:315–328.
[140] E. Eronen-Rasimus, H. Kaartokallio, C. Lyra, et al. Bacterial community dynamics and activity in relation to dissolved organic matter availability during sea-ice formation in a mesocosm experiment. MicrobiologyOpen 2014;3:139–156.
[141] J. S. Bowman, C. Larose, T. M. Vogel, J. W. Deming. Selective occurrence of Rhizobiales in frost flowers on the surface of young sea ice near Barrow, Alaska and distribution in the polar marine rare biosphere. Env Microbiol Rep 2013;5:575,82.
[142] J. S. Bowman, J. W. Deming. Elevated bacterial abundance and exopolymers in saline frost flowers and implications for atmospheric chemistry and microbial dispersal. Geophys Res Lett 2010;37:L13501.
[143] K. R. Arrigo, T. Mock, M. P. Lizotte. Primary Producers and Sea Ice. In: D. N. Thomas, G. S. Dieckmann, eds. Sea Ice: An Introduction to its Physics, Chemistry and Biology, 2nd Ed: Wiley-Blackwell, Oxford, UK; 2010:283–325.
[144] A. Boetius, S. Albrecht, K. Bakker, et al. Export of algal biomass from the melting Arctic sea ice. Science 2013;339:1430–1432.
[145] E. Y. Koh, R. O. M. Cowie, A. M. Simpson, R. O'Toole, K. G. Ryan. The origin of cyanobacteria in Antarctic sea ice: marine or freshwater? Env Microbiol Rep 2012;4:479–483.
[146] J. W. Deming. Sea Ice Bacteria and Viruses. In: D. N. Thomas, G. S. Dieckmann, eds. Sea Ice: An Introduction to its Physics, Chemistry and Biology, 2nd Ed: Wiley-Blackwell Oxford; 2010:247–282.
[147] E. Y. Koh, N. Atamna-Ismaeel, A. Martin, et al. Proteorhodopsin-bearing bacteria in Antarctic sea ice. Appl Environ Microbiol 2010;76:5918–5925.
[148] R. O. M. Cowie, E. W. Maas, K. G. Ryan. Archaeal diversity revealed in Antarctic sea ice. Ant Sci 2011;23:531.
[149] M. F. DeFlaun, J. H. Paul, W. H. Jeffrey. Distribution and molecular weight of dissolved DNA in subtropical estuarine and oceanic environments. Mar Ecol Prog Ser 1987;48:65–73.
[150] J. Sikorski, S. Graupner, M. G. Lorenz, W. Wackernagel. Natural genetic transformation of Pseudomonas stutzeri in a non-sterile soil. Microbiology 1998;144:569–576.
[151] M. Dröge, A. Pühler, W. Selbitschka. Horizontal gene transfer among bacteria in terrestrial and aquatic habitats as assessed by microcosm and field studies. Biol Fertil Soils 1999;29:221–245.
[152] M. E. Frischer, J. M. Thurmond, J. H. Paul. Natural plasmid transformation in a high-frequency-of-transformation marineVibrio strain. Appl Environ Microbiol 1990;56:3439–3444.
[153] M. E. Frischer, G. J. Stewart, J. H. Paul. Plasmid transfer to indigenous marine bacterial populations by natural transformation. FEMS Microbiol Ecol 1994;15:127–135.
[154] M. M. Gowing, B. E. Riggs, D. L. Garrison, A. H. Gibson, M. O. Jeffries. Large viruses in Ross Sea late autumn pack ice habitats. Mar Ecol Prog Ser 2002;241:1–11.
[155] M. M. Gowing, D. L. Garrison, A. H. Gibson, J. M. Krupp, M. O. Jeffries, C. H. Fritsen. Bacterial and viral abundance in Ross Sea summer pack ice communities. Mar Ecol Prog Ser 2004;279:3–12.
[156] L. E. Wells, J. W. Deming. Effects of temperature, salinity and clay particles on inactivation and decay of cold-active marine Bacteriophage 9A. Aquat Microb Ecol 2006;45:31–39.
[157] J. A. Baross, J. Liston, R. Y. Morita. Incidence of Vibrio parahaemolyticus bacteriophages and other Vibrio bacteriophages in marine samples. Appl Environ Microbiol 1978;36:492–499.
[158] H. Kobori, C. W. Sullivan, H. Shizuya. Bacterial plasmids in Antarctic natural microbial assemblages. Appl Environ Microbiol 1984;48:515–518.

[159] M. G. Lorenz, W. Wackernagel. Bacterial gene transfer by natural genetic transformation in the environment. Microbiol Rev 1994;58:563–602.

[160] C. Dahlberg, M. Bergström, M. Hermansson. In situ detection of high levels of horizontal plasmid transfer in marine bacterial communities. Appl Environ Microbiol 1998;64:2670–2675.

[161] K. L. Meibom, M. Blokesch, N. A. Dolganov, C. Y. Wu, G. K. Schoolnik. Chitin induces natural competence in Vibrio cholerae. Science 2005;310:1824–1827.

[162] H. Eicken, R. Gradinger, A. Gaylord, A. Mahoney, I. Rigor, H. Melling. Sediment transport by sea ice in the Chukchi and Beaufort Seas: Increasing importance due to changing ice conditions? Deep-Sea Res Pt II 2005;52:3281–3302.

[163] F. M. Cohan. Bacterial species and speciation. Syst Biol 2001;50:513–524.

[164] C. L. Nesbø, Y. Boucher, M. Dlutek, W. F. Doolittle. Lateral gene transfer and phylogenetic assignment of environmental fosmid clones. Environ Microbiol 2005;7:2011–2026.

[165] A. M. Cusano, E. Parrilli, G. Marino, M. L. Tutino. A novel genetic system for recombinant protein secretion in the Antarctic Pseudoalteromonas haloplanktis TAC125. Microb Cell Fact 2006;5:40.

[166] M. Borriss, E. Helmke, R. Hanschke, T. Schweder. Isolation and characterization of marine psychrophilic phage-host systems from Arctic sea ice. Extremophiles 2003;7:377–384.

[167] J. R. Colangelo-Lillis, J. W. Deming. Genomic analysis of cold-active Colwelliaphage 9A and psychrophilic phage–host interactions. Extremophiles 2013;17:99–114.

[168] D. Cortez, L. Delaye, A. Lazcano, A. Becerra. Composition-based methods to identify horizontal gene transfer. Horizontal Gene Transfer: Springer; 2009:215–225.

[169] M. S. Poptsova, J. P. Gogarten. The power of phylogenetic approaches to detect horizontally transferred genes. BMC Evol Biol 2007;7:45.

[170] D. S. Nichols, M. R. Miller, N. W. Davies, A. Goodchild, M. Raftery, R. Cavicchioli. Cold adaptation in the Antarctic archaeon Methanococcoides burtonii involves membrane lipid unsaturation. J Bacteriol 2004;186:8508–8515.

[171] S. Campanaro, T. J. Williams, D. W. Burg, et al. Temperature-dependent global gene expression in the Antarctic archaeon Methanococcoides burtonii. Method Enzymol 2011;13:2018–2038.

[172] B. A. Methé, K. E. Nelson, J. W. Deming, et al. The psychrophilic lifestyle as revealed by the genome sequence of Colwellia psychrerythraea 34H through genomic and proteomic analyses. Proc Natl Acad Sci USA 2005;102:10913–10918.

[173] R. E. Collins, J. W. Deming. An inter-order horizontal gene transfer event enables the catabolism of compatible solutes by Colwellia psychrerythraea 34H. Extremophiles 2013;17:601–610.

[174] S. Feng, S. M. Powell, R. Wilson, J. P. Bowman. Extensive gene acquisition in the extremely psychrophilic bacterial species Psychroflexus torquis and the link to sea-ice ecosystem specialism. Genome Biol Evol 2014;6:133–148.

[175] J. M. González, B. Fernández-Gómez, A. Fernàndez-Guerra, et al. Genome analysis of the proteorhodopsin-containing marine bacterium Polaribacter sp. MED152 (Flavobacteria). Proc Natl Acad Sci USA 2008;105:8724–8729.

Maximiliano J. Amenabar, Matthew R. Urschel, and Eric S. Boyd
4 Metabolic and taxonomic diversification in continental magmatic hydrothermal systems

4.1 Introduction

Hydrothermal systems integrate geological processes from the deep crust to the Earth's surface yielding an extensive array of spring types with an extraordinary diversity of geochemical compositions. Such geochemical diversity selects for unique metabolic properties expressed through novel enzymes and functional characteristics that are tailored to the specific conditions of their local environment. This dynamic interaction between geochemical variation and biology has played out over evolutionary time to engender tightly coupled and efficient biogeochemical cycles. The timescales by which these evolutionary events took place, however, are typically inaccessible for direct observation. This inaccessibility impedes experimentation aimed at understanding the causative principles of linked biological and geological change unless alternative approaches are used. A successful approach that is commonly used in geological studies involves comparative analysis of spatial variations to test ideas about temporal changes that occur over inaccessible (i.e. geological) timescales. The same approach can be used to examine the links between biology and environment with the aim of reconstructing the sequence of evolutionary events that resulted in the diversity of organisms that inhabit modern day hydrothermal environments and the mechanisms by which this sequence of events occurred. By combining molecular biological and geochemical analyses with robust phylogenetic frameworks using approaches commonly referred to as phylogenetic ecology [1, 2], it is now possible to take advantage of variation within the present – the distribution of biodiversity and metabolic strategies across geochemical gradients – to recognize the extent of diversity and the reasons that it exists.

The distribution of organisms and their metabolic functions in modern environments is rooted in the selection for physiological adaptations that allowed variant populations to radiate into new ecological niches. Such radiation events are recorded in extant organismal distribution patterns (e.g. habitat range), as well as in the genetic record of the organisms as they are distributed along environmental gradients [3, 4]. These patterns are the result of vertical descent, which is a primary means by which organisms inherit their metabolic or physiological potential from their ancestors, although horizontal gene transfer (HGT), gene loss, and gene fusions are also likely to influence extant distribution patterns. Overall, these phenomena manifest as a posi-

---

Maximiliano J. Amenabar, Matthew R. Urschel: These authors contributed equally to this work

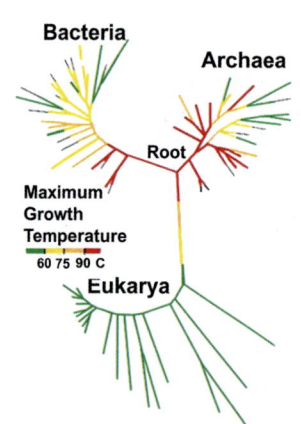

Fig. 4.1. Universal tree of life, overlain with maximum growth temperature, implies a thermophilic origin of life (adapted from Lineweaver and Schwartzman, 2004 [14]).

tive relationship between the physiological, ecological, and evolutionary relatedness of organisms [1, 2, 5, 6]. The correlation between the phylogenetic relatedness of microbial taxa and their overall ecological similarity allows an analysis of the overall phylogenetic relatedness and structure of communities of interacting organisms to be used to investigate the contemporary ecological processes that structure their composition [1]. The aforementioned attributes of biological systems offer 'a window into the past' as extant patterns in the distribution of species or metabolic function can be used to infer historical constraints imposed by the environment on the diversification of species and/or metabolic function [7–13].

Most phylogenetic studies conducted on early life have used gene or protein sequence data obtained from culture collections making it difficult to place results in an ecological context. For example, phylogenetic reconstructions of contemporary genes or proteins indicate that the first forms of life (▶ Fig. 4.1) [14] and many of the physiological functions that sustained it (e.g. hydrogen oxidation [15]) may have a hydrothermal heritage. Additional evidence in support of a thermophilic character for early forms of life for this suggestion comes from an analysis of resurrected proteins that are likely to share attributes of the last universal common ancestor (LUCA) of Archaea and Bacteria. Biochemical characterization of these proteins indicates that they function more efficiently at elevated temperatures than they do at mesophilic temperature [16]. This is consistent with the elevated temperatures predicted by silicon and oxygen isotope data for near surface environments in Archean oceans based on silicon and oxygen isotopic data (55 °C to 85 °C) [17–19]. However, more recent studies based on phosphate isotopic data suggest that Archean oceans were clement (< 40 °C) [20]. While this seems contradictory to the aforementioned phylogenetic and molecular reconstruction data, it is consistent with other phylogenetic studies of ribosomal RNAs that suggest LUCA was a mesophilic organism which then evolved to form the thermophilic ancestor of Bacteria and Archaea [21, 22]. Clearly, there is much still to be learned about the nature of early forms of life on Earth and the characteristics of the environment that drove its emergence and early evolution.

The variation in the geochemical composition of present day hydrothermal environments likely encompasses much of the geochemical compositional space that was present on early Earth [23], when key metabolic processes (e.g. methanogenesis, iron reduction, sulfur metabolism, and photosynthesis) that sustain populations inhabiting these systems are thought to have evolved. Geochemical variation in modern hydrothermal environments provides an ideal field laboratory for examination of the natural distribution of taxa, their metabolic functionalities, and for defining the range of geochemical conditions tolerated at the taxonomic and metabolic levels (i.e. habitat range or zone of habitability) [24, 25]. Approaches that integrate molecular biological and geochemical data collected simultaneously within an evolutionary framework can potentially provide new insights into the factors that drove the diversification of metabolic processes that sustain life in hydrothermal environments. The studies reviewed here are aimed at providing a better understanding of the interplay between the geological processes that fuel hydrothermal systems, the characteristics of the modern-day inhabitants of these systems, and the role that geological processes have played in shaping the diversification of ancestral populations that inhabited hydrothermal systems on ancient Earth. In doing so, this chapter will bring together topics on geology, geochemistry, microbial physiology, and phylogenetics to explore the taxonomic and metabolic evolution of microorganisms in continental magmatic systems in the context of the geochemically variable hydrothermal environments that they occupy.

4.2 Geological drivers of geochemical variation in continental hydrothermal systems

Hydrothermal systems form as the result of the interaction between magmatic fluid and a heat source. Three key components are required for the formation of a hydrothermal system: a source of water, a source of heat, and permeability of the rock strata overlying the heat source [26]. Although water sources, heat sources, and bedrock permeability may vary across hydrothermal systems, they all include the following general features: 1) a recharge area where water enters the system from the surface; 2) a subsurface network allowing this water to descend, come into contact with the heat source, and leach minerals from surrounding rock to form hydrothermal fluids; and 3) a discharge area where newly-formed hydrothermal fluids driven to the surface by heat-induced pressure or density changes emerge on the surface to form hydrothermal features such as hot springs, geysers, fumaroles, or mudpots. Magmatic gases such as hydrogen, carbon dioxide, methane, and hydrogen sulfide play an important role in the chemistry of hydrothermal fluids in both marine and continental systems. Since many of these gases drive chemotrophic microbial metabolisms [25, 27, 28], it is essential to understand the distribution and metabolic diversity of microbial life inhabiting these systems. Moreover, abiotic synthesis reactions, as recently reviewed by McCol-

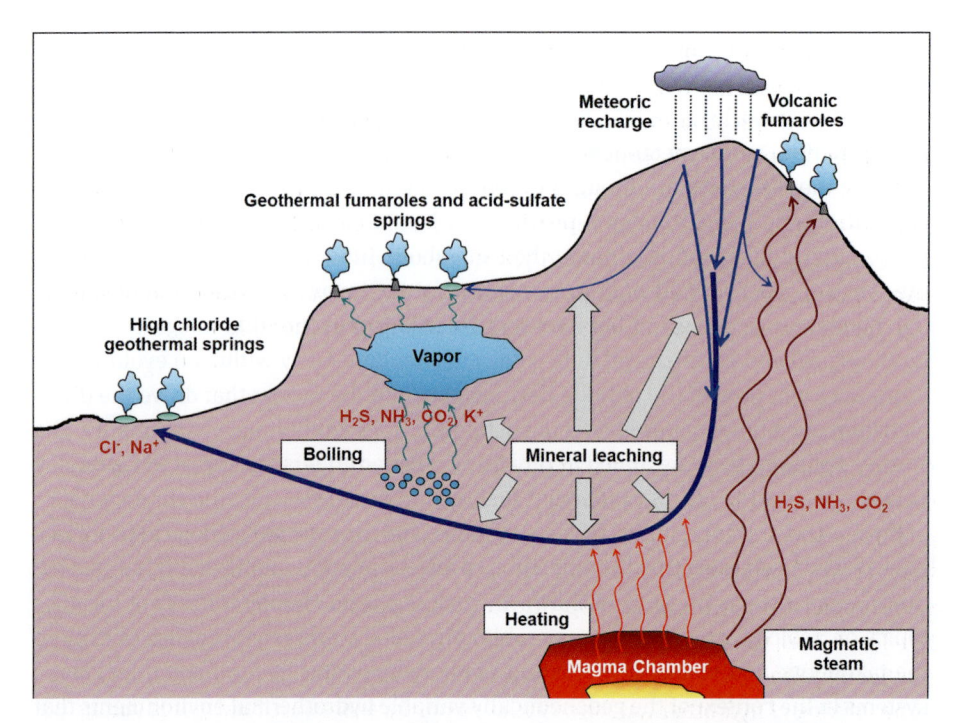

Fig. 4.2. Schematic illustrating the hypothesized functioning of a continental magmatic hydrothermal system. Meteoric water infiltrates the subsurface where it is heated by a magma chamber. Less dense heated water returns to the surface. During flow back to the surface, heated water leaches various minerals from different bedrock types. Ascending water can also undergo phase transitions leading to acidic fumaroles and acid-sulfate springs or alkaline high chloride springs.

lum and Seewald [29], are capable of producing a variety of reduced hydrocarbons (e.g. formate, carbon monoxide, methane) that likely support chemotrophic populations in hydrothermal ecosystems.

Hydrothermal systems are typically classified according to the heat source which drives them. Those systems driven by heat from volcanic activity are termed "magmatic" (e.g. Yellowstone National Park (YNP), USA; Kamchatka, Russia; Tengchong, China), while those systems driven by the natural temperature increase that occurs with increasing depth (i.e. heat emanating from the Earth's mantle, radioactive heat, etc.) are termed "non-magmatic" (e.g. hot springs in the Great Basin, USA; Hot Springs National Park, USA).

Continental magmatic hydrothermal systems are recharged by precipitation descending through cracks and fissures into the deep subsurface (▶ Fig. 4.2), where it mixes with and is heated by ion-rich magmatic fluids to temperatures as high as 500 °C [30–33]. Rising hydrothermal fluids can undergo boiling and phase separation, resulting in the partitioning and concentration of chemical species between the vapor phase (such as ammonia, hydrogen sulfide, and metal cations) and the liquid phase (non-gaseous ions, such as chloride) [34–37]. Systems that interact with deep magmatic brines tend to be alkaline and rich in sodium and chloride. Vapor continues to rise and spread through rock crevices, forming a vapor-filled subsurface area termed the "vapor zone". Some vapor cools and condenses at the surface, where it again mixes with the liquid phase hydrothermal fluids in hot springs, or forms mud pots. Remaining vapor rises via rock fissures and is ejected at the surface through fumaroles. Hydrothermal features (e.g. fumaroles or hot springs) that receive substantial vapor phase input tend to be acidic due to near surface oxidation of sulfide and the generation of protons (i.e. acidity) according to Reaction 1:

$$H_2S + 2O_2 \rightarrow SO_4^{2-} + 2H^+ \quad \text{(Reaction 1)}$$

The partitioning of sulfide in the vapor phase and its near surface oxidation result in sulfuric acid buffered systems ($pK_a \sim 2.0$ at 20 °C and 3.0 at 100 °C) whereas systems receiving low sulfide tend to be buffered by bicarbonate ($pK_a \sim 6.4$ at 20 °C and 6.6 at 100 °C) [38]. The bimodal distribution of these hot spring types can be delineated by plotting the frequency of hot springs as a function of their pH, as has been done for globally distributed hot springs [39] as well as in a focused study of YNP [40]. In ▸ Fig. 4.3 (a), the pH and temperature combinations for 7700 hot springs in YNP are plotted revealing two clusters of spring types that span the gradient of spring temperature. Mixing of hydrothermal fluids with near surface meteoric fluid yields variations in spring pH that differ from this bimodal distribution. This phenomenon is illustrated by the plot in ▸ Fig. 4.3 (c), in which springs with pH that lie between those of sulfuric acid and bicarbonate buffered springs have lower conductivity (total dissolved ions), indicating the mixing and dilution of highly conductive acidic spring water with lower conductive and near neutral meteoric water.

Protons generated through the oxidation of sulfide contribute to the weathering of subsurface bedrock (e.g. rhyolite) resulting in the high conductivities associated with the low pH springs in YNP, when compared to alkaline systems. The fact that little correlation exists between hot spring conductivity and temperature (▸ Fig. 4.3 (b)) suggests that temperature plays a secondary role to pH in dictating the ionic strength of hydrothermal fluids. The differential weathering of subsurface bedrock by vapor- and liquid-phase dominated hydrothermal systems has important ramifications for the type of hot springs formed. Vapor dominated systems tend to have little liquid and form fumaroles (steam vents), mudpots (dissolved clay or volcanic ash), and hot springs (upon mixing with meteoric fluid), whereas liquid phase dominated systems often form hot springs or geysers. While geysers can also form from acidic waters, this is a rare occurrence due to the tendency for acidic water to dissolve the constriction that is needed for pressure to develop and eject fluids from the subsurface reservoir.

In addition to chemical variation induced by subsurface boiling and phase separation, variation in subsurface hydrology results in differences in the fluid flow paths that allow for interaction with variable rock units. This interaction results in different fluid residence times, which are reflected by conductivity differences (▸ Fig. 4.3 (b)

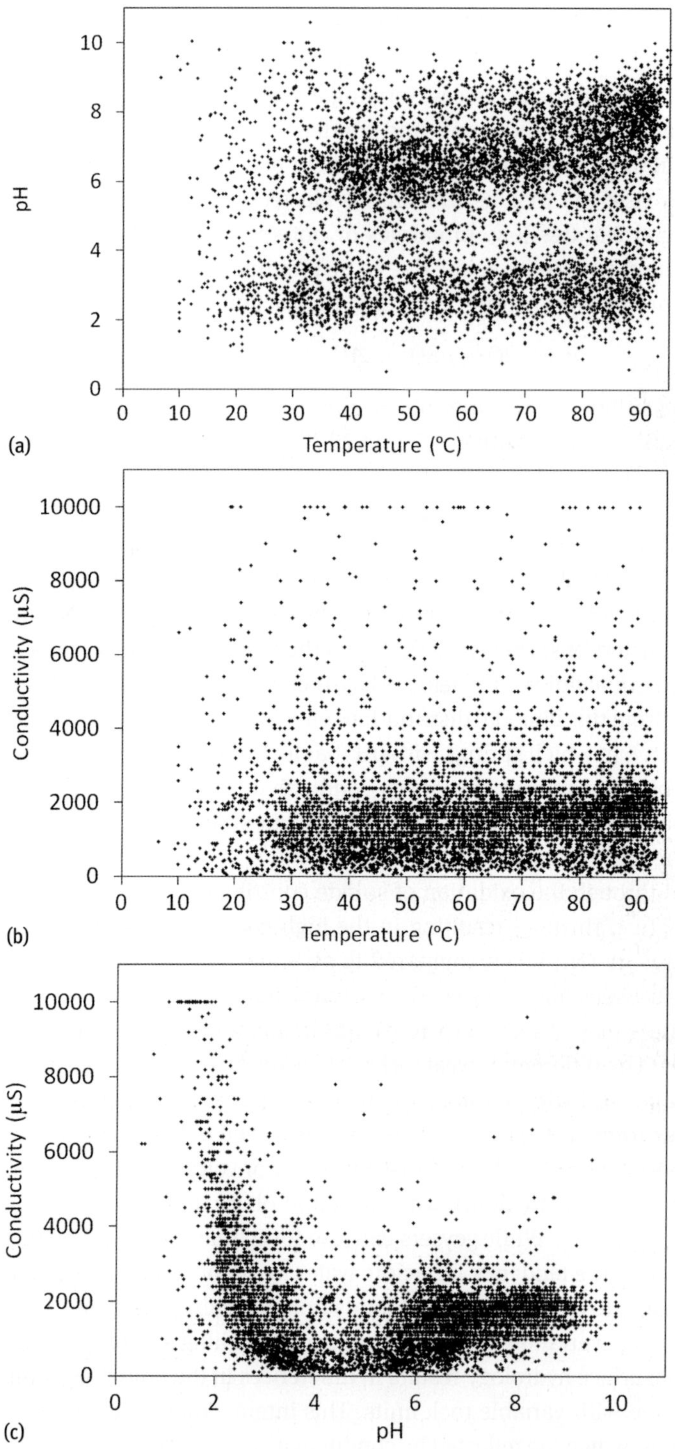
pH
Temperature (°C)
(a)
Conductivity (μS)
Temperature (°C)
(b)
Conductivity (μS)
pH
(c)

◄ **Fig. 4.3.** pH, temperature, and conductivity measurements plotted against each other for ~ 7700 hot springs in Yellowstone National Park. Data was compiled from the Yellowstone Research Coordination Network website (http://www.rcn.montana.edu/).

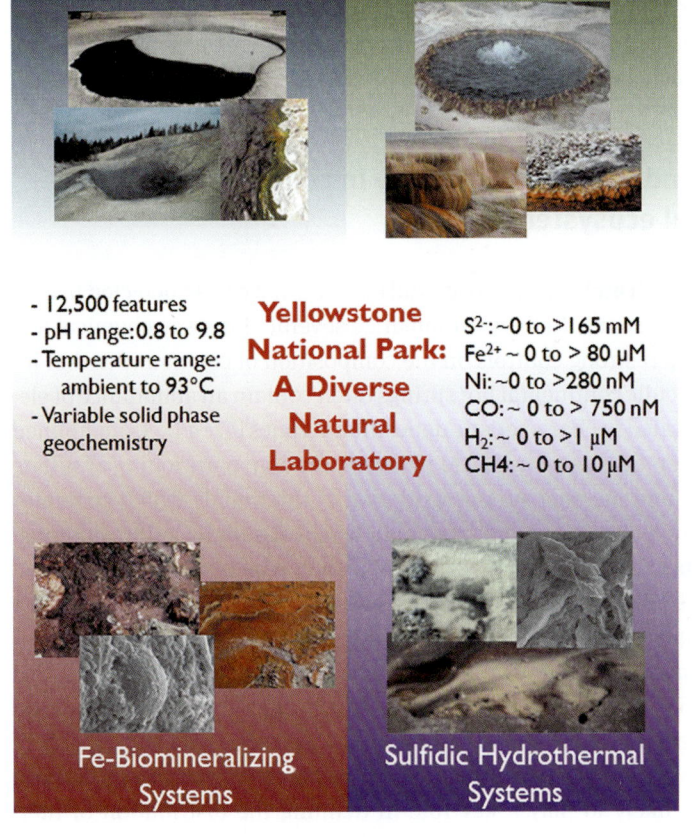

Pyritic Hydrothermal Systems

Siliceous, Carbonate Hydrothermal Systems

- 12,500 features
- pH range: 0.8 to 9.8
- Temperature range: ambient to 93°C
- Variable solid phase geochemistry

Yellowstone National Park: A Diverse Natural Laboratory

S^{2-}: ~0 to >165 mM
Fe^{2+} ~ 0 to > 80 μM
Ni: ~0 to >280 nM
CO: ~ 0 to > 750 nM
H_2: ~ 0 to >1 μM
CH_4: ~ 0 to 10 μM

Fe-Biomineralizing Systems

Sulfidic Hydrothermal Systems

Fig. 4.4. Examples of several of the predominant spring types present in Yellowstone National Park, which is a typical magmatic continental hydrothermal system. The variation in physical and chemical properties of the hot spring types present in YNP is indicated.

and (c)) and leaching of different minerals from the bedrock (► Fig. 4.2). Water-rock reactions change the composition of both the rocks and of the water as it flows through the subsurface. This results in fluid compositions that are extraordinarily diverse when compared to other natural surface waters. The chemical variability of the hydrothermal fluids is largely attributable to compositional differences between the underlying host rocks [41]. Thus, flow paths and the composition of rocks with which hydrothermal fluids interact play a key factor in determining fluid chemistry. Several different hot spring types are present in YNP hot springs. Those that the authors are most familiar with are presented in ▸ Fig. 4.4, along with the ranges of substrates present in these systems. Differences between hot spring fluid compositions may include variations in the availability of metals, such as zinc, iron, molybdenum, and manganese, as well as phosphorous compounds like hypophosphite (PO_2^{3-}) and/or phosphite (PO_3^{3-}). Decreased density resulting from heating causes these ion-enriched hydrothermal fluids to ascend toward the surface. Importantly, near surface oxidation of dissolved chemicals such as sulfide or ferrous iron can result in the deposition of different solid phases, such as elemental sulfur or iron oxyhydroxides, respectively [42].

4.3 Taxonomic and functional diversity in continental hydrothermal ecosystems

Microorganisms from both archaeal and bacterial domains have been detected in continental hydrothermal systems. Here, we summarize several of the key chemical features that appear to exert strong controls on the composition of communities that inhabit these systems. Acidic continental hot springs often contain an abundance of elemental sulfur, precipitated in characteristic depositional zones [42–44]. As mentioned previously, acidic hot springs develop due to phase separation, partitioning of sulfide into the vapor phase, and the near surface oxidation of this sulfide which generates sulfuric acid. Thus, acidic springs often have high levels of sulfide, sulfur (intermediate oxidation product), and sulfate [24, 37, 45–47], all of which can be metabolized by hyperthermophiles [48–55]. Moreover, a range of typically less stable intermediate sulfur species including thiosulfate, polysulfides, and polythionites has been detected in many hot spring environments [45–47, 56] and has been shown to be metabolized by microorganisms [56–58]. As a result, sulfurous hot springs across the globe tend to harbor similar lineages of microorganisms [59, 60], underscoring the importance of sulfur in the metabolism and diversification of thermophilic microorganisms [59, 60].

Hydrogen is also likely to play a key role in defining the distribution of thermophilic organisms in hot spring environments [28, 48]. Numerous H_2-oxidizing organisms, typically associated with the bacterial order *Aquificales* [61, 62] and members of the *Euryarchaeota* and *Crenarchaeota* [63–66], have been isolated from hot springs. In addition to H_2 oxidation, fermentative H_2 production appears to be an important process sustaining organisms inhabiting continental magmatic hydrothermal environments [7].

The discovery of ammonia oxidation by thermophilic Archaea [67] transformed the current understanding of the importance of the nitrogen cycle in hydrothermal environments. Since this initial discovery, molecular signatures of ammonia-oxidizing archaea have been shown to be widespread in continental magmatic hydrothermal ecosystems [68–70]. Moreover, the process of ammonia oxidation and the depletion of the bioavailable ammonia pool has been suggested to represent a top-down control on the structure and composition of communities inhabiting circumneutral to alkaline continental hot springs, creating a strong selective pressure for inclusion of N_2-fixing populations [71]. Biological N_2 fixation is likely to play a key role in relieving fixed N limitation in continental hot spring environments, and evidence for N_2-fixing populations has been identified by molecular approaches targeting the distribution and diversity of *nifH* genes [8, 72] and their transcripts [71, 73, 74]. Incorporation of $^{15}N_2$ into biomass in microcosm assays containing hot spring sediments indicates N_2 fixation in numerous continental hot spring environments with temperatures of up to 89 °C [72]. This temperature is close to the upper temperature limit for N_2 fixation (92 °C) as demonstrated in a marine ecosystem [75], and provides additional evidence for the role of N_2 fixation in relieving fixed N limitation. Whereas a thermophilic bacterium distantly related to *Leptospirillum* spp. was responsible for the N_2 fixation activity observed in the 89 °C continental hot spring [72], a thermophilic methanogen was responsible for this activity observed in the marine vent ecosystem [75]. Surprisingly, despite the extensive molecular-based studies of N_2 fixation potential and gene expression in continental hot springs, only bacterial N_2 fixation has been observed to date.

Several unique and deeply rooted lineages of Bacteria and Archaea have only been identified in high temperature environments. For example, the bacterial order *Aquificales* has never been observed in environments with temperatures less than 55 °C [62]. The *Aquificae* genera *Hydrogenobacter*, *Hydrogenobaculum*, *Thermocrinis*, and *Sulfurihydrogenibium* spp. are typically observed in continental hot springs and solfataric fields [58, 76–79]. The distribution of these genera in continental hot springs is driven by pH, suggesting diversification in response to acidity gradients. Whereas *Hydrogenobaculum* predominates in acidic springs, *Thermocrinis* and *Hydrogenobacter* predominate in alkaline springs. *Sulfurihydrogenibium* tends to inhabit sulfide-rich alkaline springs [62]. Unlike the above strains, members of the genera *Aquifex* and *Persephonella* are routinely isolated from marine hydrothermal systems only [58, 61, 80–82]. It is currently unknown what factors led to the unique distribution pattern of *Aquificae* strains in marine and continental hydrothermal systems but these factors likely include differences in fitness in their respective niches.

Several thermophilic bacteria have also been isolated from terrestrial hydrothermal habitats. Examples include strains of *Thermocrinis*, *Hydrogenobaculum*, *Hydrogenobacter*, and *Sulfurihydrogenibium* from the *Aquificales* order, which are important constituents of a variety of geochemically-diverse geothermal communities [62]. *Aquificales* are chemolithoautotrophic thermophilic bacteria that couple CO_2 fixation with the oxidation of reduced compounds such as hydrogen, sulfide, or elemental sulfur [58, 61, 83]. Owing to their chemolithoautotrophic mode of metabolism, they are often the dominant primary producers of biomass within high temperature ecosystems [61, 83].

Together with members of the order *Thermotogales*, these bacteria have the highest growth temperatures currently known [58].

Other thermophilic bacteria isolated from terrestrial hot springs include members of the *Desulfurella* genus. Examples of this genus include *Desulfurella multipotens*, *D. kamchatkensis*, *D. acetivorans*, and *D. propionica*. These strains were isolated from cyanobacterial mats and sediments collected from thermal areas with neutral pH (6.0–7.0) in New Zealand and Kamchatka, Russia. All are capable of chemoorganotrophic growth using elemental sulfur as a terminal electron acceptor; however, *D. multipotens* can also grow chemolithotrophically with hydrogen [84–86].

Like the distribution of *Aquificae* discussed above, the distribution of archaeal phyla appears to be dependent on environmental characteristics and selection for specific metabolic activities that can be tracked phylogenetically. Among the *Crenarchaeota*, all members of the *Sulfolobales* and *Acidolobales* orders and several representatives of the *Thermoproteales* order are acidophilic [63, 64, 87]. Among the *Euryarchaeota*, acidophily is restricted to members of the order *Thermoplasmales* [66]. Moreover, a number of metabolic properties can be tracked phylogenetically at the phylum or domain level. For example, all thermophilic methanogens and halophiles belong to the phylum *Euryarchaeota*. In contrast, thermophilic organisms capable of reducing sulfate are known to be widely distributed among the bacteria (e.g. *Deltaproteobacteria*, *Thermodesulfobacteria*, *Nitrospira*, and *Firmicutes* [49, 88]) but are narrowly distributed among archaea. The only archaeal sulfate-reducers described so far belong to the genera *Archaeoglobus* and *Caldivirga* [89, 90]. It is unclear if sulfate reduction was a character of the LUCA, but sulfate was unlikely to be an abundant anion prior to the rise of oxygenic photosynthesis [91]. An alternative hypothesis is that the genes required for sulfate reduction in either archaea or bacteria were acquired by HGT (discussed in more detail in ▸ Section 6). For example, it has been suggested that the dissimilatory sulfite reductase (*dsrAB*) genes in *Archaeoglobus* species are the result of an ancient horizontal transfer from a bacterial donor, and that microbial sulfate respiration may have a bacterial origin [92]. Moreover, it appears that the patchy distribution of sulfate respiration among bacteria is due to multiple HGT events as well as gene loss [92, 93].

Heterotrophic, thermoacidophilic archaea that couple the oxidation of organic acids, carbohydrates, and/or complex peptides to the reduction of elemental sulfur have been isolated from various acidic hot springs. Example organisms include the anaerobic, thermoacidophilic crenarchaeotes from the order *Acidilobales*. Members of this lineage include *Acidilobus aceticus* and *Acidilobus saccharovorans*, both isolated from an acidic hot spring in Kamchatka, Russia [87, 94], and *Acidilobus sulfurireducens* isolated from an acidic hot spring in YNP [54]. Molecular analyses indicate that *A. sulfurireducens* is a numerically dominant organism in acidic, high-temperature YNP springs [13, 54], and that closely related strains of *Acidilobales* reside in acidic terrestrial geothermal environments around the world, illustrating the wide distribution of this genus [60, 87, 94]. Other organotrophic thermoacidophilic crenarchaeota include *Caldisphaera lagunensis* and *Caldisphaera draconis*, which were isolated from acidic hot springs in the Philippines and YNP, respectively [54, 95]. Additional organotrophic and thermoacidophilic archaea typical of acidic hot springs include *Thermocladium modestius* [96], *Caldivirga maquilingensis* [89], *Vulcanisaeta distributa* [97], and *Vulcanisaeta souniana* [97]. Unlike members of the *Acidilobales* (*Acidilobus* and *Caldisphaera* spp.) which are strict anaerobes, these archaea are able to tolerate low levels of oxygen, suggesting a role for oxygen gradients in the diversification of thermoacidiphilic crenarchaeota.

Members of the order *Sulfolobales* represent another group of thermoacidophilic archaea that are commonly detected in continental acidic hot springs [59, 98, 99]. This order is comprised of members of the genera *Sulfolobus*, *Metallosphaera*, and *Sulfurococcus*, which are obligate aerobes. The order also includes members of the genus *Stygiolobus*, which are strict anaerobes, and facultative anaerobes from the genera *Acidianus*, *Sulfurisphaera* [64, 65], and *Sulfolobus*. During autotrophic growth, elemental sulfur or hydrogen may be oxidized to yield sulfuric acid or water, respectively, as end products with carbon dioxide serving as a carbon source [64, 65]. Sugars, yeast extract, and peptone may serve as electron and carbon sources during heterotrophic growth of isolates in the lab, but these may not be the complex sources of carbon that support these organisms *in situ* [99]. To date several members of *Sulfolobus* have been isolated from various continental and acidic hot springs around the world. Some representative species of this genus include *Sulfolobus acidocaldarius*, *S. metallicus*, *S. solfataricus*, *S. tokodaii*, *S. islandicus*, *S. shibatae*, and *S. yangmingensis*, among others [99–103]. *Metallosphaera* and *Sulfurococcus* are also facultative autotrophs capable of oxidizing elemental sulfur or organic compounds. Three species of the *Metallosphaera* genus, including *M. sedula*, *M. prunae*, and *M. hokonensis*, have been isolated from terrestrial hot springs [104–106]. *Sulfurococcus* species have also been isolated from terrestrial sites. *Sulfurococcus mirabilis* and *Sulfurococcus yellowstonensis* were isolated from Kamchatka, Russia and from a hot spring in YNP, respectively [107, 108]. In contrast, *Sulfurisphaera ohwakuensis*, the only species from the *Sulfurisphaera* genus, grows mixotrophically and was isolated from an acidic hot spring in Japan [109]. In addition to being capable of autotrophic and facultative heterotrophic growth, *Acidianus* species are physiologically versatile and grow as facultative aerobes with the ability to oxidize or reduce elemental sulfur depending on oxygen availability [110]. Some representative species of this genus include *Acidianus ambivalens*, *A. manzaensis*, *A. infernus*, and *A. brierleyi* [111–114]. *Acidianus* species have been isolated primarily from terrestrial hot springs sites, but also occur in shallow submarine hydrothermal systems [114]. The capacity to inhabit both terrestrial and marine hydrothermal systems likely reflects the broad physiological diversity of this genus. Members of the *Acidianus* genus are also metabolically diverse and can grow under a number of distinct conditions. One unique *Acidianus* species, *A. manzaensis*, can grow chemolithoautotrophically by coupling the reduction of Fe^{3+} to the oxidation of H_2 or elemental sulfur under anoxic conditions [113].

Archaeal autotrophs that couple hydrogen oxidation with the reduction of elemental sulfur have also been isolated from acidic terrestrial hot springs. One example of this group of organisms is *Stygiolobus azoricus*, which is the only strictly anaerobic chemolithotrophic thermoacidophile from the *Sulfolobales* order. The type strain was isolated from solfataric fields in Azores, Portugal [115]. Arguably the most important metabolism discovered in continental magmatic hot springs over the past decade is the oxidation of ammonia coupled to the reduction of O_2. The first cultivated aerobic ammonia oxidizer, "*Candidatus* Nitrosocaldus yellowstonii", is a chemoautotroph [67] and is widely distributed in circumneutral to alkaline hot spring environments [68, 70, 71, 116].

4.4 Application of phylogenetic approaches to map taxonomic and functional diversity on spatial geochemical landscapes

The accessibility of continental hydrothermal systems, coupled with their extensive geochemical variability and comparatively simple taxonomic diversity (when compared to non-thermal environments), makes them ideal targets for phylogenetic and community ecology approaches aimed at generating insights into the role of geochemical variation in dictating taxonomic and functional diversification. The application of phylogenetic ecology tools, while common in studies of the distribution and diversification in plant communities [1, 2, 5, 6], is relatively new in microbial studies. Phylogenetic ecology is in essence a fusion of molecular microbial ecology techniques and phylogenetics. Here, molecular sequence (gene, transcript, genomic, etc.) data are obtained across spatial geochemical or geographic gradients. This collection of sequence data is then used to generate a phylogeny of those sequences. Geochemical or geographic data obtained from the environments where the sequences were obtained is then quantitatively mapped onto the phylogeny and used to assess and rank the extent to which variation in physical or chemical measurements can explain the topology of the phylogeny. A schematic of a hypothetical phylogenetic reconstruction is presented in ▶ Fig. 4.5. Here, temperature explains the topology of the tree well, suggesting that temperature played a role in the diversification of this hypothetical sequence. Incorporation of additional geochemical variables into explanatory models, as outlined in the examples below, can be used to gauge the role of a particular parameter(s) in driving the diversification of the sequence(s) being considered.

Numerous studies of targeted single gene loci or community genomes in hydrothermal ecosystems reveal that variation in the composition of geothermal fluids strongly influences both the types of organisms present and their functional diversity [7–11]. As an example, Alsop et al., 2014 [11] compiled metagenomic sequence data from 28 continental hot springs in YNP and the Great Basin and subjected these data to a novel bioinformatics approach termed Markov clustering [117, 118]. Markov

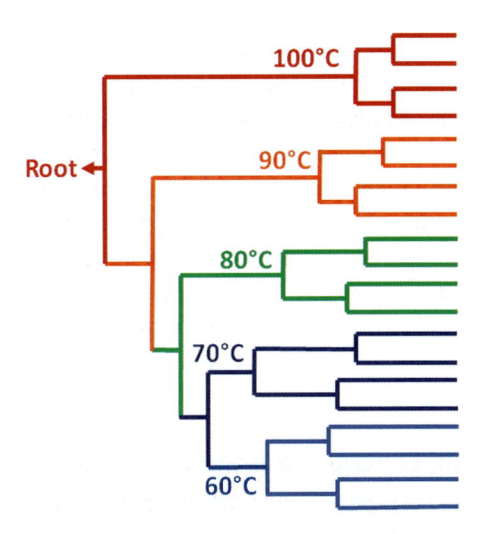

Fig. 4.5. Schematic illustrating phylogenetic approaches to ecological studies. In this case, sequences were collected across a temperature gradient ranging from 60 to 100 °C and were subjected to phylogenetic reconstruction. Using community ecology tools, it is possible to map geochemical characteristics onto the tree and quantify the extent to which they influenced the diversification of a given gene, protein, or genomic tree.

clustering is a robust mechanism to "bin" metagenomic sequence reads based on phylogenetic similarity (e.g. using BLAST expectancy values or e values). Once reads are binned, they can be subjected to a number of multivariate statistical analyses aimed at uncovering relationships between differences in the genomic and functional composition of communities and the geochemical characteristics of their local environment. The results of the work of Alsop et al., 2014 [11] suggest that high temperature ecosystems harbor lower taxonomic, phylogenetic, and functional diversity when compared to lower temperature environments. Likewise, acidic environments harbor lower taxonomic, phylogenetic, and functional diversity when compared to circumneutral to alkaline environments. Importantly, the genomic composition of the communities inhabiting hot springs characterized by extremes of acidity and temperature were substantially different from those inhabiting hot springs characterized by higher pH and lower temperature and show a strict delineation between the functional composition of communities supported by photosynthesis and those supported by chemosynthesis. In this case, communities supported by photosynthesis exhibit substantially greater taxonomic, phylogenetic, and functional diversity when compared to communities supported by chemosynthesis. These results show that hydrothermal communities are structured by interactions with the geologic processes that drive the geochemical composition of their habitats.

Targeted studies of the distribution of taxonomic and functional gene lineages across spatial and geochemical gradients in hydrothermal systems reveal more intricate details of the role of environment as well as geographic isolation in shaping their diversification. For example, studies aimed at understanding the role of environmental variation on the genomic evolution of specific groups of organisms have been conducted in hydrothermal environments. In particular, the work of Whitaker et al., 2003 [59] found genomic evidence for barriers to the dispersal of *Sulfolobus* species.

Similar studies targeting thermophilic cyanobacteria [119] and thermophilic *Aquificae* [120] also suggest that geographic barriers limit the dispersal of these lineages and allow divergence through local adaptation or random genetic drift (i.e. allopatry).

While the aforementioned studies suggest a role for geographic isolation in the diversification of thermophiles, other studies indicate that the signal for geographic isolation is diluted by environmental parameters. As an example, the phylogenetic similarity of archaeal 16S rRNA genes sampled across geochemical gradients in YNP can be best explained by variation in spring pH, suggesting that pH is the primary driver of the diversification of these 16S rRNA gene lineages [68]. Geographic distance between hot springs (as a proxy for dispersal limitation) failed to explain a significant amount of the variation in the data. This suggests that, in some cases, the probability and rate of successful dispersal is higher than the rate of local evolution. Similar studies conducted on hot springs in China found that a shift in highly structured assemblages toward less structured assemblages was associated with seasonal patterns in hot spring recharge and input of exogenous substrates (e.g. due to climatic events), suggesting that hot spring temporal dynamics are also an important contributor to the diversification of life in these systems [121].

The non-random patterns in the distribution of taxonomic lineages across spatial geochemical gradients in continental hot springs suggests a role for geochemical variation in shaping the distribution and diversification of functional components of these assemblages. By focusing on a gene required for the synthesis of (bacterio)chlorophyll as a marker for photosynthesis, it was shown that a combination of temperature and pH restrict the distribution of light-driven primary production in YNP hot springs [3, 7, 9]. While the upper temperature limit for photosynthesis of 73 °C was confirmed in circumneutral to alkaline systems [122], it was shown that in acidic systems (pH < 4.0) photosynthesis was restricted to environments with temperatures of less than 56 °C. A subsequent study that focused on the distribution of pigments across pH and temperature gradients in YNP confirmed these results and indicated an important and previously unrecognized role for sulfide in shaping the distribution of photosynthesis in the Park, with springs containing > 5 µM sulfide being devoid of photosynthetic pigments. A follow-up study revealed that the sulfide-dependent distribution of photosynthesis in acidic environments was due to its inhibitory effect on CO_2 assimilation in algae (which predominate among phototrophs in hot springs with pH < 4.0 [123]), but not in cyanobacterial dominated systems (pH > 4.0) [3]. These observations, coupled with those from metagenomic sequencing efforts indicating that phototrophic systems support greater taxonomic and functional diversity, suggest that the lack of outward radiation of phototrophs due to the inhibitory effects of sulfide may have impacted the functional diversification of populations and communities inhabiting acidic and sulfide-rich high temperature ecosystems.

Phylogenetic ecology studies focused on other functional genes as proxies for key metabolic processes in high temperature environments, including those involved in the conversion of dinitrogen gas (N_2) to bioavailable ammonia (i.e. nitrogenase, nitrogenase iron protein encoded by *nifH* [8]), the reversible oxidation of H_2 (i.e. [FeFe]-hydrogenase, large subunit encoded by *hydA* [7]), and the detoxification of mercury (i.e. mercuric reductase encoded by *merA* [10]), further implicate the role of geological processes in shaping the distribution and functional diversification of microbes in these ecosystems. For example, the availability of fixed forms of nitrogen (e.g. ammonia) required by all forms of life is not uniform in hydrothermal environments, as evinced by the generally lower availabilities of ammonia in circumneutral to high pH environments. The low concentrations of bioavailable ammonia in circumneutral to alkaline geothermal springs is due, in part, to geological and chemical factors, most notably the equilibration of $NH_4^+{}_{(aq)}$ with $NH_{3(g)}$ (pK = 7.6 at 90 °C) and the subsequent volatilization of $NH_{3(g)}$ [71, 124]. The abundance of *nifH* genes as a marker for nitrogenase, which functions to relieve fixed nitrogen limitation in ecosystems through the conversion of dinitrogen gas to ammonia, has been found to track with the availability of ammonia (Boyd et al., unpublished). Moreover, the diversification of NifH in hot springs in YNP has been shown to be driven primarily by variation in pH [8], which is consistent with the pH-dependent availability of fixed forms of nitrogen in YNP hot spring ecosystems.

The distribution of *hydA*, which encodes the large subunit of [FeFe]-hydrogenase that typically functions in the reduction of protons to generate H_2 in fermentative bacteria, was found to be pH dependent in hot spring ecosystems [7]. Here, *hydA* was more prevalent in circumneutral to alkaline environments and those with lower temperatures when compared to acidic, high temperature hot springs. The constrained distribution of *hydA* to environments with lower temperature and alkaline pH was hypothesized to reflect the unfavorable thermodynamics of organic carbon fermentation in the presence of high H_2 partial pressure [125]. Elevated concentrations of H_2 are routinely measured in springs in YNP where temperatures exceed 65 °C [126, 127]. H_2 in these systems is likely of abiotic origin, derived from subsurface iron-catalyzed water hydrolysis (radiolytic and/or serpentinization) [127]. Iron-catalyzed H_2 production is sensitive to both solution pH and reaction temperature, with a near doubling of H_2 production rates resulting from the doubling of incubation temperature [128]. Similarly, the rate of basalt-catalyzed H_2 production increases with increasing water acidity (lower pH) [128]. Coincidentally, *hydA* was not detected in YNP springs predicted to have elevated geological H_2 production (e.g. acidic pH, > 36 °C; alkaline pH, > 65 °C). Thus, environments with elevated inputs of geological H_2 might select against bacterial fermentative metabolisms which in turn might constrain the distribution of organisms dependent on this metabolism in hot spring ecosystems.

4.5 Molecular adaptation to high temperature

The patterns in the distribution of individual gene lineages in hydrothermal environments points to the presence of specific adaptations that facilitate life under the extreme temperature and pH conditions present in these systems. Thermophiles have a number of unique traits to cope with their high temperature habitats including the production of structurally more stable enzymes and proteins, different mechanisms of motility, and adjustment of membrane lipid compositions.

4.5.1 Lipids

The lipid membrane plays a fundamental role in energy conservation and in the maintenance of intercellular homeostasis by acting as a barrier between the cellular cytoplasm and the external environment. Microorganisms synthesize diverse lipid structures with widely varying biophysical properties [129–131] that have facilitated their diversification into environments with wide ranging geochemical conditions, including hydrothermal environments with extremes of temperature and pH [68, 132, 133]. The ether bonds of archaeal lipids are thought to be more resistant to acid and thermal stress as they are not broken down under conditions in which ester linkages are completely methanolyzed (5% HCl/MeOH, 100 °C) [134]. The predominant membrane lipids of thermophilic archaea are isoprenoid glycerol dialkyl glycerol tetraethers (iGDGTs) [135–138], which consist of two ether-linked C_{40} polyisoprenoid (i.e. biphytanyl) chains with zero to eight cyclopentyl rings and sometimes a cyclohexyl ring (i.e. crenarchaeol). The internal cyclopentyl rings are thought to increase the packing density of the membranes, enhancing their thermal stability [139, 140] and decreasing their permeability to ions [141]. Both pure culture [132, 142] and environmental surveys [68, 133] indicate that the number of cyclopentyl rings per iGDGT positively correlates with the temperature and the acidity of the system. This suggests that archaea acclimate to shifts in pH or temperature by adjusting the cyclopentyl ring composition of their GDGT lipids. "*Ca.* N. yellowstonii" [67] is the only archaeon identified to date that synthesizes crenarchaeol and the function and competitive advantage of synthesizing this unique lipid remains unknown.

Thermophilic archaea have also recently been shown to synthesize a variant of iGDGTs termed H-shaped GDGTs (H-GDGTs), also known as glycerol monoalkyl glycerol tetraethers (GMGT) [135, 143]. H-GDGTs, which are structurally similar to iGDGTs but contain a C–C bond between the two C_{40} biphytanyl chains, have been detected in both thermophilic *Crenarchaeota* [143] and *Euryarchaeota* [144–147] inhabiting high temperature continental hot springs and marine hydrothermal vents [148–151]. These lipids may function to further enhance the thermal stability of the lipid membrane and reduce its permeability to protons. In support of these hypotheses, the fractional abundance of H-GDGTs to total iGDGTs was found to be inversely correlated with pH but not with temperature in a survey of biomass sampled from YNP hot springs [151].

Although the predominant lipids synthesized by bacteria are glycerol fatty acyl diesters, bacteria also acclimate to changing temperature conditions by adjusting the composition of their lipid membranes in order to maintain a liquid crystalline state suitable for embedding of membrane proteins [152, 153]. The temperature of hot springs can be dynamic, and is modulated by processes that operate over time scales ranging from seconds (e.g. earthquakes), to seasonal cycles (e.g. precipitation), to decadal or even millennial cycles (e.g. caldera inflation, glacial dynamics) [154]. Bacteria can acclimate to thermal stress imposed by rapidly increasing temperatures due to the aforementioned factors and decrease the permeability of the membrane while still maintaining a liquid crystalline state by increasing the length of acyl chains, increasing the saturation of acyl lipids, and increasing the ratio of iso/anteiso composition of their fatty acids [155–158]. Interestingly, some bacterial thermophiles have been demonstrated to increase the amount of lipid produced during acclimatization to thermal stress [159].

Recently, it was shown that some thermophilic bacteria may also produce branched chain GDGT lipids (i.e. bGDGTs), which vary in the number of cyclopentyl moieties and in the degree of methylation of the alkyl chains [160]. Like GDGTs, bGDGTs have ether linkages, however, they are not isoprenoidal and differ in the stereochemical configuration of the second carbon position of the glycerol backbone [161]. Several recent studies have detected bGDGTs in biomass collected from hot springs spanning a wide temperature gradient [162–166]. However, the abundance of bGDGT lipids was shown to be lower in hot spring biomass sampled from environments with temperatures > 70 °C, suggesting that these lipids may not be synthesized by hyperthermophilic bacteria [164].

4.5.2 Protein stability

A number of complex interacting features have been identified as possible contributors to increased thermostability of proteins. These features have been the subject of extensive reviews and thus will be mentioned only briefly here [167–170]. A partial list of these features include *i)* helix dipole stabilization by negatively charged residues near their N-terminus and positively charged residues near the C-terminus, *ii)* intersubunit interaction and oligomerization, *iii)* a relatively small solvent-exposed hydrophobic surface area, *iv)* an increased packing density, *v)* an increase in core hydrophobicity, and *vi)* a deletion and/or shortening of surface loops. In general, the same forces that contribute to protein folding also may act as potential stabilization mechanisms. These include hydrophobic effects, disulfide bridges, hydrophobic interactions, aromatic interactions, hydrogen bonding, and ionic interactions. However, as outlined briefly below, different thermophilic enzymes appear to utilize different combinations of the aforementioned strategies to tolerate high temperature.

Numerous studies have highlighted the importance of ionic interactions as key structural determinants of thermal stability at high temperatures. For example, Russell et al., reported an increase in the number and extent of ion-pair networks in the thermostable enzymes glutamate dehydrogenase and citrate synthase from the hyperthermophilic archaeon *Pyrococcus furiosus* [171, 172]. Citrate synthase from *P. furiosus* also exhibits more intimate association of the subunits, an increase in intersubunit interactions, and a reduction in thermolabile residues relative to non-thermophilic enzymes [171].

The relationship between an increase in ionic interactions and thermostability is not universal among thermophile proteins. For example, the glutamate dehydrogenase from *Thermotoga maritima* has fewer intersubunit ion pairs and an increased number of hydrophobic interactions when compared with the enzyme from *P. furiosus* [173]. In contrast, adenylate kinase from *Methanococcus jannaschii* has a larger and more hydrophobic core, an increased number of hydrophobic and aromatic interactions, shorter loops, and helix dipole stabilization which together are thought to confer thermal stability [174].

The mechanism of thermal adaptation has also been studied in the two component enzyme nitrogenase (Fe protein and MoFe protein) from the thermophile *Methanobacter thermoautotrophicus*. The Fe protein component of this enzyme possesses shorter loop regions and lower random coil content than its mesophilic counterpart from *Azotobacter vinelandii*. Both of these features have also been observed in several other thermostable enzymes. Moreover, there are also a greater number of ion-pairing interactions between the two components of this enzyme, suggesting the existence of dynamic protein-protein interactions at higher temperatures [175]. Other protein features that could enhance thermostability include extrinsic parameters such as the presence of high concentrations of intracellular compatible solutes (discussed in ▸ Section 4.5.3), elevated concentrations of proteins, the use of molecular chaperones, or variations in environmental factors such as pressure [167, 176].

4.5.3 Cytoplasmic osmolytes

Many microorganisms accumulate compatible solutes in response to increases in the levels of salts or sugars in the environment. Thermophilic and hyperthermophilic organisms are exposed to dynamically changing geochemical and physical conditions, such as changes in temperature, pH, and solute concentrations due to differences in mixing of hydrothermal and near surface waters. Mechanisms for adjusting and accumulating osmolytes are therefore necessary to survive such dramatic shifts. A survey of osmolytes produced in several species of archaea found that *Pyrobaculum aerophilum*, *Thermoproteus tenax*, *Thermoplasma acidophilum*, and a suite of members of the order *Sulfolobales* accumulated the common osmolyte trehalose [177]. However, other thermophilic taxa were found to produce several unique osmolytes that have only been identified in thermophiles, suggesting that the biosynthesis of these osmolytes represents an adaptation to high temperature environments. For example, the accumulation of mannosylglycerate has been reported in the bacterium *Rhodothermus marinus* [178, 179] and the archaeon *P. furiosus* [177, 180] while *Pyrodictium occultum* accumulates di-*myo*-inositol-1,1'(3,3')-phosphate [177]. The unusual osmolyte cyclic-2,3-bisphosphoglycerate was found to accumulate to high concentrations in cells of the hyperthermophilic methanogen *Methanopyrus* while *Archaeoglobus fulgidus* accumulates diglycerolphosphate in response to increasing salinity of growth medium or incubation temperature [177]. Additionally, both *P. furiosus* [180] and *Methanococcus igneus* [181] produce higher concentrations of the unusual phosphorous-containing solute di-myo-1,1'-inositol-phosphate when grown at elevated temperatures. Likewise, *R. marinus* has been shown to accumulate higher concentrations of mannosylglycerate with increasing incubation temperature [182]. Interestingly, a characterization of intracellular organic solutes in several members of the *Thermotogales* revealed large differences in the solutes detected in each strain when grown under optimal conditions [183], indicating that osmolyte production can vary even between closely related strains. This difference may be related to selective pressures imposed by the environment from which they were originally isolated. Similar findings were reported in the two closely related strains *Rhodothermus marinus* and *Rhodothermus obamensis*. At several growth temperatures and salinities, the major compatible solutes in *R. marinus* were alpha-mannosylglycerate and alpha-mannosylglyceramide, whereas *R. obamensis* only accumulated alpha-mannosylglycerate. The absence of the amide solute in *R. obamensis* at the growth conditions examined was the most pronounced difference in the compatible solute accumulation profile of the two strains [182].

The observations outlined above indicate that specific compatible solutes may have been selected for due to their ability to also protect intracellular molecules against thermal denaturation [184]. Indeed, a comparison of the thermostabilizing effects of a number of organic osmolytes found that mannosylglycerate and trehalose increased the thermotolerance of the enzymes lactate dehydrogenase and glucose oxidase *in vitro* [185]. Likewise, the thermolabile enzymes glyceraldehyde-3-phosphate dehydrogenase and malate dehydrogenase from *Methanothermus fervidus* were found to be stabilized by the potassium salt of the osmolyte cyclic diphosphoglycerate *in vitro* [186]. Diglycerol phosphate, the predominant organic osmolyte synthesized by *A. fulgidus* [177], was found to stabilize lactate dehydrogenase purified from rabbit muscle, alcohol dehydrogenase purified from yeast, and glutamate dehydrogenase purified from *Thermococcus litoralis* at elevated temperatures. This same study found that diglycerol phosphate stabilized purified rubredoxin, a low-molecular-weight iron-containing protein, from a number of anaerobic bacterial strains [187]. For example, the intracellular concentrations of cyclic 2,3-diphosphoglycerate and potassium in the methanogen *Methanopyrus kandleri* have been reported to be as high as

1.1 M and 3 M, respectively [188]. These salts have been shown to play an important role in the thermostability of cyclohydrolase and formyltransferase from *M. kandleri* [189], which is the most thermotolerant methanogen currently known [190].

4.5.4 Motility

Thermophiles have evolved a number of lifestyle strategies to cope with stress imposed by dynamic chemical and physical conditions. For example, to survive the geochemical variations that occur with the mixing of hydrothermal and near surface waters, many hyperthermophiles have been shown to develop surface-adherent, biofilm communities. The production of extracellular polymeric substances (EPS) not only facilitates attachment to surfaces as the first step in biofilm formation, but is also associated with enhanced survival during periods of environmental stress [191]. Cultivated members of the *Thermococcales* are capable of forming copious amounts of capsular polysaccharides and EPS in laboratory cultures, often under conditions of chemical or physical stress [192], suggesting that this might also be a predominant mode of growth for these organisms in natural environments.

Biofilms are particularly common in areas of hot springs with geochemical conditions that allow for photosynthesis [3, 4]. The structure and composition of naturally formed and laminated biofilms inhabiting alkaline siliceous hot springs has been subject to decades of study by a number of researchers, perhaps most notably Ward, Brock, and Castenholz (as reviewed in [193]). Using glass rods suspended at the air-water interface in the runoff channel of a photosynthetic alkaline hot spring, Boomer and colleagues studied the successional development of multilayered mat (i.e. biofilm) communities in a photosynthetic alkaline hot spring (60–70 °C) [194]. This study showed that the pioneer populations in the formation of the photosynthetic biofilm were cyanobacteria closely affiliated with *Synechococcus* spp. as well as *Thermus* spp. With additional incubation time, a red layer developed deeper in the mat (under the green layer) and was attributed to the establishment of the anoxygenic phototroph *Roseiflexus* in the biofilm community. In this same study, the communities that formed the biofilm were compared with the planktonic populations emanating from the geyser vent, revealing similar taxa. This finding is consistent with the idea that many thermophilic populations have evolved to attach to surfaces in order to proliferate, concentrate nutrients, and resist environmental stress [191].

Biofilms also appear to provide an adaptive advantage for chemosynthetic microbial communities that inhabit high temperature (> 70 °C) transects of continental hot springs. By placing sterile glass slides or cover slips in the outflow channels of several hot springs in YNP with temperatures that range from 80 to 90 °C, attachment of microbial cells and growth of filamentous structures was observed [195–197]. Remarkably, the generation times of these surface-associated filaments (estimated by normalizing cells produced along a filament structure to incubation time) were found to range from 2 to 7 hrs, which is similar to the generation times observed in populations inhabiting non-thermal environments [195]. The substantial numbers of cells that accumulate on the surfaces when compared to the near absence of microbes in spring waters emanating from the source of the hot spring suggests that surface attachment is pervasive in high temperature continental hot springs and may represent an evolutionary strategy for populations to distribute themselves at specific thermal and/or geochemical transects [197]. More recently, glass slides have been incubated in hot spring effluent channels and pools at high temperature to promote colonization and growth of hyperthermophiles for use in molecular-based studies [28]. Intriguingly, the populations that attached to these surfaces were similar in composition to those present in sediments, supporting the notion that hyperthermophiles preferentially attach to surfaces. In a separate study, the composition of native elemental sulfur flocs in an acidic hot spring (73 °C) were shown to harbor similar communities to those that emanated from the hot spring source, suggesting a role for these organisms in the formation and transformation of the substrate [83].

Motility is another important mechanism used by thermophiles to respond to dynamic conditions. Chemotaxis refers to the movement toward or away from chemicals and has been observed in a wide variety of bacteria and archaea [198, 199]. Chemotaxis represents a survival mechanism used by microorganisms to search for local environments favorable for colonization. The basis of chemotaxis is a two-component signal transduction pathway whereby the phosphorylation of a histidine autokinase that senses environmental parameters signals the subsequent phosphorylation of a response regulator [200]. The response regulator then controls diverse processes such as chemotaxis. While the majority of chemotactic research has been performed on the model bacterium *Escherichia coli*, some research exists on thermophile chemotaxis. For example, cells of an uncharacterized thermophilic bacterium termed "PS-3", which grows optimally between 60 and 70 °C, were shown to be attracted to a variety of amino acids and carbohydrates [201]. However, the absence of genes required for chemotactic behavior in the genomes of other representative hyperthermophiles such as *P. furiosus*, *Pyrococcus horikoshii*, *Methanocaldococcus jannachii*, and *Aquifex aeolicus* [202–204] suggests that chemotactic behavior may be more restricted in hyperthermophiles than in their mesophilic counterparts.

Motility allows thermophiles to respond and reposition themselves in response to changing geochemical or thermal gradients. However, it is possible that chemotaxis is less important to hyperthermophiles than their positioning in the thermal gradient [205]. Thermotaxis, or the sensing and motile response to thermal gradients, has been demonstrated in nemotodes [206] and *E. coli* [207] but has not been robustly studied in thermophiles. Nevertheless, thermotaxis may be of equal importance to chemotaxis for thermophiles, because maintaining a position in a thermal gradient is likely critical to the functionality of their biomolecules [205]. Indeed, the distribution of numerous thermophiles in natural transects has been shown to closely correspond to their optimal growth temperatures. For example, while it is not known if

Synechococcus are thermotactic, ecotypes of closely-related and thermophilic strains have been shown to have remarkably different cardinal growth temperatures [208] and distributions along a hot spring thermal gradient [209]. Likewise, the maximum rate of acetylene reduction activity (a proxy for N_2 fixation) observed in an enrichment of a thermoacidiphilic diazotroph was shown to correspond to the temperature of the environment where it was isolated, a dynamic and poorly mixed pool containing diffuse elemental sulfur flocs [72]. Such conditions might select for thermotactic behavior in hyperthermophiles.

4.6 Mechanisms of evolution in high temperature environments

Differences in the genomic compositions and organization of closely-related taxa indicate that vertical descent alone cannot account for the evolution of these lineages. For example, analysis of the genome sequences of the three closely related thermophilic archaea *Pyrococcus abyssi*, *P. horikoshii*, and *P. furiosus* revealed different levels of conservation among four regions of their chromosomes containing genetic elements that likely mediated chromosomal reorganization, along with a substantial degree of divergence [210]. Divergence of *P. furiosus* from the ancestral *Pyrococcus* strain might have occurred through Darwinian evolutionary processes (i.e. positive selection) acting on genes involved in translation and/or due to loss of some genes involved in signal transduction or cell motility. In the case of *P. horikoshii* and *P. abyssi*, positive selection was found to operate primarily on the transcription machinery and on the genes involved in inorganic ion transport [211]. At the proteomic level, the comparison of the three *Pyrococcus* species revealed substantial differences in their proteolytic enzymes [210], implying that these closely-related archaea may have different fitness in diverse hydrothermal environments depending on carbon source availability. This may suggest that genetically related but physiologically distinct *Thermococcales* spp. evolved to occupy slightly different niches [205]. The resulting polymorphism is probably linked to an adaptation of these thermophiles to differential environmental constraints. Gunbin et al., 2009 suggested hydrostatic pressure as one of the environmental factors that played a key role in evolutionary divergence of *P. furiosus*, *P. abyssi*, and *P. horikoshii* from their common ancestor given differences in cardinal pressures for growth of the three strains [211]. However, as pointed out by Gunbin et al., 2009, adaptation to pressure is not the sole causative factor, as there are also differences in the metabolisms of these strains, in particular at the level of carbohydrate utilization and amino acid auxotrophy. For example, while *P. furiosus* and *P. abyssi* can utilize maltose and pyruvate for growth, growth of *P. horikoshii* is inhibited by these substrates. Likewise, whereas *P. furiosus* and *P. abyssi* can synthesize tryptophan, *P. horikoshii* is auxotrophic for this amino acid. For these reasons, nutrient partitioning is suggested to be a secondary driver in the diversification of *Pyrococcus* strains, allowing them to occupy different trophic levels within the food web.

HGT associated with clustered regularly interspaced short palindrome repeat (CRISPR) elements has also been observed in *P. furiosus* [212]. The CRISPR system constitutes a microbial immune system that functions to target and neutralize foreign DNA in a manner similar to eukaryotic RNA interference [213, 214]. HGT associated with CRISPR elements likely involves the insertion of DNA fragments from other organisms into archaeal genomes, thus contributing to genomic and physiological diversification of hyperthermophiles [212].

Although HGT events are difficult to prove unambiguously, one potential method for doing so involves placing genes or proteins within a phylogenetic context and comparing them to taxonomic trees that are constructed through analysis of ribosomal RNA genes. Phylogenetic analyses of genes or proteins that are differentially represented in the genomes of closely related strains can be used to determine whether these genetic differences are due to gene loss or acquisition through the process of HGT. For example, genomic analysis of bacteria belonging to the thermophilic order *Thermotogales* has revealed extensive evidence for HGT with other thermophilic archaea likely occupying partially overlapping niches [215–218]. Many of the genes that have been horizontally transferred within archaea encode ATP binding cassette (ABC) transporters. Analysis of the substrate binding affinities of the proteins encoded by these operons indicate that oligosaccharides are their likely substrates and not oligopeptides as was originally suggested [219].

A recent screening of two representative *Aquificales* genomes (*Thermocrinis albus* [220] and *Hydrogenobacter thermophilus* [221]) revealed the presence of nitrogenase gene clusters [222]. Maximum likelihood-based phylogenetic reconstructions of a concatenation of the nitrogenase structural protein sequences, NifHDK, indicates that Nif proteins in these two *Aquificales* genomes were acquired recently through HGT with a more recently evolved thermophilic member of the bacterial phylum *Deferribacteres* (e.g. *Calditerrivibrio nitroreducens* or *Denitrovibrio acetiphilus*). This suggests that *Aquificales* acquired *nif* comparatively recently from an exchange with a bacterial partner in a thermal environment. In further support of this hypothesis, numerous *Aquificales* genera (e.g. *Hydrogenobaculum*) do not encode *nif* [223], despite branching more basally than *Thermocrinis* and *Hydrogenobacter* in taxonomic trees [58]. *Hydrogenobaculum* spp. tend to populate acidic geothermal environments where NH_4^+ produced from magmatic degassing is in much higher supply [124], whereas *Thermocrinis* and *Hydrogenobacter* tend to populate circumneutral to alkaline environments that are N limited [62]. Thus, the recent diversification of *Aquificales* into N-limited environments may have been facilitated by acquisition of *nif*. Together, these findings add to a growing body of evidence suggesting that HGT has played a significant role in expanding the taxonomic and ecological distribution of N_2 fixation [224–226]. Moreover, this analysis illustrates both how the nuances of an environment (e.g. N limitation) can select for HGT events that increase fitness and how environmental characteristics shape the co-distribution of lineages (e.g. *Thermocrinis*/*Hydrogenobacter* and *Deferribacteres*), allowing HGT between these species [227].

This same phenomenon also likely explains "phylogenetic barriers" to HGT that probably reflect different evolutionary paths by which specific lineages come to occupy different and non-overlapping environmental niches [228]. For example, the genomes of the hyperthermophilic bacteria *A. aeolicus* and *T. maritima* were shown to contain a considerably greater fraction of archaeal genes than any non-thermophilic bacterial genomes. This suggests a connection between the similarity in growth conditions of donor/recipient strains, their ecological niches, and the apparent rate of horizontal gene exchange between them. *A. aeolicus* and *T. maritima* likely acquired their genes by HGT from hyperthermophilic archaea, perhaps the euryarchaeote *P. horikoshii*, which co-inhabit similar high temperature environmental niches [215, 229, 230]. Conversely, the thermoacidophilic archaeon *T. acidophilum* and the moderately thermophilic *Halobacterium* sp. appear to possess a significantly greater number of genes that were horizontally transferred from bacteria, when compared to other archaeal species [228, 231].

The probability of HGT varies among genes in a given genome. Generally, informational genes (those involved in transcription, translation, and related processes) are less prone to horizontal transfer than are other categories of genes, such as operational genes (those involved in housekeeping). This difference may be due to the fact that informational genes are typically members of large and complex systems, thereby making their horizontal transfer less probable [232]. Interesting exceptions include aminoacyl-tRNA synthetases, which are key components of the protein translation machinery whose evolution involves HGT [233–235]. Nearly all of the 20 aminoacyl-tRNA synthetases are ubiquitous and essential for cell growth in all living organisms (i.e. informational genes). However, unlike most informational genes, they do not have multiple protein partners and their interactions are limited to their contacts with their cognate tRNA, ATP, and amino acid. Furthermore, athough aminoacyl-tRNA synthetases belong to the translational (informational) machinery, they function in isolation from this complex [235]) and due to their central role in protein translation, there is strong purifying selective pressure acting on the evolution of aminoacyl-tRNA synthetases making it unlikely that they will be lost during cell replication [236, 237].

Another unexpected case of a likely HGT between eukaryotes and archaea is the Trp-tRNA synthetase from the hyperthermophilic archaeal genus *Pyrococcus* [234]. Given the high temperatures at which *Pyrococcus* species grow, the presence of a eukaryotic aminoacyl-tRNA synthetase in these hyperthermophilic *Archaea* is surprising and may suggest that *Pyrococcus* acquired this gene from a thermophilic eukaryote such as a polychaete annelid [228, 238]. This apparent horizontal transfer of eukaryotic Trp-tRNA synthetase gene into *Pyrococcus* most likely involved a xenologous gene displacement, where the ancestral Trp-tRNA synthetase from *Pyrococcus* is replaced by the eukaryotic version [234].

Other evolutionary mechanisms that can account for divergences between genomes are gene duplications, gene loss, and gene fusions. As previously mentioned, analysis of the genomes of the closely related hyperthermophiles *P. furiosus* and

P. horikoshii revealed extensive differences in their composition. Both genomes differ considerably in gene order, displaying displacements and inversions. Gene composition also differed between genomes, suggesting genomic rearrangements and gene loss. On the other hand, the occurrence of two paralogous families of ferredoxin oxidoreductases in both *Pyrococcus* genomes provides evidence for gene duplication preceding the divergence of the *Pyrococcus* species [202].

Gene duplications, losses, and fusions were prevalent in the evolution of molybdenum (Mo)-nitrogenase, which has its origins in hyperthermophilic methanogens and which provides the majority of biologically fixed N on the planet [239]. While *nif* emerged in hydrogenotrophic methanogens and was likely acquired vertically in other members of the *Methanobacteriales*, *Methanomicrobiales*, and *Methanococcales* [142, 239], not all members of these orders encode Mo-nitrogenase indicating gene loss among these members. Moreover, several of the structural proteins of Mo-nitrogenase encoded by *nifDK* are homologous to *nifEN*, which encodes proteins required to synthesize the Mo-based cofactor of this enzyme [240–243]. Phylogenetic evidence indicates that *nifEN* are clearly the result of a tandem duplication of *nifDK* [224, 239, 244, 245]. NifB, another protein required for the synthesis of the active site cofactor of Mo-nitrogenase [246, 247], is a fusion protein consisting of an amino-terminal domain belonging to the radical S-adenosyl methionine (SAM) family of proteins and a carboxy-terminal domain belonging to the NifX/NafY family of proteins in most organisms [248]. However, in early evolving thermophilic methanogens, NifB is not fused [239], indicating that the fusion event took place later in the diversification of this metabolic process.

4.7 Concluding remarks

The early evolutionary events that led to the spectrum of functional diversity present in contemporary microorganisms may trace back to a hydrothermal heritage [14, 16]. While this appears to be the case for several processes that are considered by many to have been key to the origin of life, such as hydrogen oxidation [249], as well as processes that were key to the emergence of higher forms of life, such as nitrogen fixation [222, 239, 250], the events leading to taxonomic diversification are less clear. In particular, phylogenetic reconstructions of taxonomic gene samples across environmental gradients often fail to show high temperature environments as harboring the most basal branching members [68]. This may suggest that evidence for a high temperature origin of life has been obscured by subsequent diversification in response to other environmental drivers such as pH and oxygen. Alternatively, such results may point to a mesophilic origin of life [21, 22]. Additional studies that incorporate more hot spring environments and more robust genetic datasets (e.g. whole genomes or concatenations of house-keeping genes) may provide a more comprehensive picture of the characteristics of the environment that most likely supported early evolving lineages.

The taxonomic and functional diversity of microbial life in modern hydrothermal systems, in particular those that are too hot for photosynthesis (> 73 °C), may have been the result of the underlying geological processes that create extensive geochemical variation in these systems [25, 38]. The integration of phylogenetic tools and molecular microbial ecology approaches (i.e. phylogenetic ecology) now permit these hypotheses to be quantitatively evaluated. Significant questions remain, however, including the extent to which interspecies interactions (facilitative, competitive) influence the evolutionary trajectory of key biological processes and whether these interactions overwhelm the geochemical influences on the diversification of these processes. Indeed, attempts to explain the diversification of key proteins in hydrothermal environments based on geochemical data using phylogenetic tools often fail to explain the majority of the variation in the sequence dataset [7–9, 11]. This suggests that other unaccounted variables, such as interspecies interactions, or additional geochemical variables need to be included in such analyses. Omics approaches provide a potential mechanism for unraveling the nature and extent of such interactions and may provide a path forward to furthering our understanding of the processes that have influenced the evolution of thermophilic microorganisms.

Thermophiles have a number of unique adaptations that enable their persistence in high temperature environments. However, it is likely that adaptations allowing hyperthermophiles to persist in high temperature environments remain to be discovered. The development and application of next generation sequencing technologies continue to reveal new protein encoding genes that cannot be classified based on homology to biochemically characterized proteins. As an example, annotations of thermophile genome sequences typically result in functional assignment of only a small fraction (< 40%) of the encoded open reading frames (ORFs). The function of the remaining ORFs and their role in thermophile physiology remain in the frontier of our understanding of the evolution of thermophiles. Despite the success of applying molecular tools to characterize the taxonomic and functional diversity of microbial communities in hydrothermal systems, much of this diversity is only reported in sequence datasets and is not represented in culture collections preventing physiological and biochemical studies. Continued effort to bring representatives of ecologically-relevant (dominant) lineages into culture, in conjunction with development of new genetic systems suitable for heterologous expression of thermophile genes, is needed to begin to close the gap in our understanding of the functional role of poorly characterized genes identified through genome sequencing efforts.

The development and application of next generation sequencing tools also permit analyses of the rate of evolutionary change in thermophiles under the natural conditions of their environment. Similar approaches have been applied in studies of rates of evolutionary change in other extreme environments, such as acid mine drainage [251]. Application of such techniques to communities that span the geochemical gradients present in continental hydrothermal systems offers the unique opportunity to examine questions related to the role of environmental variation, both spatial and temporal, in constraining or promoting evolutionary change.

References

[1] C. O. Webb, D. D. Ackerly, M. A. McPeek, M. J. Donoghue. Phylogenies and community ecology. Annu Rev of Ecol Syst 2002,33,475–505.
[2] M. Westoby. Phylogenetic ecology at world scale, a new fusion between ecology and evolution. Ecology 2006,87,S163–S5.
[3] E. S. Boyd, K. M. Fecteau, J. R. Havig, E. L. Shock, J. W. Peters. Modeling the habitat range of phototrophs in Yellowstone National Park: toward the development of a comprehensive fitness landscape. Front Microbiol 2012,3,221.
[4] A. Cox, E. L. Shock, J. R. Havig. The transition to microbial photosynthesis in hot spring ecosystems. Chem Geol 2011,280,344–351.
[5] J. J. Wiens. Speciation and ecology revisited: Phylogenetic niche conservatism and the origin of species. Evolution 2004,58,193–197.
[6] J. J. Wiens, C. H. Graham. Niche conservatism: Integrating evolution, ecology, and conservation biology. Annu Rev Ecol Evol Syst 2005,36,519–539.
[7] E. S. Boyd, T. L. Hamilton, J. R. Spear, M. Lavin, J. W. Peters. [FeFe]-hydrogenase in Yellowstone National Park: evidence for dispersal limitation and phylogenetic niche conservatism. ISME J 2010,4,1485–1495.
[8] T. L. Hamilton, E. S. Boyd, J. W. Peters. Environmental constraints underpin the distribution and phylogenetic diversity of *nifH* in the Yellowstone geothermal complex. Microb Ecol 2011,61,860–870.
[9] T. L. Hamilton, K. Vogl, D. A. Bryant, E. S. Boyd, J. W. Peters. Environmental constraints defining the distribution, composition, and evolution of chlorophototrophs in thermal features of Yellowstone National Park. Geobiology 2011,10,236–249.
[10] Y. Wang, E. Boyd, S. Crane, et al. Environmental conditions constrain the distribution and diversity of archaeal *merA* in Yellowstone National Park, Wyoming, USA Microb Ecol 2011,62,739–752.
[11] E. B. Alsop, E. S. Boyd, J. Raymond. Merging metagenomics and geochemistry reveals environmental controls on biological diversity and evolution. BMC Genomics 2014,In press.
[12] J. R. Havig, J. Raymond, D. R. Meyer-Dombard, N. Zolotova, E. L. Shock. Merging isotopes and community genomics in a siliceous sinter-depositing hot spring. Geochim Cosmochim Acta 2010.
[13] W. P. Inskeep, D. B. Rusch, Z. Jay, et al. Metagenomes from high-temperature chemotrophic systems reveal geochemical controls on microbial community structure and function. PLoS ONE 2010,5,e9773.
[14] C. H. Lineweaver, D. Schwartzman. Cosmic Thermobiology, thermal constraints on the origin and evolution of life in the universe. In: J. Seckbach, ed. Origins:Genesis, evolution and biodiversity of microbial life in the Universe: Kluwer, 2004,233–248.
[15] E. S. Boyd, G. J. Schut, M. W. W. Adams, J. W. Peters. Hydrogen metabolism and the evolution of biological respiration. Microbe 2014,In press.
[16] E. A. Gaucher, J. M. Thomson, M. F. Burgan, S. A. Benner. Inferring the palaeoenvironment of ancient bacteria on the basis of resurrected proteins. Nature 2003,425,285–288.
[17] F. Robert, M. Chaussidon. A palaeotemperature curve for the Precambrian oceans based on silicon isotopes in cherts. Nature 2006,443,969–972.

[18] J. Karhu, S. Epstein. The implication of the oxygen isotope records in coexisting cherts and phosphates. Geochimica et Cosmochimica Acta 1986,50,1745–1756.
[19] L. P. Knauth, D. R. Lowe. High Archean climatic temperature inferred from oxygen isotope geochemistry of cherts in the 3.5 Ga Swaziland Supergroup, South Africa. Geological Society of America Bulletin 2003,115,566–580.
[20] R. E. Blake, S. J. Chang, A. Lepland. Phosphate oxygen isotopic evidence for a temperate and biologically active Archaean ocean. Nature 2010,464,1029–1032.
[21] B. Boussau, S. Blanquart, A. Necsulea, N. Lartillot, M. Gouy. Parallel adaptations to high temperatures in the Archaean eon. Nature 2008,456,942–945.
[22] M. Groussin, B. Boussau, S. Charles, S. Blanquart, M. Gouy. The molecular signal for the adaptation to cold temperature during early life on Earth. Biol Lett 2013,9,20130608.
[23] E. L. Shock, M. E. Holland. Quantitative habitability. Astrobiology 2007,7,839–851.
[24] D. K. Nordstrom, J. W. Ball, R. B. McClesky. Ground water to surface water: chemistry of thermal outflows in Yellowstone National Park. In: W. P. Inskeep, T. R. McDermott, eds. Geothermal biology and geochemistry in Yellowstone National Park. Bozeman: Montana State University, 2005,143–162.
[25] E. L. Shock, M. Holland, D. Meyer-Dombard, J. P. Amend, G. R. Osburn, T. P. Fischer. Quantifying inorganic sources of geochemical energy in hydrothermal ecosystems, Yellowstone National Park, USA. Geochim Cosmochim Acta 2010,74,4005–4043.
[26] H. P. Heasler, C. Jaworowski, D. Foley. Geothermal systems and monitoring hydrothermal features. In: R. Young, L. Norby, eds. Geological Monitoring. Boulder, Colorado: Geological Society of America, 2009,105–140.
[27] W. J. Brazelton, M. O. Schrenk, D. S. Kelley, J. A. Baross. Methane- and sulfur-metabolizing microbial communities dominate the Lost City Hydrothermal Field ecosystem. Appl Environ Microbiol 2006,72,6257–6270.
[28] J. R. Spear, J. J. Walker, T. M. McCollom, N. R. Pace. Hydrogen and bioenergetics in the Yellowstone geothermal ecosystem. Proc Natl Acad Sci USA 2005,102,2555–2560.
[29] T. M. McCollom, J. S. Seewald. Abiotic synthesis of organic compounds in deep-sea hydrothermal environments. Chem Rev 2007,107,382–401.
[30] J. W. Hedenquist, J. B. Lowenstern. The role of magmas in the formation of hydrothermal ore-deposits. Nature 1994,370,519–527.
[31] A. E. Williams-Jones, C. A. Heinrich. Vapor transport of metals and the formation of magmatic-hydrothermal ore deposits. Econ Geol 2005,100,1287–1312.
[32] V. Duchi, A. Minissale, M. Manganelli. Chemical composition of natural deep and shallow hydrothermal fluids in the Larderello geothermal field. J Volcanol Geotherm Res 1992,49,313–328.
[33] M. A. Clynne, C. J. Janik, L. J. P. Muffer. "Hot Water" in Lassen Volcanic National Park – Fumaroles, Steaming Ground, and Boiling Mudpots Menlo Park, CA1993.
[34] R. O. Fournier. Geochemistry and dynamics of the Yellowstone National Park hydrothermal system. Ann Rev Earth Plan Sci 1989,17,13–53.
[35] A. H. Truesdell, R. O. Fournier. Conditions in the deeper parts of the hot spring systems of Yellowstone National Park, Wyoming.1976.
[36] D. White, L. Muffler, A. Truesdell. Vapor-dominated hydrothermal systems compared with hot-water systems. Econ Geol 1971,66,75–97.
[37] D. Nordstrom, R. Blaine McCleskey, J. W. Ball. Sulfur geochemistry of hydrothermal waters in Yellowstone National Park: IV Acid–sulfate waters. Appl Geochem 2009,24,191–207.
[38] J. P. Amend, E. L. Shock. Energetics of overall metabolic reactions of thermophilic and hyperthermophilic Archaea and Bacteria. FEMS Microbiology Reviews 2001,25,175–243.

[39] T. Brock. Bimodal distribution of pH values of thermal springs of the world. Geol Soc Amer Bull 1971,82,1393–1394.
[40] W. P. Inskeep, Z. J. Jay, S. G. Tringe, M. J. Herrgard, D. B. Rusch. The YNP metagenome project: environmental parameters responsible for microbial distribution in the Yellowstone geothermal ecosystem. Front Microbiol 2013,4,67.
[41] J. P. Amend, T. M. McCollom, M. Hentscher, W. Bach. Catabolic and anabolic energy for chemolithoautotrophs in deep-sea hydrothermal systems hosted in different rock types. Geochim Cosmochim Acta 2011,75,5736–5748.
[42] H. W. Langner, C. R. Jackson, T. R. McDermott, W. P. Inskeep. Rapid oxidation of arsenite in a hot spring ecosystem, Yellowstone National Park. Environ Sci Technol 2001,35,3302–3309.
[43] R. E. Macur, H. W. Langner, B. D. Kocar, W. P. Inskeep. Linking geochemical processes with microbial community analysis: successional dynamics in an arsenic-rich, acid-sulphate-chloride geothermal spring. Geobiology 2004,2,163–177.
[44] C. R. Jackson, H. W. Langner, J. Donahoe-Christiansen, W. P. Inskeep, T. R. McDermott. Molecular analysis of microbial community structure in an arsenite-oxidizing acidic thermal spring. Environ Microbiol 2001,3,532–542.
[45] Xu Y, A. A. Schoonen, D. K. Nordstrom, K. M. Cunningham, J. W. Ball. Sulfur geochemistry of hydrothermal waters in Yellowstone National Park: I. The origin of thiosulfate in hot spring waters. Geochim Cosmochim Acta 1998,62,3729–3743.
[46] G. Lorenson. Application of in situ Au-Amalgam microelectrodes in Yellowstone National Park to guide microbial sampling: An investigation into arsenite and polysulfide detection to define microbial habitats. Burlington: Vermont, 2006.
[47] A. Kamyshny, G. Druschel, Z. F. Mansaray, J. Farquhar. Multiple sulfur isotopes fractionations associated with abiotic sulfur transformations in Yellowstone National Park geothermal springs. Geochem Trans 2014,15,7.
[48] S. D'Imperio, C. R. Lehr, H. Oduro, G. Druschel, M. Kühl, T. R. McDermott. Relative importance of H_2 and H_2S as energy sources for primary production in geothermal springs. Appl Environ Microbiol 2008,74,5802–5808.
[49] R. Rabus, T. A. Hansen, F. Widdel, eds. Dissimilatory Sulfate- and Sulfur-Reducing Prokaryotes. New York, NY, USA: Springer-Verlag, 2013.
[50] C. G. Friedrich, D. Rother, F. Bardischewsky, A. Quentmeier, J. Fischer. Oxidation of reduced inorganic sulfur compounds by Bacteria: Emergence of a common mechanism? Appl Environ Microbiol 2001,67,2873–2882.
[51] G. Muyzer, A. J. M. Stams. The ecology and biotechnology of sulphate-reducing Bacteria. Nat Rev Microbiol 2008,6,441–454.
[52] E. A. Bonch-Osmolovskaya. Bacterial sulfur reduction in hot vents. FEMS Microbiol Rev 1994,15,65–77.
[53] A. Kletzin, T. Urich, F. Müller, T. M. Bandeiras, C. M. Gomes. Dissimilatory oxidation and reduction of elemental sulfur in thermophilic Archaea. J Bioenerg Biomemb 2004,36,77–91.
[54] E. S. Boyd, R. A. Jackson, G. Encarnacion, et al. Isolation, characterization, and ecology of sulfur-respiring *Crenarchaea* inhabiting acid-sulfate-chloride-containing geothermal springs in Yellowstone National Park. Appl Environ Microbiol 2007,73,6669–6677.
[55] S. Fishbain, J. G. Dillon, H. L. Gough, D. A. Stahl. Linkage of high rates of sulfate reduction in Yellowstone Hot Springs to unique sequence types in the dissimilatory sulfate respiration pathway. Appl Environ Microbiol 2003,69,3663–3667.
[56] G. Ravot, B. Ollivier, M. Magot, et al. Thiosulfate reduction, an important physiological feature shared by members of the order Thermotogales. Appl Environ Microbiol 1995,61,2053–2055.

[57] E. S. Boyd, G. K. Druschel. Involvement of intermediate sulfur species in biological reduction of elemental sulfur under acidic, hydrothermal conditions. Appl Environ Microbiol 2013,79,2061–2068.

[58] W. Eder, R. Huber. New isolates and physiological properties of the *Aquificales* and description of *Thermocrinis albus* sp. nov. Extremophiles 2002,6,309–318.

[59] R. J. Whitaker, D. W. Grogan, J. W. Taylor. Geographic barriers isolate endemic populations of hyperthermophilic Archaea. Science 2003,301,976–978.

[60] A. A. Perevalova, T. V. Kolganova, N.-K. Birkeland, C. Schleper, E. A. Bonch-Osmolovskaya, A. V. Lebedinsky. Distribution of Crenarchaeota representatives in terrestrial hot springs of Russia and Iceland. Appl Environ Microbiol 2008,74,7620–7628.

[61] H. Huber, W. Eder, eds. *Aquificales*. New York, NY, USA: Springer-Verlag, 2006.

[62] A.-L. Reysenbach, A. Banta, S. Civello, et al. *Aquificales* in Yellowstone National Park. Bozeman: Montana State University, 2005.

[63] H. Huber, R. Huber, K. O. Stetter, eds. *Thermoproteales*. New York, NY, USA: Springer-Verlag, 2006.

[64] H. Huber, D. Prangishvili, eds. *Sulfolobales*. New York, NY USA: Springer-Verlag, 2006.

[65] H. Huber, K. O. Stetter, eds. Order III: *Sulfolobales*. 2nd edn. New York, NY, USA: Springer-Verlag, 2001.

[66] H. Huber, K. O. Stetter, eds. *Thermoplasmatales*2006.

[67] J. R. de la Torre, C. B. Walker, A. E. Ingalls, M. Könneke, D. A. Stahl. Cultivation of a thermophilic ammonia oxidizing archaeon synthesizing crenarchaeol. Environ Microbiol 2008,10,810–818.

[68] E. S. Boyd, T. L. Hamilton, J. Wang, He L, C. L. Zhang. The role of tetraether lipid composition in the adaptation of thermophilic Archaea to acidity. Front Microbiol 2013,4,62.

[69] W. Hou, S. Wang, H. Dong, et al. A comprehensive census of microbial diversity in hot springs of Tengchong, Yunnan Province China using 16S rRNA gene pyrosequencing. PLoS ONE 2013,8,e53350.

[70] H. Jiang, Q. Huang, H. Dong, et al. RNA-based investigation of ammonia-oxidizing Archaea in hot springs of Yunnan Province, China. Appl Environ Microbiol 2010,76,4538–4541.

[71] T. L. Hamilton, E. Koonce, A. Howells, et al. Competition for ammonia influences the structure of chemotrophic communities in geothermal springs. Appl Environ Microbiol 2014,80,653–661.

[72] T. L. Hamilton, R. K. Lange, E. S. Boyd, J. W. Peters. Biological nitrogen fixation in acidic high-temperature geothermal springs in Yellowstone National Park, Wyoming. Environ Microbiol 2011,13,2204–2215.

[73] S. T. Loiacono, D. A. R. Meyer-Dombard, J. R. Havig, A. T. Poret-Peterson, H. E. Hartnett, E. L. Shock. Evidence for high-temperature in situ *nifH* transcription in an alkaline hot spring of Lower Geyser Basin, Yellowstone National Park. Environ Microbiol 2012,14,1272–1283.

[74] A.-S. Steunou, S. I. Jensen, E. Brecht, et al. Regulation of nif gene expression and the energetics of N2 fixation over the diel cycle in a hot spring microbial mat. ISME J 2008,2,364–378.

[75] M. P. Mehta, J. A. Baross. Nitrogen fixation at 92 °C by a hydrothermal vent archaeon. Science 2006,314,1783–1786.

[76] S. Nakagawa, Z. Shtaih, A. Banta, T. J. Beveridge, Y. Sako, A.-L. Reysenbach. *Sulfurihydrogenibium yellowstonense* sp. nov., an extremely thermophilic, facultatively heterotrophic, sulfur-oxidizing bacterium from Yellowstone National Park, and emended descriptions of the genus *Sulfurihydrogenibium*, *Sulfurihydrogenibium subterraneum* and *Sulfurihydrogenibium azorense*. Int J Syst Evol Microbiol 2005.

[77] T. Kawasumi, Y. Igarashi, T. Kodama, Y. Minoda. *Hydrogenobacter thermophilus* gen. nov., sp. nov., an extremely thermophilic, aerobic, hydrogen-oxidizing bacterium. Int J Syst Bact 1984,34,5–10.

[78] J. Donahoe-Christiansen, S. D'Imperio, C. R. Jackson, W. P. Inskeep, T. R. McDermott. Arsenite-oxidizing *Hydrogenobaculum* strain isolated from an Acid-Sulfate-Chloride geothermal spring in Yellowstone National Park. Appl Environ Microbiol 2004,70,1865–1868.

[79] C. Romano, S. D'Imperio, T. Woyke, et al. Comparative genomic analysis of phylogenetically closely related *Hydrogenobaculum* sp. isolates from Yellowstone National Park. Appl Environ Microbiol 2013,79,2932–2943.

[80] S. Nakagawa, K. Takai, K. Horikoshi, Y. Sako. *Persephonella hydrogeniphila* sp. nov., a novel thermophilic, hydrogen-oxidizing bacterium from a deep-sea hydrothermal vent chimney. Int J Syst Evol Microbiol 2003,53,863–869.

[81] D. Götz, A. Banta, T. J. Beveridge, A. I. Rushdi, B. R. T. Simoneit, A. L. Reysenbach. *Persephonella marina* gen. nov., sp. nov. and *Persephonella guaymasensis* sp. nov., two novel, thermophilic, hydrogen-oxidizing microaerophiles from deep-sea hydrothermal vents. Int J Syst Evol Microbiol 2002,52,1349–1359.

[82] S. Mino, H. Maikita, T. Toki, et al. Biogeography of *Persephonella* in deep-sea hydrothermal vents of the Western Pacific. Front Microbiol 2013,4.

[83] E. S. Boyd, W. D. Leavitt, G. G. Geesey. CO_2 uptake and fixation by a thermoacidophilic microbial community attached to precipitated sulfur in a geothermal spring. Appl Environ Microbiol 2009,75,4289–4296.

[84] E. A. Bonch-Osmolovskaya, T. G. Sokolova, N. A. Kostrikina, G. A. Zavarzin. *Desulfurella acetivorans* gen. nov. and sp. nov. – a new thermophilic sulfurreducing eubacterium. Arch Microbiol 1990,153,151–155.

[85] M. L. Miroshnichenko, F. A. Rainey, H. Hippe, N. A. Chernyh, N. A. Kostrikina, E. A. Bonch-Osmolovskaya. *Desulfurella kamchatkensis* sp. nov. and *Desulfurella propionica* sp. nov., new sulfur-respiring thermophilic Bacteria from Kamchatka thermal environments. Int J Syst Bacteriol 1998,2,475–479.

[86] M. L. Miroshnichenko, G. A. Gongadze, A. M. Lysenko, E. A. Bonch-Osmolovskaya. *Desulfurella multipotens* sp. nov., a new sulfur-respiring thermophilic eubacterium from Raoul Island (Kermadec archipelago, New Zealand). Arch Microbiol 1994,161,88–93.

[87] M. I. Prokofeva, N. A. Kostrikina, T. V. Kolganova, et al. Isolation of the anaerobic thermoacidophilic crenarchaeote *Acidilobus saccharovorans* sp. nov. and proposal of *Acidilobales* ord. nov., including *Acidilobaceae* fam. nov. and *Caldisphaeraceae* fam. nov. Int J Syst Evol Microbiol 2009,59,3116–3122.

[88] K. Mori, H. Kim, T. Kakegawa, S. Hanada. A novel lineage of sulfate-reducing microorganisms: *Thermodesulfobiaceae* fam. nov., *Thermodesulfobium narugense*, gen. nov., sp. nov., a new thermophilic isolate from a hot spring. Extremophiles 2003,7,283–290.

[89] T. Itoh, K. Suzuki, P. C. Sanchez, T. Nakase. *Caldivirga maquilingensis* gen. nov., sp. nov., a new genus of rod-shaped crenarchaeote isolated from a hot spring in the Philippines. Int J Syst Bacteriol 1999,49,1157–1163.

[90] S. Burggraf, H. W. Jannasch, B. Nicolaus, K. O. Stetter. *Archaeoglobus profundus* sp. nov., represents a new species within the sulfate-reducing Archaeabacteria. Syst Appl Microbiol 1990,13,24–28.

[91] K. S. Habicht, D. E. Canfield. Sulphur isotope fractionation in modern microbial mats and the evolution of the sulphur cycle. Nature 1996,382,342–343.

[92] M. Klein, M. Friedrich, A. J. Roger, et al. Multiple lateral transfers of dissimilatory sulfite reductase genes between major lineages of sulfate-reducing prokaryotes. J Bacteriol 2001,183,6028–6035.

[93]   V. Zverlov, M. Klein, S. Lücker, et al. Lateral gene transfer of dissimilatory (bi)sulfite reductase revisited. J Bacteriol 2005,187,2203–2208.
[94]   M. I. Prokofeva, M. L. Miroshnichenko, N. A. Kostrikina, et al. *Acidilobus aceticus* gen. nov., sp. nov., a novel anaerobic thermoacidophilic archaeon from continental hot vents in Kamchatka. Int J Syst Evol Microbiol 2000,50,2001–2008.
[95]   T. Itoh, K. Suzuki, P. Sanchez, T. Nakase. *Caldisphaera lagunensis* gen. nov., sp. nov., a novel thermoacidophilic crenarchaeote isolated from a hot spring at Mt Maquiling, Philippines. Int J Syst Evol Microbiol 2003,53,1149–1154.
[96]   T. Itoh, K. Suzuki, T. Nakase. *Thermocladium modestius* gen. nov., sp. nov., a new genus of rod-shaped, extremely thermophilic crenarchaeote. Int J Syst Bacteriol 1998,3,879–887.
[97]   T. Itoh, K. Suzuki, T. Nakase. *Vulcanisaeta distributa* gen. nov., sp. nov., and *Vulcanisaeta souniana* sp. nov., novel hyperthermophilic, rod-shaped crenarchaeotes isolated from hot springs in Japan. Int J Syst Evol Microbiol 2002,52,1097–1104.
[98]   W. Hou, S. Wang, H. Dong, et al. A comprehensive census of microbial diversity in hot springs of Tengchong, Yunnan Province China. PLoS ONE 2013,8,e53350.
[99]   T. D. Brock, K. M. Brock, R. T. Belly, R. L. Weiss. *Sulfolobus*: A new genus of sulfur-oxidizing bacteria living at low pH and high temperature. Arch Microbiol 1972,84,54–68.
[100] D. Grogan, P. Palm, W. Zillig. Isolate B12, which harbours a virus-like element, represents a new species of the archaebacterial genus *Sulfolobus*, *Sulfolobus shibatae*, sp. nov. Arch Microbiol 1990,154,594–599.
[101] G. Huber, K. O. Stetter. *Sulfolobus metallicus* sp. nov., a novel strictly chemolithotrophic thermophilic archaeal species of metal-mobilizers. Syst Appl Microbiol 1991,14.
[102] R. L. Jan, Wu J, S. M. Chaw, C. W. Tsai, S. D. Tsen. A novel species of thermoacidophilic archaeon, *Sulfolobus yangmingensis* sp. nov. Int J Syst Bacteriol 1999,4,1809–1816.
[103] T. Suzuki, Y. Iwasaki, T. Uzawa, et al. *Sulfolobus tokodaii* sp. nov. (f. *Sulfolobus* sp. strain 7), a new member of the genus *Sulfolobus* isolated from Beppu Hot Springs, Japan. Extremophiles 2002,6,39–44.
[104] T. Fuchs, H. Huber, K. Teiner, S. Burggraf, K. O. Stetter. *Metallosphaera prunae*, sp. nov., a novel metal mobilizing, thermoacidophilic Archaeum, isolated from a uranium mine in Germany. Syst Appl Microbiol 1995,18,560–566.
[105] G. Huber, C. Spinnler, A. Gambacorta, K. O. Stetter. *Metallosphaera sedula* gen. and sp. nov. represents a new genus of aerobic, metal-mobilizing, thermoacidophilic Archaebacteria. Syst Appl Microbiol 1989, 12, 38–47.
[106] N. Kurosawa, Y. H. Itoh, T. Itoh. Reclassification of *Sulfolobus hakonensis* Takayanagi et al. 1996 as *Metallosphaera hakonensis* comb. nov. based on phylogenetic evidence and DNA G+C content. Int J Syst Evol Microbiol 2003,53,1607–1608.
[107] R. S. Golovacheva, K. M. Valieho-Roman, A. V. Troitskii. *Sulfurococcus mirabilis* gen. nov., sp. nov., a new thermophilic archaebacterium with the ability to oxidize sulfur. Mikrobiologiya 1987,56,100–107.
[108] G. I. Karavaiko, O. V. Golyshina, A. V. Troitskii, K. M. Valieho-Roman, R. S. Golovacheva, T. A. Pivovarova. *Sulfurococcus yellowstonii* sp. nov., a new species of iron- and sulphur-oxidizing thermoacidophilic Archaebacteria. Microbiol Mol Biol Rev 1994,63,379–387.
[109] N. Kurosawa, Y. H. Itoh, T. Iwai, et al. *Sulfurisphaera ohwakuensis* gen. nov., sp. nov., a novel extremely thermophilic acidophile of the order Sulfolobales. Int J Syst Bacteriol 1998,48,451–456.
[110] M. A. Giaveno, M. S. Urbieta, J. R. Ulloa, E. G. Toril, E. R. Donati. Physiologic versatility and growth flexibility as the main characteristics of a novel thermoacidophilic *Acidianus* strain isolated from Copahue geothermal area in Argentina. Microb Ecol 2013,65, 336–346.

[111] W. Zillig, S. Yeats, I. Holz, et al. *Desulfurolobus ambivalens*, gen. nov., sp. nov., an autotrophic archaebacterium facultatively oxidizing or reducing sulfur. Syst Appl Microbiol 1986,8,197–203.

[112] T. Fuchs, H. Huber, S. Burggraf, K. O. Stetter. 16S rDNA-based phylogeny of the archaeal order *Sulfolobales* and reclassification of *Desulfurolobus ambivalens* as *Acidianus ambivalens* comb. nov. Syst Appl Microbiol 1996,19,56–60.

[113] N. Yoshida, M. Nakasato, N. Ohmura, et al. *Acidianus manzaensis* sp. nov., a novel thermoacidophilic archaeon growing autotrophically by the oxidation of H_2 with the reduction of Fe^{3+}. Curr Microbiol 2006,53,406–411.

[114] A. Segerer, A. Neuner, J. K. Kristjansson, K. O. Stetter. *Acidianus infernus* gen. nov., sp. nov., and *Acidianus brierleyi* comb. nov.: facultatively aerobic, extremely acidophilic thermophilic sulfur-metabolizing Archaebacteria. Int J Syst Bacteriol 1986,36,559–564.

[115] A. H. Segerer, A. Trincone, M. Gahrtz, K. O. Stetter. *Stygiolobus azoricus* gen. nov., sp. nov. represents a novel genus of anaerobic, extremely thermoacidophilic Archaebacteria of the order *Sulfolobales*. Int J Syst Bacteriol 1991,41,495–501.

[116] L. J. Reigstad, A. Richter, H. Daims, T. Urich, L. Schwark, C. Schleper. Nitrification in terrestrial hot springs of Iceland and Kamchatka. FEMS Microbiology Ecology 2008,64,167–174.

[117] A. J. Enright, S. Van Dongen, C. A. Ouzounis. An efficient algorithm for large-scale detection of protein families. Nuc Acids Res 2002,30,1575–1584.

[118] W. D. Swingley, R. E. Blankenship, J. Raymond. Integrating Markov clustering and molecular phylogenetics to reconstruct the cyanobacterial species tree from conserved protein families. Mol Biol Evol 2008,25,643–654.

[119] R. T. Papke, N. B. Ramsing, M. M. Bateson, D. M. Ward. Geographical isolation in hot spring cyanobacteria. Environmental Microbiology 2003,5,650–659.

[120] C. Takacs-Vesbach, K. Mitchell, O. Jackson-Weaver, A.-L. Reysenbach. Volcanic calderas delineate biogeographic provinces among Yellowstone thermophiles. Environmental Microbiology 2008,10,1681–1689.

[121] B. R. Briggs, E. L. Brodie, L. M. Tom, et al. Seasonal patterns in microbial communities inhabiting the hot springs of Tengchong, Yunnan Province, China. Environ Microbiol 2013,n/a-n/a.

[122] T. D. Brock. Micro-organisms adapted to high temperatures. Nature 1967,214,882–885.

[123] T. D. Brock. Lower pH limit for the existence o fblue-green algae: evolutionary and ecological implications. Science 1973,179,480–483.

[124] J. M. Holloway, D. K. Nordstrom, J. K. Böhlke, R. B. McCleskey, J. W. Ball. Ammonium in thermal waters of Yellowstone National Park: Processes affecting speciation and isotope fractionation. Geochim Cosmochim Acta 2011,75,4611–4636.

[125] B. Schink. Energetics of syntrophic cooperation in methanogenic degradation. Microbiol Mol Biol Rev 1997,61,262–280.

[126] W. P. Inskeep, T. R. McDermott. Geomicrobiology of acid-sulfate-chloride springs in Yellowstone National Park. In: W. P. Inskeep, T. R. McDermott, eds. Geothermal biology and geochemistry in Yellowstone National Park. Bozeman: Montana State University, 2005,143–162.

[127] J. R. Spear, J. J. Walker, T. M. McCollom, N. R. Pace. Hydrogen and bioenergetics in the Yellowstone geothermal ecosystem. Proc Natl Acad Sci USA 2005,102,2555–2560.

[128] T. O. Stevens, J. P. McKinley. Abiotic controls on H_2 production from basalt-water reactions and iImplications for aquifer biogeochemistry. Environ Sci Technol 2000,34,826–831.

[129] Y. Koga, H. Morii. Biosynthesis of ether-type polar lipids in Archaea and evolutionary considerations. Microbiol Mol Biol Rev 2007,71,97–120.

[130] P. L.-G. Chong, U. Ayesa, V. Prakash Daswani, E. C. Hur. On physical properties of tetraether lipid membranes: effects of cyclopentane rings. Archaea 2012,138439.

[131] J. B. Parsons, C. O. Rock. Bacterial lipids: Metabolism and membrane homeostasis. Prog Lip Res 2013,52,249–276.
[132] J. L. Macalady, M. M. Vestling, D. Baumler, N. Boekelheide, C. W. Kasper, J. F. Banfield. Tetraether-linked membrane monolayers in *Ferroplasma* spp: a key to survival in acid. Extremophiles 2004,8,411–419.
[133] A. Pearson, Pi Y, W. Zhao, et al. Factors controlling the distribution of archaeal tetraethers in terrestrial hot springs. Appl Environ Microbiol 2008,74,3523–3532.
[134] Y. Koga. Thermal adaptation of the Archaeal and Bacterial lipid membranes. Archaea 2012,789652.
[135] S. Schouten, E. C. Hopmans, J. S. Sinninghe Damsté. The organic geochemistry of glycerol dialkyl glycerol tetraether lipids: a review. Org Geochem 2013,54,19–61.
[136] S. Schouten, E. C. Hopmans, R. D. Pancost, J. S. Sinninghe Damsté. Widespread occurence of structurally diverse tetraether membrane lipids: evidence for the ubiquitous presence of low-temperature relatives of hyperthermophiles. Proc Natl Acad Sci USA 2000,97,14421–14426.
[137] A. Pearson, Z. Huang, A. E. Ingalls, et al. Nonmarine Crenarchaeol in Nevada hot springs. Appl Environ Microbiol 2004,70,5229–5237.
[138] M. De Rosa, A. Gambacorta. The lipids of Archaebacteria. Prog Lipid Res 1988,27,153–175.
[139] J. L. Gabriel, P. L. G. Chong. Molecular modeling of archaebacterial bipolar tetraether lipid membranes. Chem Phys Lipids 2000,105,193–200.
[140] A. Gliozzi, G. Paoli, M. de Rosa, A. Gambacorta. Effect of isoprenoid cyclization on the transition temperature of lipids in thermophilic Archaebacteria. Biochim Biophys Acta 1983,735,234–242.
[141] M. G. L. Elferink, J. G. de Wit, A. J. M. Driessen, W. N. Konings. Stability and proton-permeability of liposomes composed of archaeal tetraether lipids. Biochim Biophys Acta 1994,1193,247–254.
[142] E. S. Boyd, A. Pearson, Pi Y, et al. Temperature and pH controls on glycerol dibiphytanyl glycerol tetraether lipid composition in the hyperthermophilic crenarchaeon *Acidilobus sulfurireducens*. Extremophiles 2011,15,59–65.
[143] C. S. Knappy, C. E. M. Nunn, H. W. Morgan, B. J. Keely. The major lipid cores of the archaeon *Ignisphaera aggregans*: implications for the phylogeny and biosynthesis of glycerol monoalkyl glycerol tetraether isoprenoid lipids. Extremophiles 2011,15,517–528.
[144] A. Sugai, Y. Masuchi, I. Uda, T. Itoh, Y. Itoh. Core lipids of hyperthermophilic archaeon, *Pyrococcus horikoshii* OT 3. Japan Oil Chem Soc 2000,49,695–700.
[145] A. Sugai, I. Uda, Y. Itoh, H. Itoh. The core lipid composition of the 17 Strains of hyperthermophilic Archaea, *Thermococcales*. J Oleo Science 2004,53,41–44.
[146] S. Schouten, M. Baas, E. C. Hopmans, A.-L. Reysenbach, J. S. Sinninghe Damsté. Tetraether membrane lipids of Candidatus "Aciduliprofundum boonei", a cultivated obligate thermoacidophilic euryarchaeote from deep-sea hydrothermal vents. Extremophiles 2008,12,119–124.
[147] H. Morii, T. Eguchi, M. Nishihara, K. Kakinuma, H. König, Y. Koga. A novel ether core lipid with H-shaped C_{80}-isoprenoid hydrocarbon chain from the hyperthermophilic methanogen *Methanothermus fervidus*. Biochim Biophys Acta 1998,1390,339–345.
[148] A. Jaeschke, S. L. Jørgensen, S. M. Bernasconi, R. B. Pedersen, I. H. Thorseth, G. L. Früh-Green. Microbial diversity of Loki's Castle black smokers at the Arctic Mid-Ocean Ridge. Geobiol 2012,10,548–561.
[149] S. A. Lincoln, A. S. Bradley, S. A. Newman, R. E. Summons. Archaeal and bacterial glycerol dialkyl glycerol tetraether lipids in chimneys of the Lost City Hydrothermal Field. Org Geochem 2013,60,45–53.
[150] R. A. Gibson, M. T. J. van der Meer, E. C. Hopmans, A.-L. Reysenbach, S. Schouten, J. S. Sinninghe Damsté. Comparison of intact polar lipid with microbial community composition of vent deposits of the Rainbow and Lucky Strike hydrothermal fields. Geobiol 2013,11,72–85.

[151] J. Jia, C. L. Zhang, W. Xie, et al. Differential temperature and pH controls on the abundance and composition of H-GDGTs in terrestrial hot springs. Org Geochem 2014,*In press*.
[152] E. R. L. Gaughran. The saturation of bacterial lipids as a function of temperature. J Bacteriol 1947,53,506.
[153] N. J. Russell, N. Fukunaga. A comparison of thermal adaptation of membrane lipids in psychrophilic and thermophilic Bacteria. FEMS Microbiol Rev 1990,75,171–182.
[154] S. Hurwitz, J. B. Lowenstern. Dynamics of the Yellowstone hydrothermal system. Reviews of Geophysics 2014,2014RG000452.
[155] S. Aihara, M. Kato, M. Ishinaga, M. Kito. Changes in positional distribution of fatty acids in the phospholipids of *Escherichia coli* after shift-down in temperature. Biochim Biophys Acta 1972,270,301–306.
[156] J. L. C. M. van de Vossenberg, A. J. M. Driessen, W. N. Konings. The essence of being extremophilic: the role of the unique archaeal membrane lipids. Extremophiles 1998,2,163–170.
[157] M. C. Mansilla, L. E. Cybulski, D. Albanesi, D. De Mendoza. Control of membrane lipid fluidity by molecular thermosensors. J Bacteriol 2004,186,6681–6688.
[158] J. Reizer, N. Grossowicz, Y. Barenholz. The effect of growth temperature on the thermotropic behavior of the membranes of thermophilic Bacillus. Composition-structure-function relationships. Biochim Biophys Acta 1985,815,268–280.
[159] J. M. Aerts, A. M. Lauwers, W. Heinen. Temperature-dependent lipid content and fatty acid composition of three thermophilic Bacteria. Antonie Van Leeuwenhoek 1985,51,155–165.
[160] J. S. Sinninghe Damsté, E. C. Hopmans, R. D. Pancost, S. Schouten, J. A. J. Geenevasen. Newly discovered non-isoprenoid dialkyl diglycerol tetraether lipids in sediments. J Chem Soc Chem Comm 2000,23,1683–1684.
[161] J. W. H. Weijers, S. Schouten, E. C. Hopmans, et al. Membrane lipids of mesophilic anaerobic Bacteria thriving in peats have typical archaeal traits. Appl Environ Microbiol 2006,8,648–657.
[162] S. Schouten, M. T. J. van der Meer, E. C. Hopmans, et al. Archaeal and bacterial glycerol dialkyl glycerol tetraether lipids in hot springs of Yellowstone National Park. Appl Environ Microbiol 2007,73,6181–6191.
[163] F. Peterse, S. Schouten, J. van der Meer, M. T. J. van der Meer, J. S. Sinninghe Damsté. Distribution of branched tetraether lipids in geothermally heated soils: implications for the MBT/CBT temperature proxy. Org Geochem 2009,40,201–205.
[164] He L, C. L. Zhang, H. Dong, B. Fang, G. Wang. Distribution of glycerol dialkyl glycerol tetraethers in Tibetan hot springs. Geosci Front 2012,3,289–300.
[165] B. P. Hedlund, J. J. Paraiso, A. J. Williams, et al. Wide distribution of autochthonous branched glycerol dialkyl glycerol tetraethers (bGDGTs) in US Great Basin hot springs. Front Microbiol 2013,4.
[166] C. Zhang, J. Wang, J. Dodsworth, et al. *In situ* production of branched glycerol dialkyl glycerol tetraethers in a Great Basin Hot Spring (USA). Front Microbiol 2013,4.
[167] C. Vieille, G. J. Zeikus. Hyperthermophilic enzymes: sources, uses, and molecular mechanisms for thermostability. Microbiol Mol Biol Rev 2001,65,1–43.
[168] R. Sterner, W. Liebl. Thermophilic adaptation of proteins. Crit Rev Biochem Mol Biol 2001,36,39–106.
[169] C. Vetriani, D. L. Maeder, N. Tolliday, et al. Protein thermostability above 100 C: a key role for ionic interactions. Proc Natl Acad Sci USA 1998,95,12300–12305.
[170] R. Scandurra, V. Consalvi, R. Chiaraluce, L. Politi, P. C. Engel. Protein stability in extremophilic Archaea. Front Biosci 2000,5,D787-D95.
[171] R. J. M. Russell, J. M. C. Ferguson, D. W. Hough, M. J. Danson, G. L. Taylor. The crystal structure of citrate synthase from the hyperthermophilic Archaeon *Pyrococcus furiosus* at 1.9 Å resolution. Biochemistry 1997,36,9983–9994.

[172] K. S. P. Yip, K. L. Britton, T. J. Stillman, et al. Insights into the molecular basis of thermal stability from the analysis of ion-pair networks in the glutamate dehydrogenase family. Eur J Biochem 1998,255,336–346.

[173] S. Knapp, W. M. de Vos, D. Rice, R. Ladenstein. Crystal structure of glutamate dehydrogenase from the hyperthermophilic eubacterium *Thermotoga maritima* at 3.0 Å resolution. J Mol Biol 1997, 267, 916–932.

[174] P. Haney, J. Konisky, K. K. Koretke, Z. Luthey-Schulten, P. G. Wolynes. Structural basis for thermostability and identification of potential active site residues for adenylate kinases from the archaeal genus *Methanococcus*. Proteins 1997,1997,117–130.

[175] S. Sen, J. W. Peters. The thermal adaptation of the nitrogenase Fe protein from thermophilic *Methanobacter thermoautotrophicus*. Proteins 2006, 62, 450–460.

[176] R. Sterner, W. Liebl. Thermophilic adaptation of proteins. Crit Rev Biochem Mol Biol 2001,36,39–106.

[177] L. O. Martins, R. Huber, H. Huber, K. O. Stetter, M. S. Da Costa, H. Santos. Organic solutes in hyperthermophilic Archaea. Appl Environ Microbiol 1997,63,896–902.

[178] L. O. Martins, N. Empadinhas, J. D. Marugg, et al. Biosynthesis of mannosylglycerate in the thermophilic bacterium *Rhodothermus marinus*: Biochemical and genetic characterization of a mannosylglycerate synthase. J Biol Chem 1999,274,35407–35414.

[179] O. C. Nunes, C. M. Manaia, M. S. Da Costa, H. Santos. Compatible solutes in the thermophilic bacteria *Rhodothermus marinus* and "Thermus thermophilus". Appl Environ Microbiol 1995,61,2351–2357.

[180] L. O. Martins, H. Santos. Accumulation of mannosylglycerate and di-myo-inositol-phosphate by *Pyrococcus furiosus* in response to salinity and temperature. Appl Environ Microbiol 1995,61,3299–3303.

[181] R. A. Ciulla, S. Burggraf, K. O. Stetter, M. F. Roberts. Occurrence and role of di-myo-Inositol-1,1'-phosphate in *Methanococcus igneus*. Appl Environ Microbiol 1994,60,3660–3664.

[182] Z. Silva, N. Borges., L. O. Martins, R. Wait, M. S. da Costa, H. Santos. Combined effect of the growth temperature and salinity of the medium on the accumulation of compatible solutes by *Rhodothermus marinus* and *Rhodothermus obamensis*. Extremophiles 1999,3,163–172.

[183] L. O. Martins, L. S. Carreto, M. S. Da Costa, H. Santos. New compatible solutes related to Di-myo-inositol-phosphate in members of the order *Thermotogales*. J Bacteriol 1996,178,5644–5651.

[184] M. S. da Costa, H. Santos. Extremophiles: Compatible solutes in microorganisms that grow at high temperature. In: C. Gerday, N. Glansdorff, eds. in Encyclopedia of Life Support Systems (EOLSS). Paris, France: Eolss Publishers, 2009,265–281.

[185] N. Borges, A. Ramos, N. D. H. Raven, R. J. Sharp, H. Santos. Comparative study on the thermostabilizing effect of mannosylglycerate and other compatible solutes on model enzymes. Extremophiles 2002,6,209–216.

[186] R. Hensel, H. König. Thermoadaptation of methanogenic Bacteria by intracellular ion concentration. FEMS Micobiol Lett 1988,49,75–79.

[187] P. Lamosa, A. Burke, R. Peist, et al. Thermostabilization of proteins by diglycerol phosphate, a new compatible solute from the hyperthermophile *Archaeoglobus fulgidus*. Appl Environ Microbiol 2000,66,1974–1979.

[188] S. Shima, R. K. Thauer, U. Ermler. Hyperthermophilic and salt-dependent formyltransferase from *Methanopyrus kandleri*. Biochem Soc Trans 2004,32,269–272.

[189] S. Shima, D. A. Hérault, A. Berkessel, R. K. Thauer. Activation and thermostabilization effects of cyclic 2, 3-diphosphoglycerate on enzymes from the hyperthermophilic *Methanopyrus kandleri*. Arch Microbiol 1998,170,469–472.

[190] M. Kurr, R. Huber, H. König, et al. *Methanopyrus kandleri*, gen. and sp. nov. represents a novel group of hyperthermophilic methanogens, growing at 110 °C. Arch Microbiol 1991,156,239–247.
[191] J. W. Costerton, Z. Lewandowski, D. E. Caldwell, D. R. Korber, H. M. Lappin-Scott. Microbial biofilms. Ann Rev Microbiol 1995,49,711–745.
[192] K. D. Rinker, R. M. Kelly. Effect of carbon and nitrogen sources on growth dynamics and exopolysaccharide production for the hyperthermophilic archaeon *Thermococcus litoralis* and bacterium *Thermotoga maritima*. Biotechnol Bioeng 2000,69,537–547.
[193] D. M. Ward, M. J. Ferris, S. C. Nold, M. M. Bateson. A natural view of microbial biodiversity within hot spring cyanobacterial mat communities. Microb Mol Biol Rev 1998,62,1353–1370.
[194] S. M. Boomer, K. L. Noll, G. G. Geesey, B. E. Dutton. Formation of multilayered photosynthetic biofilms in an alkaline thermal spring in Yellowstone National Park, Wyoming. Appl Environ Microbiol 2009,75,2464–2475.
[195] T. L. Bott, T. D. Brock. Bacterial growth rates above 90 degrees C in Yellowstone hot springs. Science 1969,164,1411–1412.
[196] T. D. Brock, M. L. Brock, T. L. Bott, M. R. Edwards. Microbial life at 90 C: the sulfur Bacteria of Boulder Spring. J Bacteriol 1971,107,303–314.
[197] T. D. Brock. Life at high temperatures. Science 1967,158,1012–1019.
[198] G. H. Wadhams, J. P. Armitage. Making sense of it all: bacterial chemotaxis. Nature Rev Mol Cell Biol 2004,5,1024–1037.
[199] H. Szurmant, G. W. Ordal. Diversity in chemotaxis mechanisms among the Bacteria and Archaea. Microbiol Mol Biol Rev 2004,68,301–319.
[200] E. C. Kofoid, J. S. Parkinson. Transmitter and receiver modules in bacterial signaling proteins. Proc Natl Acad Sci USA 1988,85,4981–4985.
[201] N. Hirota. Chemotaxis in thermophilic bacterium PS-3. J Biochem 1984,96,645–650.
[202] D. L. Maeder, R. B. Weiss, D. M. Dunn, et al. Divergence of the hyperthermophilic Archaea *Pyrococcus furiosus* and *P. horikoshii* inferred from complete genomic sequences. Genetics 1999,152,1299–1305.
[203] C. J. Bult, O. White, G. J. Olsen, et al. Complete genome sequence of the methanogenic archaeon, *Methanococcus jannaschii*. Science 1996,273,1058–1073.
[204] G. Deckert, P. V. Warren, T. Gaasterland, et al. The complete genome of the hyperthermophilic bacterium *Aquifex aeolicus*. Nature 1998,392,353–358.
[205] A.-L. Reysenbach, E. Shock. Merging genomes with geochemistry in hydrothermal ecosystems. Science 2002,296,1077–1082.
[206] W. S. Ryu, A. D. T. Samuel. Thermotaxis in *Caenorhabditis elegans* analyzed by measuring responses to defined thermal stimuli. J Neuroscience 2002,22,5727–5733.
[207] T. Nara, I. Kawagishi, S. Nishiyama-i, M. Homma, Y. Imae. Modulation of the thermosensing profile of the *Escherichia coli* aspartate receptor Tar by covalent modification of its methyl-accepting sites. J Biol Chem 1996,271,17932–17936.
[208] J. P. Allewalt, M. M. Bateson, N. P. Revsbech, K. Slack, D. M. Ward. Effect of temperature and light on growth of and photosynthesis by *Synechococcus* isolates typical of those predominating in the Octopus Spring microbial mat community of Yellowstone National Park. Appl Environ Microbiol 2006,72,544–550.
[209] A. L. Ruff-Roberts, J. G. Kuenen, D. M. Ward. Distribution of cultivated and uncultivated cyanobacteria and *Chloroflexus*-like bacteria in hot spring microbial mats. Appl Environ Microbiol 1994,60,697–704.
[210] O. Lecompte, R. Ripp, V. Puzos-Barbe, et al. Genome evolution at the genus level: comparison of three complete genomes of hyperthermophilic Archaea. Genome Res 2001,11,981–993.

[211] K. V. Gunbin, D. A. Afonnikov, N. A. Kolchanov. Molecular evolution of the hyperthermophilic Archaea of the *Pyrococcus* genus: analysis of adaptation to different environmental conditions. BMC Genomics 2009,10,639.
[212] M. C. Portillo, J. M. Gonzalez. CRISPR elements in the *Thermococcales*: evidence for associated horizontal gene transfer in *Pyrococcus furiosus*. J Appl Genet 2009,50,421–430.
[213] K. S. Makarova, N. V. Grishin, S. A. Shabalina, Y. I. Wolf, E. V. Koonin. A putative RNA-interference based immune system in prokaryotes: computational analysis of the predicted enzymatic machinery, functional analogies with eukaryotic RNAi, and hypothetical mechanisms of action. Biol Direct 2006,7,1–26.
[214] R. Sorek, V. Kunin, P. Hugenholtz. CRISPR – a widespread system that provides acquired resistance against phages in Bacteria and Archaea. Nat Rev Microbiol 2008,6,181–186.
[215] K. E. Nelson, R. A. Clayton, S. R. Gill, et al. Evidence for lateral gene transfer between Archaea and Bacteria from genome sequence of *Thermotoga maritima*. Nature 1999,399,323–329.
[216] K. E. Nelson, J. A. Eisen, C. M. Fraser. Genome of *Thermotoga maritima* MSB8. Methods Enzymol 2001,330,169–180.
[217] C. L. Nesbo, S. L'Haridon, K. O. Stetter, W. F. Doolittle. Phylogenetic analyses of two "archaeal" genes in *Thermotoga maritima* reveal multiple transfers between Archaea and Bacteria. Mol Biol Evol 2001,18,362–375.
[218] C. L. Nesbo, K. E. Nelson, W. F. Doolittle. Suppressive subtractive hybridization detects extensive genomic diversity in *Thermotoga maritima*. J Bacteriol 2002,184,4475–4488.
[219] D. M. Nanavati, K. Thirangoon, K. M. Noll. Several archaeal homologs of putative oligopeptide-binding proteins encoded by *Thermotoga maritima* bind sugars. Appl Environ Microbiol 2006,72,1336–1345.
[220] R. Wirth, J. Sikorski, E. Brambilla, et al. Complete genome sequence of *Thermocrinis albus* type strain (HI 11/12 T). Stand Genomic Sci 2010, 2, 194–202.
[221] A. Zeytun, J. Sikorski, M. Nolan, et al. Complete genome sequence of *Hydrogenobacter thermophilus* type strain (TK-6 T). Stand Genomic Sci 2011,4,131–143.
[222] E. S. Boyd, T. L. Hamilton, J. W. Peters. An alternative path for the evolution of biological nitrogen fixation. Front Microbiol 2011,2,205.
[223] C. Romano, S. D'Imperio, T. Woyke, et al. Comparative genomic analysis of phylogenetically closely related *Hydrogenobaculum* sp. isolates from Yellowstone National Park. Appl Environ Microbiol 2013,79,2932–2943.
[224] J. Raymond, J. L. Siefert, C. R. Staples, R. E. Blankenship. The natural history of nitrogen fixation. Mol Biol Evol 2004,21,541–554.
[225] H. Bolhuis, I. Severin, V. Confurius-Guns, U. I. A. Wollenzien, L. J. Stal. Horizontal transfer of the nitrogen fixation gene cluster in the cyanobacterium *Microcoleus chthonoplastes*. ISME J 2009,4,121–130.
[226] K. J. Kechris, J. C. Lin, P. J. Bickel, A. N. Glazer. Quantitative exploration of the occurrence of lateral gene transfer by using nitrogen fixation genes as a case study. Proc Natl Acad Sci U S A 2006,103,9584–9589.
[227] C. M. Thomas, K. M. Nielsen. Mechanisms of, and barriers to, horizontal gene transfer between Bacteria. Nat Rev Microbiol 2005,3,711–721.
[228] E. V. Koonin, K. S. Makarova, L. Aravind. Horizontal gene transfer in prokaryotes: Quantification and classification. Annu Rev Microbiol 2001,55,709–742.
[229] L. Aravind, R. L. Tatusov, Y. I. Wolf, D. R. Walker, E. V. Koonin. Evidence for massive gene exchange between archaeal and bacterial hyperthermophiles. Trends Genet 1998,14,442–444.
[230] P. Worning, L. J. Jensen, K. E. Nelson, S. Brunak, D. W. Ussery. Structural analysis of DNA sequence: evidence for lateral gene transfer in *Thermotoga maritima*. Nuc Acids Res 2000,28,706–709.

[231] A. Kanhere, M. Vingron. Horizontal gene transfers in prokaryotes show differential preferences for metabolic and translational genes. BMC Evol Biol 2009,9,9.
[232] R. Jain, M. C. Rivera, J. A. Lake. Horizontal gene transfer among genomes: the complexity hypothesis. Proc Natl Acad Sci USA 1999,96,3801–3806.
[233] R. F. Doolittle, J. Handy. Evolutionary anomalies among the aminoacyl-tRNA synthetases. Curr Opin Genet Dev 1998,8,630–636.
[234] Y. I. Wolf, L. Aravind, N. V. Grishin, E. V. Koonin. Evolution of aminoacyl-tRNA synthetases– analysis of unique domain architectures and phylogenetic trees reveals a complex history of horizontal gene transfer events. Genome Res 1999,9,689–710.
[235] C. R. Woese, G. J. Olsen, M. Ibba, D. Soll. Aminoacyl-tRNA synthetases, the genetic code, and the evolutionary process. Microbiol Mol Biol Rev 2000,64,202–236.
[236] S. F. Ataide, M. Ibba. Small molecules: big players in the evolution of protein synthesis. ACS Chemical Biology 2006,1,285–297.
[237] Y. I. Wolf, L. Aravind, N. V. Grishin, E. V. Koonin. Evolution of aminoacyl-tRNA synthetases– analysis of unique domain architectures and phylogenetic trees reveals a complex history of horizontal gene transfer events. Genome Res 1999,9,689–710.
[238] F. X. Sicot, M. Mesnage, M. Masselot, et al. Molecular adaptation to an extreme environment: origin of the thermal stability of the pompeii worm collagen. J Mol Biol 2000,302,811–820.
[239] E. S. Boyd, A. D. Anbar, S. Miller, T. L. Hamilton, M. Lavin, J. W. Peters. A late methanogen origin for molybdenum-dependent nitrogenase. Geobiol 2011,9,221–232.
[240] J. T. Roll, V. K. Shah, D. R. Dean, G. P. Roberts. Characteristics of NIFNE in *Azotobacter vinelandii* strains. J Biol Chem 1995,270,4432–4437.
[241] R. A. Ugalde, J. Imperial, V. K. Shah, W. J. Brill. Biosynthesis of iron-molybdenum cofactor in the absence of nitrogenase. J Bacteriol 1984,159,888–893.
[242] Hu Y, A. W. Fay, M. W. Ribbe. Identification of a nitrogenase FeMo cofactor precursor on NifEN complex. Proc Natl Acad Sci U S A 2005,102,3236–3241.
[243] M. R. Jacobson, K. E. Brigle, L. T. Bennett, et al. Physical and genetic map of the major nif gene cluster from *Azotobacter vinelandii*. J Bacteriol 1989,171,1017–1027.
[244] R. Fani, R. Gallo, Liò P. Molecular evolution of nitrogen fixation: the evolutionary history of the *nifD*, *nifK*, *nifE*, and *nifN* genes. J Mol Evol 2000,51,1–11.
[245] B. Soboh, E. S. Boyd, D. Zhao, J. W. Peters, L. M. Rubio. Substrate specificity and evolutionary implications of a NifDK enzyme carrying NifB-co at its active site. FEBS Letters 2010,584,1487–1492.
[246] L. Curatti, P. W. Ludden, L. M. Rubio. NifB-dependent *in vitro* synthesis of the iron-molybdenum cofactor of nitrogenase. Proc Natl Acad Sci USA 2006,103,5297–5301
[247] J. Christiansen, P. J. Goodwin, W. N. Lanzilotta, L. C. Seefeldt, D. R. Dean. Catalytic and biophysical properties of a nitrogenase Apo-MoFe protein produced by a *nifB*-deletion mutant of *Azotobacter vinelandii*. Biochem 1998,37,12611–12623.
[248] L. M. Rubio, P. W. Ludden. Biosynthesis of the iron-molybdenum cofactor of nitrogenase. Annu Rev Microbiol 2008,62,93–111.
[249] E. S. Boyd, G. Schut, M. W. W. Adams, W. John. J. W. Peters. Hydrogen metabolism and the evolution of biological respiration. Microbe 2014,In press.
[250] E. Boyd, J. W. Peters. New insights into the evolutionary history of biological nitrogen fixation. Frontiers in Microbiology 2013,4.
[251] V. J. Denef, J. F. Banfield. In situ evolutionary rate measurements show ecological success of recently emerged bacterial hybrids. Science 2012,336,462–466.

Aharon Oren

5 Halophilic microorganisms and adaptation to life at high salt concentrations – evolutionary aspects

5.1 Phylogenetic and physiological diversity of halophilic microorganisms

In a paper published in 1998 [1], Ian Dundas speculated about the question: "Was the environment for primordial life hypersaline?" Based on the fact that the cells of extremely halophilic Archaea such as *Halobacterium* contain molar concentrations of KCl, the author wrote: "One can envisage, in the prebiotic world, an aqueous hypersaline environment enriched in dissolved potassium chloride and from which most of the sodium chloride had been lost by evaporation." The finding that many representatives of the *Halobacteriaceae* can also grow in the absence of molecular oxygen [2–4] was presented to support the idea. It is interesting to note that in the Miller/Urey type spark discharge experiments there is a strong bias for the formation of acidic amino acids and that the foldable polypeptides derived from available prebiotic α-amino acids may have been dependent on salt for stability [5, 6]. But there is little evidence in favor of a hypersaline origin of life. Still, halophiles are found nearly everywhere in the tree of life as based on small subunit rRNA sequence comparions. We find microorganisms growing above 100–150 g/l salt and often up to NaCl saturation in all three domains of life (▶ Tab. 5.1). The *Halobacteriaceae* [2, 7] are not the only archaeal halophiles: some methanogens require and tolerate high salt, and the recent discovery of the 'Nanohaloarchaea', a group of extremely small members of the Euryarchaeota not closely related to the *Halobacteriaceae*, shows that more groups of archaeal halophiles may be discovered. These 'Nanohaloarchaea' are widespread in saltern ponds and natural hypersaline lakes, and they have small (~ 1.2 Mb) genomes [8, 9]. No representatives of the group have yet been cultured.

Within the Domain Bacteria we find halophiles (here operationally defined as organisms able to grow at relatively high rates at salt concentrations above 100 g/l) in many branches of the phylogenetic tree, and they show very different life styles. They include aerobic and anoxygenic phototrophic *Gammaproteobacteria*, dissimilatory sulfate reducing *Deltaproteobacteria*, members of the *Firmicutes*, the *Actinobacteria*, the *Spirochaetes*, and the *Bacteroidetes* [10, 11]. The extremely halophilic aerobic red *Salinibacter* (*Bacteroidetes*) [12–15] and the order *Halanaerobiales* (*Firmicutes*), consisting of anaerobic, mostly fermentative species [16] will be discussed below in further depth. Also intriguing are the polyextremophilic anaerobic halothermoalkaliphiles of the order *Natranaerobiales* [17].

Table 5.1. Phylogenetic affiliation of halophilic microorganisms mentioned in this chapter.

Domain	Phylum/Kingdom	Class	Order	Family	Genus
Archaea	Euryarchaeota	Halobacteria	Halobacteriales	Halobacteriaceae	*Halobacterium*
					Haloferax
					Halohasta
					Haloquadratum
					Halorubrum
Bacteria	Proteobacteria	Gammaproteobacteria	Chromatiales	Ectothiorhodospiraceae	*Halorhodospira*
			Oceanospirillales	Halomonadaceae	*Chromohalobacter*
					Halomonas
	Bacteroidetes	Cytophagia	Cytophagales	Rhodothermaceae	*Salinibacter*
	Firmicutes	Clostridia	Halanaerobiales	Halanaerobiaceae	*Halanaerobium*
					Halothermothrix
				Halobacteroidaceae	*Halobacteroides*
					Natroniella
					Orenia
					Salinibacter
			Natranaerobiales	Natranaerobiaceae	*Natranaerobius*
		Bacilli	Bacillales	Bacillaceae	*Halobacillus*
	Deferribacteres	Deferribacteres	Deferribacterales	Deferribacteraceae	*Flexistipes*
Eukaryota	Chlorophyta	Chlorophyceae	Chlamydomonadales	Dunaliellaceae	*Dunaliella*
	Fungi	Dothideomycetes	Capnodiales	Teratosphaeriaceae	*Hortaea*
		Wallemiomycetes	Wallemiales	Wallemiaceae	*Wallemia*

The best known halophilic or highly halotolerant eukaryote is the green alga *Dunaliella salina*, the main or sole primary producer in brines approaching salt saturation. Its cells are often colored orange-red due to massive accumulation of β-carotene granules within the chloroplast. Also some fungi can grow at extremely high salinities, e.g. the black yeast *Hortaea werneckii* (NaCl range from near-zero to almost saturation) and *Wallemia ichthyophaga*, a true halophile requiring > 100 g/l salt. It is interesting to compare the genomes of these two fungi: *H. werneckii* has a large genome (51.6 Mb) and possesses two highly identical gene copies for almost every protein. The increase in genomic DNA may have been triggered by salt stress [18]. *W. ichthyophaga* has a much smaller genome (9.6 Mb) lacking obvious salt-related features [19].

5.2 What adaptations are necessary to become a halophile?

Biological membranes are permeable to water. Therefore halophilic or halotolerant microorganisms must maintain a cytoplasm that is osmotically at least equivalent to the osmotic pressure of the surrounding brines. Osmotic balance can be achieved by two fundamentally different strategies. One involves the accumulation of inorganic ions and adaptation of the entire intracellular machinery to presence of molar concentrations of ions (mainly K^+ and Cl^-); the second is based on exclusion of salts and synthesis of organic 'compatible' solutes or their accumulation from outside when present in their medium.

The 'salt-in' strategy is best known from *Halobacterium* and other members of the archaeal family *Halobacteriaceae*. Examination of their proteome shows a large excess of acidic amino acids (glutamate, aspartate) over basic amino acids (lysine, arginine) in nearly every protein. In addition, the content of hydrophobic amino acids tends to be low [20–22]. The high negative surface charge makes the proteins more soluble and renders them more flexible at high salt. Such proteins require the constant presence of salts to maintain their structural integrity, and they denature in the absence of salt [20, 23, 24]. As a result, the 'salt-in' strategy does not allow a high degree of adaptability to changing salinity. The 'salt-in' strategy, combined with a highly acidic proteome, is also used by *Salinibacter*, which phylogenetically belongs to the *Bacteroidetes* branch of the Bacteria. Convergent evolution may have led to highly similar physiological properties of *Halobacterium* and relatives and *Salinibacter*, organisms that occur together in the same habitats [12]. High concentrations of inorganic ions were documented in a few other members of the Bacteria: species of the order *Halanaerobiales* (*Firmicutes*) [16] and the anoxygenic phototrophic gammaproteobacterium *Halorhodospira halophila* [25]. Evolutionary aspects relating to these two groups are further discussed below.

The alternative to the 'salt-in' strategy is the use of organic osmotic solutes, generally uncharged or zwitterionic low-molecular-weight compounds that are accumulated inside the cell while excluding salt to a large extent. *Dunaliella* and halophilic fungi use glycerol for this purpose, and compounds such as glycine betaine and ectoine (1,4,5,6-tetrahydro-2-methyl-pyrimidine-4-carboxylic acid) and hydroxyectoine are used by many halophilic and halotolerant members of the Bacteria. More than twenty different compounds were identified thus far as osmotic solutes, and often cocktails of different compounds are used [10, 26]. Halophilic methanogenic Archaea also use organic solutes for osmotic balance. The advantage of the use of organic solutes is the possibility to rapidly adapt to salinity changes by adjusting the solute concentrations within the cells. No special modifications of the intracellular enzymes are needed and no highly acidic proteome is expected (see however, further discussions below). Possible exceptions are the periplasmic proteins that are exposed to the high salinity of the medium, as shown in an analysis of selected proteins of *Chromohalobacter salexigens* (*Gammaproteobacteria*) [27].

In many cases there still is considerable ambiguity as to the exact cytoplasmic concentration of sodium and other inorganic salts and of organic osmotic solutes. The precise determination of sodium inside the cells (against a high concentration outside) is experimentally challenging. And often the estimated cytoplasmic concentrations of organic osmolytes are far from sufficient to balance the osmotic pressure of the medium. The possibility cannot be excluded that organic osmotic solutes may work in conjunction with inorganic salts. As discussed in the final section of this chapter, there indeed have been reports of cases in which both strategies can be combined.

5.3 Is an acidic (meta)proteome indeed indicative for halophily and high intracellular ionic concentrations?

As halophilic behavior based on high intracellular ion concentrations is linked to a highly acidic proteome both in the *Halobacteriaceae* and (probably by convergent evolution) in *Salinibacter* [12], organisms that dominate the biota at the highest salt concentrations, it is interesting to examine the correlation between salinity and the acidic nature of the proteins along salinity gradients up to salt saturation. Indeed a strong correlation was found between the ratio of acidic to basic amino acids in the metaproteome and salinity, the highest values (~1.46) being reported for the Dead Sea and for saltern crystallizer ponds. Marine samples gave values of 0.86–0.95, and the benthic microbial mats in the 9% salt lagoons of Guerrero Negro, Mexico yielded an intermediate value of ~1.01 [28, 29]. The acidic nature of the metaproteome in saltern ponds is also reflected in the isoelectric point (pI) profiles calculated from metagenomic data [8].

The finding of an acidic proteins-enriched metaproteome in the 9% salt Guerrero Negro microbial mats was somewhat surprising. No 'salt-in' strategists are known to grow at this salinity, and microorganisms that use organic osmotic solutes do not need a far-reaching adaptation of most of their proteins to salt. Still, in spite of the strong vertical stratification of the types of microorganisms present in the mat [30], a strikingly similar pI pattern was found at all depths with major peaks at pI 4.5–4.9 and median pI values of ~6.8. This pattern was attributed to species-independent molecular convergence within the microbial community because of the increased salinity [28]. A similar pI distribution was found for the metagenome of a hypersaline (6% salt) lagoon on the Galapagos Islands [8].

However, reevaluation of the data based on published genomes of moderately halophilic members of the Bacteria and of common marine bacteria shows that the large peak of low-pI proteins in the Guerrero Negro metaproteome is by no means unusual. Similar peaks were found for the moderately halophilic *Chromohalobacter salexigens* and *Halomonas elongata* (*Gammaproteobacteria*) (maximum at pI 4.4–5.0 and 4.5–5.1; median pI values 6.60 and 6.32, respectively) [31]. These are organisms that produce ectoine or accumulate glycine betaine for osmotic balance [10, 26]. Similar pI distributions were also found for typical marine *Proteobacteria* such as *Alteromonas macleodii* and *Aliivibrio fischeri* (peaks at pI 4.6–4.8; median pI values 6.46 and 6.52, respectively) [31]. This pattern very much resembles that of the 9% salt Guerrero Negro mat metaproteome [28], and there is thus little evidence for convergent evolution of the communities in the mat as an adaptation to elevated salinity.

5.4 Genetic variation and horizontal gene transfer in communities of halophilic Archaea

The *Halobacteriaceae* are an interesting group to study evolution, genetic processes, and speciation. Numbers of halophilic Archaea in hypersaline brines are typically in the order of 10^7–10^8 per milliliter, i.e. two orders of magnitude higher than most marine and freshwater ecosystems. Thus, opportunities for exchange of genetic material are plentiful.

Comparison of the genomes of *Halobacterium salinarum* strain NRC-1 and strain R1 provides an interesting glimpse of the potentially rapid evolution of a halophilic archaeon in the laboratory. Although full documentation of the history of these strains is lacking, they probably do not represent independent isolates but originate from the same ancestral isolate. The chromosomes are completely co-linear and nearly identical: except for differences related to insertion sequence elements there are only twelve differences: four point mutations (three of which alter protein structure), five single-base frameshifts (three of which have a major effect on protein sequence), and three insertion/deletion events. However, strain NRC-1 has two megaplasmids while R1 has four. Large-scale duplications in the plasmids occur in both strains, but the duplication patterns are highly dissimilar and regions of co-linearity in the plasmids are short. Moreover, 210 kb of sequence is present only in strain R1 [32].

Except for the genus *Haloferax* nothing is known about the ways in which genetic material may be transferred between cells. In *Hfx. volcanii*, DNA can be transferred in a mating process during which up to 2 μm long cytoplasmic bridges are formed between cells. Each parental type can serve as a donor or as a recipient. Under certain conditions even cell fusion is possible [33]. Mating of *Hfx. volcanii* and *Hfx. mediterranei* leads to the formation of recombinant hybrids. Long DNA fragments can be transferred: 310–530 kb, equivalent to 10–17% of the total chromosome length. Thus within the genus *Haloferax* only low barriers exist between species [34].

Haloquadratum walsbyi, the unusual flat square species that dominates the archaeal community in many saltern crystallizer ponds, is difficult to grow in the laboratory, but information from the sequenced genome [35] and from metagenomic data from natural brine samples provides some insight into the rapid evolutionary processes that may be active in this species. At the 16S rRNA level, *Haloquadratum* phylotypes recovered from geographically remote sites are very similar [36]. However, metagenomic analysis of the *Hqr. walsbyi* population in a single Spanish saltern crystallizer pond showed a large 'pan-genome' of accessory genes, including many transposition and phage-related genes. Thus, numerous genetically different variants coexist in the same small ecosystem [37]. How genes move between *Haloquadratum* cells is still unknown.

Transformation of *Halobacteriaceae* by free DNA was never documented, but transformation may be an interesting option for gene transfer and evolutionary processes in halophilic Archaea. Hypersaline environments inhabited by *Halobacteriaceae* are often located in topographical minima where cellular debris can collect. This, together with the stability of DNA at high salt, implies that halophiles may become exposed to a greater quantity and diversity of extracellular DNA than many other types of prokaryotes [38].

No mechanism for horizontal gene transfer was yet discovered in the genus *Halorubrum*, but multilocus sequence analysis (16S rRNA, *atpB*, EF-2, *radA* and *secY*) of large numbers of strains isolated from salterns in Spain and in Algeria provided evidence for promiscuous exchange of genetic information, comparable to that of a sexual population and leading to highly mosaic allelic profiles within a population. Clusters could be defined by concatenation of the marker sequences, but barriers to exchange between them are leaky. It was concluded that no non-arbitrary way to circumscribe 'species' is likely to emerge for this genus, and the same may well be true for other genera of *Halobacteriaceae* and even for prokaryotes in general [39, 40].

A recent indication that large sections of DNA can be exchanged between members of different genera of *Halobacteriaceae* came from studies in Deep Lake, Antarctica, a hypersaline (210–280 g/l salts) lake that because of the high salinity never freezes, not even at the minimum water temperature of −20 °C. The water temperature rises above zero only in the top few meters in the summer months, and the higest recorded temperature is +11.5 °C. Growth of halophilic Archaea *in situ* is slow, and it was estimated that annually no more than six generations develop. Metagenomic analysis enabled reconstruction of the complete genomes of four types of Archaea, belonging to distinct and phylogenetically disparate genera, and these together account for ~ 72% of the community: *Halohasta* sp. (1 replicon; 3.33 Mb), *Halorubrum lacusprofundi* (3 replicons; 3.69 Mb), *Halobacterium* sp. (2 replicons, 3.16 Mb), and a type designated DL31, not belonging to any of the currently recognzied genera (3 replicons, 3.64 Mb). Large identical regions up to 35 kb in length are shared between all four taxa: 30 regions of > 5 kb were common to 7 of the 9 replicons, and 13 regions of > 10 kb were shared between 6 replicons. The same high-identity regions were not found in metagenomic data from other hypersaline environments. Thus, in Deep Lake a large extent of interspecies gene exchange occurs, but still the ecotypes maintain their identity as separate species belonging to diverse genera [41].

Many members of the *Halobacteriaceae* can be considered polyploid. Thus, exponentially growing cells of *Hbt. salinarum* contain ~ 25 copies of the genome, and stationary phase cells have ~ 15 copies; exponential and stationary cells of *Hfx. volcanii* have ~ 15 and ~ 10 genome copies, respectively. Also *Hfx. mediterranei* contains mutiple genome copies. Little is known yet about the possible impact of this polyploidy on evolutionary processes in *Haloferax* spp., but having many copies of the chromosome may result in low apparent mutation rates, high radiation and desiccation resistance, a reservoir of phosphorus, and survival over geological times [42, 43].

5.5 *Salinibacter*: convergent evolution and the 'salt-in' strategy of haloadaptation

Salinibacter is a relatively recently discovered member of the community of aerobic prokaryotes that live at the highest salinities. It requires at least 150 g/l salt. It uses KCl for osmotic adaptation, and can use light as an energy source. Xanthorhodopsin, a bacteriorhodopsin-like light-driven proton pump, can accept photons absorbed by the carotenoid salinixanthin. Halorhodopsin- and sensory rhodopsin-like proteins are also present. *Salinibacter* thus shares many key properties with the *Halobacteriaceae* with which it shares its habitat. *Salinibacter* may well have acquired the retinal protein genes from haloarchaea by horizontal gene transfer. Analysis of its genome shows a highly acidic nature of the encoded proteins (median pI 5.92, somewhat higher than the value of 5.03 for *Halobacterium* strain NRC-1). Several genes and gene clusters probably derive from horizontal gene transfer from halophilic Archaea (or alternatively, may have been laterally transferred to the Archaea). Of special interest is the presence of a 'hypersalinity island': a cluster of 19 genes coding for K^+ uptake/efflux systems and other transporters important for a hyperhalophilic lifestyle. This 'hypersalinity island' is a mosaic of genes probably derived from different sources: *Halobacteriaceae*, methanogenic Archaea, cyanobacteria, and *Firmicutes* [12].

Although *Salinibacter* isolates from different locations worldwide are phenotypically very similar, there are differences at the genomic level. Comparison of strains M31 (the type strain of the species) and strain M8, isolated at the same time from the same saltern pond on the Island of Mallorca, Spain, shows a mosaic structure with conserved and hypervariable regions. Ten percent of the genes are strain-specific [13]. In a study of microevolution within the genus, 35 *Salinibacter* strains were isolated from one milliliter of saltern crystallizer pond brine. No two genomes were identical and the pan-genome appears to be very large. Small differences were also observed in the metabolome [14, 15]. Nothing is yet known about possible mechanisms of transfer of genetic information within the genus *Salinibacter* or between *Salinibacter* and members of the *Halobacteriaceae* that live in the same habitat.

5.6 High intracellular K^+ concentrations but no acidic proteome? The case of the *Halanaerobiales*

Although nothing is known about genetic processes in the *Halanaerobiales*, a discussion of their mode of haloadaptation in view of recent genomic data is relevant in the framework of this review. Organic osmotic solutes were never proven to be present in this group of mostly fermentative anaerobes that phylogenetically belong to the low G+C *Firmicutes*, except for the presence of glycine betaine in *Orenia salinaria* grown in media containing yeast extract, a source of the compound [44]. High concentrations of Na^+, K^+ and Cl^- were measured inside the cells of *Halanaerobium praevalens*, *Halanaerobium acetethylicum*, and *Halobacteroides halobius*. Selected enzymes tested function well in the presence of molar concentrations of salt, and many enzymes require salt for optimal activity. Moreover, analysis of acid hydrolysates of *Halanaerobium praevalens*, *Halanaerobium saccharolyticum*, *Natroniella acetigena*, *Halobacteroides halobius*, and *Sporohalobacter lortetii* suggested a strongly acidic nature of the bulk protein [16]. Therefore it was assumed that the *Halanaerobiales* use the 'salt-in' strategy and have a salt-adapted, acidic proteome such as found in the *Halobacteriaceae*.

However, analysis of the genomes of *Halanaerobium praevalens*, the alkaliphilic "Halanaerobium hydrogeniformans", and the thermophilic *Halothermothrix orenii* did not confirm unusually high contents of acidic amino acids or low contents of basic amino acids in the predicted proteomes. Bimodal distribution of the pI values with peaks around 4.6–4.8 and 9.8–10.2 were found in a pattern similar to that of many non-halophiles. The excess of acidic amino acids in *Halanaerobium praevalens* and *Halothermothrix orenii* is only 0.4 mol%, as compared to 7.5 mol% in *Halobacterium* and 4.1 mol% in *Salinibacter*. The apparent excess of acidic amino acids reported earlier for members of the group is due to the high content of glutamine and asparagine in their proteins, which yield glutamate and aspartate upon acid hydrolysis.

The *Halanaerobiales* may thus use a 'salt-in' strategy of osmotic adaptation without the acidic proteomes associated with this strategy in the *Halobacteriaceae* and in *Salinibacter* [45]. Further details about this potentially new strategy are still lacking. Another, phylogenetically unrelated anaerobic halophile, *Flexistipes sinusarabici* (*Deferribacteres*), isolated from a deep-sea brine pool on the bottom of the Red Sea and growing between 30 and at least 100 g/l salt, also does not show an acidic proteome (median pI 7.47, compared to 7.42 for *Halanaerobium praevalens*) [31]. The only anaerobic halophile with a markedly acidic proteome reported (median pI 6.27) [31] is *Natranaerobius thermophilus* from an Egyptian soda lake, and classified in a newly proposed order, the *Natranaerobiales*, affiliated with the *Clostridia*. It is a halothermoalkaliphile and grows optimally at 3.1–4.9 M Na^+, 53 °C and pH 9.5 [17].

5.7 Different modes of haloadaptation in closely related *Halorhodospira* species

The recent analysis of the proteins and of the intracellular ionic concentrations in two species of the genus *Halorhodospira* shows that the evolutionary aspects of salt adaptation are far more complex than often assumed. *Halorhodospira* is a genus of anoxygenic phototrophic neutrophilic or alkaliphilic sulfur bacteria, and currently encompasses four species: *H. halophila*, *H. neutriphila*, *H. halochloris* and *H. abdelmalekii* [11]. Phylogenetically they form a coherent cluster, not only on the basis of 16S rRNA gene sequences but also of two other genes tested: *cbbL* (the large RuBisCo subunit) and *nifH* (a nitrogenase subunit) [46]. All tolerate at least 25% NaCl. *H. halophila* and *H. neutriphila* contain bacteriochlorophyll *a* and spirilloxanthin carotenoids, while *H. halochloris* and *H. abdelmalekii* have bacteriochlorophyll *b* and rhopdopsin. *H. halophila*, *H. halochloris* and *H. abdelmalekii* all produce glycine betaine as osmotic solute, with minor amounts of ectoine and trehalose.

H. halochloris contains low K^+ concentrations and does not have an acidic proteome, as expected for a microorganism that accumulates organic osmotic solutes. However, the bulk protein of the closely related *H. halophila* is highly acidic, and the cells contain high KCl when grown at high salt (35%) but not at low salt (5%). Deolle et al. [25] concluded that "obligate protein halophilicity is a non-adaptive property resulting from genetic drift in which constructive neutral evolution progressively incorporates weak K^+-binding sites on an increasingly acidic protein surface".

5.8 Final comments

Adaptation to life at high salt concentrations has evolved in many groups of microorganisms in all three domains of life: Bacteria, Archaea and Eucarya. While some properties such as the need to keep their intracellular environment in osmotic equilibrium with the surroundings are common to all halophiles, there are fundamental differences in the mechanisms that enable the cells to live at high salt: high or low ionic concentrations in the cytoplasm, the possibility of using organic osmotic solutes, and the adaptation of the proteins to function in the presence of high salt concentrations.

Halophilic behavior of proteins is generally based on a large excess of acidic residues in conjunction with the occurrence of smaller hydrophobic amino acids (less pronounced hydrophobicity). However, the significance of acidic amino acids in haloadaptation of proteins has recently been challenged: a decrease in lysine residues and an increase in aspartate marks the most prominent change when comparing halophilic with non-halophilic proteins [47]. One of the key mechanisms for haloadpatation of proteins may be a decrease in their solvent-accessible surface. Using site-directed mutagenesis on DNA ligase N (of both *Hfx. volcanii* and *Escherichia coli*) it was shown that replacement of charged residues such as glutamate and aspartate by their amidated counterparts had no effect on the enzymes' stability in salt solution. An excess negative charge may be important for solubility rather than stability in salt solution [48]. Hence an increase of acidic residues on the protein surface may be just one (of many) possible ways to adapt to high salt concentration or low water activity.

It is convenient to divide the halophiles into two groups according to their way to cope with high salt: the strategy used e.g. by *Halobacterium* and *Salinibacter* based on accumulation of KCl and the evolution of an acidic proteome, and the strategy used e.g. by the alga *Dunaliella*, based on exclusion of salts and accumulation of organic osmotic solutes, without the need for special salt-adapted proteins. Genomic analysis combined with physiological studies has shown that there are cases such as the *Halanaerobiales* [45] and the species of the genus *Halorhodospira* [25] that do not simply fall into one of these two categories. Another, well-documented case of an organism that can use both intracellular ions and organic osmotic solutes to cope with high salt concentrations in its environment is the Gram-positive endospore-forming *Halobacillus halophilus* [49].

Genomic and metagenomic studies demonstrate that extensive horizontal transfer may occur between species of the same genus (e.g. *Halorubrum* [39, 40]) and also between phylogenetically unrelated organisms (e.g. the case of the transfer of large DNA fragments between members of the ecosystem of Deep Lake, Antarctica [41] and the mosaic 'salinity island' in the genome of *Salinibacter* [24]). However, except for the genus *Haloferax* [33, 34] our understanding of the mechanisms enabling horizontal gene transfer remains very limited.

Comparative genomic analysis led to the conclusion that the *Halobacteriaceae* may have evolved from a methanogen following acquisition of ~1000 genes derived from the Bacteria domain [50]. How and when the extremely halophilic properties originated is still far from clear. Future comparative genomic analyses combined with laboratory studies of the mechanisms enabling horizontal transfer between different types of halophilic microorganisms will undoubtedly provide deeper insights in the evolution of halophilic behavior among different groups of salt-tolerant and salt-requiring microorganisms.

Acknowledgments

Some of the studies described in this chapter were supported by grants no. 1103/10 and 343/13 from the Israel Science Foundation.

References

[1] I. Dundas. Was the environment for primordial life hypersaline? Extremophiles 1998, 2, 375–377.
[2] A. Oren. The family *Halobacteriaceae*. In: E. Rosenberg, E. F. DeLong, F. Thompson, S. Lory, E. Stackebrandt, eds. The Prokaryotes. A handbook on the biology of bacteria: Ecophysiology and biochemistry, 4th ed. New York, Springer, 2014, in press.
[3] A. Antunes, M. Taborda, R. Huber, C. Moissl, M. F. Nobre, M. S. da Costa. *Halorhabdus tiamatea* sp. nov., a non-pigmented, extremely halophilic archaeon from a deep-sea, hypersaline anoxic basin of the Red Sea, and emended description of the genus *Halorhabdus*. Int J Syst Evol Microbiol 2008, 58, 215–220.
[4] R. Hartmann, H.-D. Sickinger, D. Oesterhelt. Anaerobic growth of halobacteria. Proc Natl Acad Sci USA 1980, 77, 3821–3825.
[5] L. M Longo, J. Lee, M. Blaber. Simplified protein design biased for prebiotic amino acids yield a foldable, halophilic protein. Proc Natl Acad Sci USA 2013, 110, 2135–2139.
[6] L. M. Longo, M. Blaber. Prebiotic protein design supports a halophile origin of foldable proteins. Frontiers Microbiol 2014, 4, 418.
[7] A. Oren. Taxonomy of halophilic Archaea: current status and future challenges. Extremophiles 2014, in press.
[8] R. Ghai, A. B. Fernández, A.-B. Martin-Cuadrado, C. Megumi. Mizuno, K. D. McMahon, R. T. Papke, R. Stepanauskas, B. Rodriguez-Brito, F. Rohwer, C. Sánchez-Porro, A. Ventosa, F. Rodríguez-Valera. New abundant microbial groups in aquatic hypersaline environments. Sci Rep 2011, 1, 135.
[9] P. Narasingarao, S. Podell, J. A. Ugalde, C. Brochier-Armanet, J. B. Emerson, J. J. Brocks, K. B. Heidelberg, J. F. Banfield, E. E. Allen. De novo assembly reveals abundant novel major lineage of Archaea in hypersaline microbial communities. ISME J 2012, 6, 81–93.
[10] A. Ventosa, J. J. Nieto, A. Oren. Biology of aerobic moderately halophilic bacteria. Microbiol. Mol Biol Rev 1998, 62, 504–544.
[11] A. Oren. Family *Ectothiorhodospiraceae*. In: E. Rosenberg, E. F. DeLong, F. Thompson, S. Lory, E. Stackebrandt, eds. The Prokaryotes. A handbook on the biology of bacteria: Ecophysiology and biochemistry, 4th ed. New York, Springer, 2014, in press.
[12] M. E. F. Mongodin, K. E. Nelson, S. Duagherty, R. T. DeBoy, J. Wister, H. Khouri, J. Weidman, D. A. Balsh, R. T. Papke, G. Sanchez Perez, A. K. Sharma, C. L. Nesbo, D. MacLeod, E. Bapteste, W. F. Doolittle, R. L. Charlebois, B. Legault, F. Rodríguez-Valera. The genome of *Salinibacter ruber*: convergence and gene exchange among hyperhalophilic bacteria and archaea. Proc Natl Acad Sci USA 2005, 102, 18147–18152.

[13] Peña A, H. Teeling, J. Huerta-Cepas, F. Santos, P. Yarza, J. Brito-Echeverría, M. Lucio, P. Schmitt-Kopplin, I. Meseguer, C. Schenowitz, C. Dossat, V. Barbe, J. Dopazo, R. Rosselló-Mora, M. Schüler, F. O. Glöckner, R. Amann, T. Gabaldón, J. Antón. Fine-scale evolution: genomic, phenotypic and ecological differentiation in two coexisting *Salinibacter ruber* strains. ISME J 2010, 4, 882–895.

[14] Peña A, H. Teeling, J. Huerta-Cepas, F. Santos, I. Meseguer, M. Lucio, P. Schmitt-Kopplin, J. Dopazo, R. Rosselló-Móra, M. Schüler, F. O. Glöckner, R. Amann, T. Gabaldön, J. Antón. From genomics to microevolution and ecology: the case of *Salinibacter ruber*. In: A. Ventosa, A. Oren, Y. Ma, eds. Halophiles and hypersaline environments. Berlin, Springer-Verlag, 2011, 109–122.

[15] Peña A, M. Gomariz, M. Lucio, P. González-Torres, J. Huertas-Cepa, M. Martínez-Garcia, F. Santos, P. Schmitt-Kopplin, T. Gabaldón, R. Rosselló-Móra, J. Antón, *Salinibacter ruber*: the never ending microdiversity? In: R. T. Papke, A. Oren, eds. Halophiles: genetics and genomics. Caister Press, 2014, 37–56.

[16] A. Oren. The order *Halanaerobiales*, families *Halanaerobiaceae* and *Halobacteroidaceae*. In: E. Rosenberg, E. F. DeLong, F. Thompson, S. Lory, E. Stackebrandt, eds. The Prokaryotes. A handbook on the biology of bacteria: Ecophysiology and biochemistry, 4th ed. New York, Springer, 2014, in press.

[17] N. M. Mesbah, D. B. Hedrick, A. D. Peacock, M. Rohde, J. Wiegel. *Natranaerobius thermophilus* gen. nov., sp. nov., a halophilic alkalithermophilic bacterium from soda lakes of the Wadi An Natrun, Egypt, and proposal of *Natranaerobiaceae* fam. nov. and *Natranaerobiales* ord. nov. Int J Syst Evol Microbiol 2007, 57, 2507–2512.

[18] M. Lenassi, Gostinčar C, S. Jackmann, M. Turk, I. Sadowski, C. Nislow, S. Jones, I. Birol, N. Gunde-Cimerman, Plemenitaš A. Whole genome duplication and enrichment of metal cation transporters revealed by de novo genome sequencing of extremely halotolerant black yeast *Hortaea werneckii*. PLoS One 2013, 8, e71328.

[19] J. Zajc, Y. Liu, W. Dai, Z. Yang, Hu J, Gostinčar C, N. Gunde-Cimerman. Genome and transcriptome sequencing of the halophilic fungus *Wallemia ichthyophaga*: haloadaptations present and absent. BMC Genomics 2013, 14, 617.

[20] J. K. Lanyi. Salt-dependent properties of proteins from extremely halophilic bacteria. Bacteriol Rev 38, 1974, 272–290

[21] Ng WV, S. P. Kennedy, G. G. Mahairas, B. Berquist, M. Pan, H. D. Shukla, S. R. Lasky, N. S. Baliga, V. Thorsson, J. Sbrogna, S. Swartzell, D. Weir, J. Hall, T. A. Dahl, R. Welti, Y. A. Goo, B. Leithauser, K. Keller, R. Cruz, M. J. Danson, D. W. Hough, D. G. Maddocks, P. E. Jablonski, M. P. Krebs, C. M. Angevine, H. Dale, T. A. Isenberger, R. F. Peck, M. Pohlschroder, J. L. Spudich, K.-H. Jong, M. Alam, T. Freitas, S. Hou, C. J. Daniels, P. P. Dennis, A. D. Omer, H. Ebhardt, T. M. Lowe, P. Liang, M. Riley, L. Hood, S. DasSarma. Genome sequence of *Halobacterium* species NRC-1. Proc Natl Acad Sci USA 2000, 97, 12176–12181.

[22] P. P. Dennis, L. C. Shimmin. Evolutionary divergence and salinity-mediated selection in halophilic archaea. Microbiol Mol Biol Rev 1997, 61, 90–104.

[23] D. J. Kushner. Life in high salt and solute concentrations: halophilic bacteria. In: D. J. Kushner, ed. Microbial life in extreme environments. London, Academic Press, 1978, 317–368.

[24] M. Mevarech, F. Frolow, L. M. Gloss. Halophilic enzymes: proteins with a grain of salt. Biophys Chem 2000, 86, 155–164.

[25] R. Deole, J. Challacombe, D. W. Raiford, W. D. Hoff. An extremely halophilic proteobacterium combines a highly acidic proteome with a low cytoplasmic potassium content. J Biol Chem 2013, 288, 581–588.

[26] E. A. Galinski. Osmoadaptation in bacteria. Adv Microb Physiol 1995, 37, 273–328.

[27] A. Oren, F. Larimer, P. Richardson, A. Lapidus, L. N. Csonka. How to be moderately halophilic with a broad salt tolerance: clues from the genome of *Chromohalobacter salexigens*. Extremophiles 2005, 9, 275–279.
[28] V. Kunin, J. Raes, J. K. Harris, J. R. Spear, J. J. Walker, N. Ivanova, C. von Mering, B. M. Bebout, N. R. Pace, P. Bork, P. Hugenholtz. Millimeter scale genetic gradients and community-level molecular convergence in a hypersaline microbial mat. Mol Systems Biol 2008, 4, 198.
[29] M. E. Rhodes, S. Fitz-Gibbon, A. Oren, C. H. House. Amino acid signatures of salinity on an environmental scale with a focus on the Dead Sea. Environ Microbiol 2010, 12, 2613–2623.
[30] J. K. Harris, J. G. Caporaso, J. J. Walker, J. R. Spear, N. J. Gold, C. E. Robertson, P. Hugenholtz, J. Goodrich, D. McDonald, D. Knights, P. Marshall, H. Tufo, R. Knight, N. R. Pace. Phylogenetic stratigraphy in the Guerrero Negro hypersaline microbial mat. ISME J 2013, 7, 50–60.
[31] Elevi R. Bardavid, A. Oren. Acid-shifted isoelectric point profiles of the proteins in a hypersaline microbial mat – an adaptation to life at high salt concentrations? Extremophiles 2012, 16, 787–792.
[32] F. Pfeiffer, S. C. Schuster, A. Broicher, M. Falb, P. Palm, K. Rodewald, A. Ruepp, J. Soppa, J. Tittor, D. Oesterhelt. Evolution in the laboratory: The genome of *Halobacterium salinarum* strain R1 compared to that of strain NRC-1. Genomics 2008, 9, 335–346.
[33] I. Rosenshine, R. Tchelet, M. Mevarech. The mechanism of DNA transfer in the mating system of an archaebacterium. Science 1989, 245, 1387–1389.
[34] A. Naor, P. Lapierre, M. Mevarech, R. T. Papke, U. Gophna. Low species barriers in halophilic archaea and the formation of recombinant hybrids. Curr Biol 2012, 22:1444–1448
[35] H. Bolhuis, P. Palm, A. Wende, M. Falb, M. Rampp, F. Rodriguez-Valera, F. Pfeiffer, D. Oesterhelt. The genome of the square archaeon *Haloquadratum walsbyi*: life at the limits of water activity. BMC Genomics 2006, 7, 169.
[36] Oh D, K. Porter, B. Russ, D. Burns, M. Dyall-Smith. Diversity of *Haloquadratum* and other haloarchaea in three, geographically distant, Australian saltern crystallizer ponds. Extremophiles 2012, 14, 161–169.
[37] B. A. Legault, A. Lopez-Lopez, J. C. Alba-Casado, W. F. Doolittle, H. Bolhuis, F. Rodríguez-Valera, R. T. Papke. Environmental genomics of "*Haloquadratum walsbyi*" in a saltern crystallizer indicates a large pool of accessory genes in an otherwise coherent species. BMC Genomics 2006, 7, 171.
[38] M. E. Rhodes, J. R. Spear, A. Oren, C. H. House. Differences in lateral gene transfer in hypersaline versus thermal environments. BMC Evol Biol 2011, 11, 199.
[39] R. T. Papke, J. E. Koenig, F. Rodríguez-Valera, W. F. Doolittle. Frequent recombination in a saltern population of *Halorubrum*. Science 2004, 306, 1928–1929.
[40] R. T. Papke, O. Zhaxybayeva, E. J. Feil, K. Sommerfeld, D. Muise, W. F. Doolittle. Searching for species in haloarchaea. Proc Natl Acad Sci USA 2007, 104, 14092–14097.
[41] M. Z. DeMaere, T. J. Williams, M. A. Allen, M. V. Brown, J. A. E. Gibson, J. Rich, F. M. Lauro, M. Dyall-Smith, K. W. Davenport, T. Woyke, N. C. Kyrpides, S. G. Tringe, R. Cavicchioli. High level of intergenera gene exchange shapes the evolution of haloarchaea in an isolated Antarctic lake. Proc Natl Acad Sci USA 2013, 110, 16939–16944.
[42] J. Soppa. Evolutionary advantages of polyploidy in halophilic archaea. Biochem Soc Trans 2013, 41, 339–343.
[43] K. Zerulla, S. Chimileski, D. Näther, U. Gophna, R. T. Papke, J. Soppa. DNA as phosphate storage polymer and the alternative advantages of polyploidy for growth and survival. Plos One 2014, 9: e94819.
[44] S. Mouné, C. Eatock, R. Matheron, J. C. Willison, A. Hirschler, R. Herbert, P. Caumette. *Orenia salinaria* sp. nov., a fermentative bacterium isolated from anaerobic sediments of Mediterranean salterns. Int J Syst Evol Microbiol 2000, 50, 721–729.

[45] Elevi R. Bardavid, A. Oren. The amino acid composition of proteins from anaerobic halophilic bacteria of the order Halanaerobiales. Extremophiles 2012, 16, 567–572.
[46] T. P. Tourova, E. M. Spiridonova, I. A. Berg, N. V. Slobodova, E. S. Boulygina, D. Y. Sorokin. Phylogeny and evolution of the family *Ectothiorhodospiraceae* based on comparison of 16S rRNA, *cbbL* and *nifH* gene sequences. Int J Syst Evol Microbiol 2007, 57, 2387–2398.
[47] X. Tadeo, B. López-Méndez, T. Trigueros, A. Laín, Castaño D, O. Millet. Structural basis for the aminoacid composition of proteins from halophilic Archaea. PLoS Biology 2009, 7, e1000257.
[48] S. Fukuchi, K. Yoshimune, M. Wakayama, M. Moriguchi, K. Nishikawa. Unique amino acid composition of proteins in halophilic bacteria. J Mol Biol 2003, 327, 347–357.
[49] S. H. Saum, F. Pfeiffer, P. Palm, M. Rampp, S. C. Schuster, V. Müller, D. Oesterhelt. Chloride and organic osmolytes: a hybrid strategy to cope with elevated salinities by the moderately halophilic, chloride-dependent bacterium Halobacillus halophilus. Environ Microbiol 2013, 15, 1619–1633.
[50] S. Nelson-Sathi, T. Dagan, G. Landan, A. Janssen, M. Steel, J. O. McInerney, U. Deppenmeier, W. F. Martin. Acquisition of 1,000 eubacterial genes physiologically transformed a methanogen at the origin of Haloarchaea. Proc Natl Acad Sci 2012, 109, 20537–20542.

John R. Battista
6 The origin of extreme ionizing radiation resistance

6.1 Introduction and background

The evolution of extreme ionizing radiation resistance is difficult to explain. One cannot argue that terrestrial exposure to ionizing radiation provided the selective pressure that built this phenotype. Average yearly exposures to ionizing radiation from cosmic rays and radioactive decay are extremely low and those values have not changed dramatically throughout Earth's history. Despite this fact, at least seven phyla of Bacteria and Archaea display this phenotype, individual species within these phyla tolerating exposures in the laboratory that are as much as six orders of magnitude higher than the annual average exposure on Earth. Identifying the selective pressure(s) responsible for building ionizing radiation resistance in microorganisms is the key to understanding the origin of this phenotype.

6.1.1 Ionizing radiation

Ionizing radiation refers to any form of emitted wave or particle possessing sufficient energy to eject orbital electrons from a molecule following its impact with those electrons [1–3]. Each ionization event has the potential to significantly modify biological molecules, releasing sufficient energy to break covalent bonds. Ionizing radiation is broadly defined as either electromagnetic or particulate. Electromagnetic radiations are photons of light and include ionizing and non-ionizing forms. A photon's energy is inversely related to its wavelength, the shorter the wavelength the higher the energy and the more likely it is to ionize biological material. X-rays and y-rays, which exhibit low linear energy transfer (LET) in living cells, are the most commonly studied high-energy electromagnetic radiations. LET is a property that reflects the rate of energy absorption by the absorbing material. High energy electromagnetic radiations transfer less energy per unit length along their path than particulate radiations. This lower rate of transfer allows this form of ionizing radiation to penetrate more deeply into a cell, but the pattern of energy deposition events is less dense. Particulate radiations include alpha particles, electrons, protons, neutrons, and heavy charged nuclei. Particles produce ionization by depositing energy along relatively short tracks as they pass through matter. Particulate radiations are considered high LET radiation, dissipating their energy over much shorter distances than high energy electromagnetic radiations. The majority of experimental studies that examine the effects of ionizing radiation on bacterial or archaeal species have utilized low LET radiations. Low LET sources are more amenable to the study of microbes in culture; these sources can irradiate larger sample volumes than particle producing devices.

6.1.2 Biological damage caused by electromagnetic radiations

Electromagnetic radiations do not cause biological damage directly [4]. Instead irradiated material absorbs the energy of the photons and transfers that energy to orbital electrons [1, 2]. When high-energy photons – such as those generated by the decay of radionuclides – interact with an electron, part of their energy is transferred to that electron, ejecting it from its orbital. These accelerated electrons are called fast electrons. Fast electrons have direct and indirect effects when generated in biological systems. During direct actions, fast electrons collide with biological macromolecules causing an ionization event that modifies and possibly inactivates that macromolecule. Through indirect action, the fast electron modifies other molecules in the cell, making them chemically reactive and potentially toxic to cellular macromolecules. In biological systems, indirect actions predominate; it is estimated that approximately 75% of the biological damage caused by high-energy ionizing radiation results from indirect effects. Fast electrons produce free radicals from water in cells, initiating the homolytic cleavage of water, which creates highly reactive hydroxyl radicals. Hydroxyl radicals are electrophilic and capable of reacting with most biologically important molecules. Numerous detailed studies document the effect of hydroxyl radicals on carbohydrates, nitrogenous bases and nucleic acids, amino acids and proteins, and lipids [3, 4]. If the macromolecule affected is critical to a cellular function, and that function is lost there is a biological effect. While exposing a bacterium to ionizing radiation will result in damage to all cellular macromolecules, there is little evidence that damage to carbohydrates [5] (with the exception of the ribose components of nucleic acids), or lipids [6] significantly affect cell viability; in general, species are inactivated before the level of damage to carbohydrates or lipids becomes lethal.

It is generally accepted that inactivation of a bacterial cell following exposure to ionizing radiation results from the irreversible destruction of the nucleic acid making up the genome of that cell. Ionizing radiation results in many types of DNA damage, including base damage, single strand breaks, double strand breaks, and DNA-DNA or DNA-protein crosslinks. Hydroxyl radicals readily add to C–C and C–N double bonds and can easily abstract protons modifying a large number of potential sites on the base and sugar moieties of the nucleic acid [7, 8]. Many of these forms of damage can be repaired, provided the genome has not suffered excessive damage. However, damage to the ribose sugar of DNA by hydroxyl radical modifies the molecule in a manner that can lead to single-strand breaks [4, 9–11], and when single-strand breaks occur in close proximity on both strands of the DNA they can result in double-strand breaks [12]. If not repaired, DNA double-strand breaks will prevent replication of the prokaryotic genome, and because the free ends generated are substrates for intracellular nucleases double-strand breaks can result in significant loss of genetic information even if the break is repaired.

While much of the literature suggests that destruction of genomic DNA is responsible for loss of cell viability, evidence is accumulating that suggests focusing on genome damage as the only reason for cell death may be misleading [13]. This work postulates that proteins are the most critical intracellular targets of ionizing radiation, arguing that in the presence of certain proteins DNA is protected from the actions of ionizing radiation, and that only when these proteins are inactivated does DNA damage accumulate to lethal levels [14, 15]. Hydroxyl radicals oxidize the amino acid residues of proteins [16–19], resulting in protein inactivation by multiple mechanisms. Proteins irradiated *in vitro* show evidence of covalently cross-linked peptides, hydrolysis of the peptide backbone, and many types of amino acid modifications at active site residues. Clearly, all of these processes have the potential to adversely affect protein function; it is assumed that similar processes occur when proteins are irradiated *in vivo*.

6.1.3 Exposure to ionizing radiation selects for ionizing radiation resistant bacteria

Generating an ionizing radiation resistant strain of a bacterium in the laboratory is not difficult [20]. The process starts by exposing a bacterial population to doses of ionizing radiation sufficiently high enough to kill most of that population. The survivors are allowed to recover and grow to dense cultures that presumably include strains that are incrementally more resistant than the population that gave rise to them. If cycles of irradiation and outgrowth are repeated, successive populations will become enriched for radioresistant strains, and there is a gradual shift toward higher numbers of individuals surviving that dose of radiation. Increasing the dose administered as the population adapts results in increased resistance among individuals within the population. Using this protocol, Harris et al. were able to isolate ionizing radiation resistant derivatives of *Escherichia coli* K12 strain MG1655 [21] that were between 300–3000 fold more resistant than the parent strain when exposed to 3000 Gy gamma radiation. These strains were phenotypically stable, maintaining their resistance to ionizing radiation after 100 generations of growth without exposure to selective pressure.

As a DNA damaging agent, ionizing radiation is inherently mutagenic, and it is assumed that mutations resulting from exposure provide sufficient genetic change to permit selection of ionizing radiation tolerant strains. This general paradigm had been used in earlier studies to increase the ionizing radiation resistance of several other ionizing radiation sensitive species, including *E. coli* B, *Salmonella enterica* serovar Typhimurium, and *Bacillus pumilis* [22–24], suggesting that many bacteria may have a latent capacity to become ionizing radiation resistant given appropriate circumstances.

6.1.4 The occurrence of extreme ionizing radiation resistance within the Bacteria and Archaea

The radioresistance of microorganisms is typically compared using either the D_{37} or the D_{10} doses of the irradiated population; these values are the absorbed dose at which 37% or 10% of the irradiated population survive [1, 2]. The D_{37} dose for almost all actively growing, aerobic, vegetative microorganisms lies between 0.5 and 10 kGy [25]. Since the D_{37} dose represents the average amount of ionizing radiation required for each cell to experience a lethal event, comparison of this value can be used to rank species with respect to their capacity to resist the lethal effects of ionizing radiation. As indicated by the range of inactivating doses reported [25], microorganisms vary greatly in their response to ionizing radiation. The vast majority of microbial species found in nature are quite sensitive to ionizing radiation, making exposure to a high energy source an effective means of sterilization. However, there are a number of vegetative microorganisms that are substantially more resistant to the ionizing radiation-induced inactivation when compared to the majority of other species. The first species identified as expressing an ionizing radiation resistant phenotype was *Micrococcus radiodurans* [26]. (The species was latter re-named *Deinococcus radiodurans* [27].) *D. radiodurans*' D_{37} dose for ionizing radiation is greater than 6500 Gy [28] – a truly remarkable tolerance given that the D_{37} dose for a typical bacterium is approximately 500 Gy. The remainder of this discussion will focus only on those species exhibiting extreme ionizing radiation resistance, which is arbitrarily defined here as species that exhibit a D_{10} value of greater than or equal to 2000 Gy during vegetative growth. Based on this restriction, there are eleven genera found within seven phyla that exhibit extreme ionizing radiation resistance within the domains Bacteria and Archaea [29] (▶ Tab. 6.1).

Table 6.1. Ionizing radiation resistant genera in the domains Bacteria and Archaea.

Domain	Phylum	Family	Genus
Bacteria	Proteobacteria	Methylobacteriaceae	Methylobacterium
		Moraxellaceae	Acinetobacter
	Cyanobacteria	Xenococcaceae	Chroococcidiopsis
	Deinococcus-Thermus	Deinococcaceae	Deinococcus
		Trueperaceae	Truepera
	Planctomycetes	Planctomycetaceae	Gemmata
	Flexibacter-Cytophaga-Bacteroides	Flexibacteraceae	Hymenobacter
	Actinobacteria	Kineosporiineae	Kineococcus
		Rubrobacteraceae	Rubrobacter
Archaea	Euryarchaeota	Thermococcaceae	Thermococcus
			Pyrococcus

6.1.5 Natural sources of ionizing radiation

On Earth ionizing radiation results from two natural sources: cosmic rays and the decay of naturally-occurring radionuclides [30]. Cosmic rays arise from extraterrestrial sources with solar flares being the primary source within our solar system. The primordial radionuclides, ^{40}K, ^{87}Rb, ^{238}U, and ^{232}Th, are the major sources of radioactivity emitted from the Earth's crust. The average absorbed doses due to cosmic radiation and radioactive decay worldwide in air at ground level are estimated to be 3×10^{-8} Gy/h and 5×10^{-8} Gy/h, respectively, but these values grossly underestimate the radiation flux in specific areas of the planet's surface. With the exception of ^{40}K, which is distributed over most of the Earth, concentrations of primordial radionuclides tend to be localized to specific areas of the Earth's surface. As a consequence, dose rates vary in areas where higher concentrations of primordial radionuclides are found at the Earth's surface. For example, there is a beach near Guarapari, Brazil that is made up of ^{232}Th-rich monazite sands that generate an average dose rate of 2×10^{-5} Gy/h, approximately three orders of magnitude higher than the worldwide average. It must be noted that even this dose rate is exceptionally low; delivering a total dose of 4.8×10^{-4} Gy in a 24 hour period. To place this level of exposure in perspective, an individual receives a dose of approximately 5×10^{-4} Gy during a routine dental x-ray (International Atomic Energy Agency: https://rpop.iaea.org/RPOP/RPoP/Content/InformationFor/HealthProfessionals/6_OtherClinicalSpecialities/Dental/concept-radiation-dose.htm).

6.2 The existence of extreme ionizing radiation resistance is difficult to reconcile with the natural history of the Earth

Assuming that extreme ionizing radiation resistance evolved as part of Earth's natural history (ideas of an extraterrestrial origin for the phenotype are described later in this text), we must assume that this phenotype is an adaptation acquired through natural selection. Natural selection is the evolutionary process that results in the spread of favorable variations within a population [31]. Superficially, one might hypothesize that extreme ionizing radiation resistance evolved when species were exposed to naturally occurring sources of ionizing radiation, in a manner analogous to what occurred during the laboratory studies discussed above where iterative application of ionizing radiation increased resistance [21]. It is not unreasonable to assume that individuals with specific genetic variations will have a selective advantage that improves their reproductive success in the radioactive environment.

Given the degree of radioresistance observed in some ionizing radiation resistant species, the scenario outlined above seems unlikely. The annual accumulated dose at the most radioactive sites on Earth is about 0.2 Gy [30]. This dose is four orders of magnitude lower than the D_{37} value of *Deinococcus radiodurans*. If one assumes that an organism remains in one location (i.e., does not move to a region of lower radiation flux) and that ionizing radiation-induced cellular damage is not repaired, it would take at least 150 years to accumulate a 30 Gy exposure at a site where the radiation flux was equivalent to that in the monazite sands of Guarapari, Brazil [30].

The preceding discussion assumes that microorganisms have always encountered the same levels of ionizing radiation exposure that cells face on modern day Earth. Strictly speaking, this assumption is not accurate. Prokaryotic life is thought to have evolved 3.5 to 4 Ga, and Earth's radiological history has changed during that time. Some of the primordial radionuclides have half-lives less than the age of the Earth; the half-lives of ^{40}K and ^{238}U are 1.3 and 4.5 Ga, respectively. In the distant past, Earth's microorganisms would have been subjected to a slightly higher exposure to ionizing radiation; the dose rate one billion years ago would have been approximately twice that experienced during the present day [32]. While a doubling of the dose rate might have implications for the survival of radiosensitive species, the rate remains too low to realistically suggest that this exposure influenced the selection of extreme ionizing radiation resistance.

There is also evidence that Earth once supported spontaneous fission reactions and fifteen "fossil" nuclear reactors can be found near present day Oklo, Gabon [33]. These ancient reactors contain uranium deposits that are characterized by an unusual distribution of uranium isotopes (particularly ^{235}U). ^{235}Uranium is consumed during a fission reaction and the levels of this isotope are approximately half of what they are elsewhere on Earth, and there is general agreement among geologists and physicists that a change this large could only result from a self-sustaining nuclear chain reaction. The Oklo reactors formed 1.7 Ga and were active for several hundred thousand years [34, 35]. At their peak the Oklo reactors are believed to have emitted a dose of approximately 5 Gy per day at their core [36], but the core temperature is estimated to have been between 150 and 360 °C. The low dose rate and high reactor temperatures argue against involvement of the Oklo reactors in the evolution and natural history of ionizing radiation resistant species.

6.3 Proposed explanations for the existence of ionizing radiation resistance

6.3.1 Panspermia: the exchange of bacteria between planets

Panspermia is a hypothesis explaining how life spread throughout the universe [37]. This hypothesis assumes that life could have originated anywhere in the universe and that over time that life traveled between planetary bodies. Planetary impacts by asteroids or meteoroids may occur with sufficient force to send ejecta into space and once there this material will drift until the gravitational pull of another planetary body brings that material to its surface [38, 39]. The idea of panspermia is built around the possibility that the ejecta produced by impacts contain microbial life and that these microbes survive the trip through space until they are deposited on a different planet's surface. Experimental work suggests that terrestrial microorganisms can survive the shock and temperatures associated with ejection [40], the impact upon transfer to another planet [40], and extended transit times in space [41–43], provided they are protected from solar radiation [42]. The potential existence of panspermia has led to the suggestion that at least some life on Earth may have originated elsewhere in the solar system and then found its way to this planet, and this possibility has been invoked to explain why ionizing radiation resistant species exist on Earth [44]. Proponents of this hypothesis suggest that it is inappropriate to discuss the evolution of ionizing radiation resistance based on parameters that are limited to what occurs or has occurred on Earth, as I did in the preceding sections. Instead they argue that there are other planets exposed to much higher fluxes of cosmic radiation and that on such planets there is appropriate selective pressure to build an ionizing radiation resistant species. Those extraterrestrial microbial species would then be transferred to Earth, expressing an exotic phenotypic characteristic (ionizing radiation resistance) of limited use to them on this planet. There are a number of well-studied samples of meteoroites from Mars that have been found on Earth, providing strong evidence that transfer events do occur. (NASA's Jet Propulsion Laboratory maintains a website listing all of Mars meteorites that are known at www2.jpl.nasa.gov/snc/index.html.)

The mean transit time for transfer of material between planets in our solar system ranges from tens of thousands to millions of years. While under some conditions bacterial endospores can survive simulated trips in space [42, 45, 46], there is at present nothing to indicate that any species can persist under the conditions encountered for this length of time. Modeling studies indicate that transit times for transfer between planets can be shorter [47], remaining within the experimentally established six year survival of *Bacillus subtilis* endospores in space [42]. These models suggest that if radioresistant species existed on another planet, and the transit time between planets is reasonably rapid, panspermia could account for the origin of ionizing radiation resistance.

There is no evidence that phyla containing ionizing radiation resistant genera (▶ Tab. 6.1) are related by vertical inheritance [29]. In other words, there is no evidence that a single phylogenetic group gave rise to all other ionizing radiation resistant prokaryotes during evolution. For this reason, panspermia as an origin for ionizing radiation resistance would require that one of the following conditions be met: a) all genera that are radioresistant were transferred to Earth simultaneously, b) that multiple material transfer events occurred during Earth's natural history, or c) that panspermia occurred a minimal number of times and the genes for the biochemical processes that mediate ionizing radiation resistance were horizontally transferred to unrelated terrestrial species. The first two circumstances listed above are improbable. However, if panspermia resulted in the transmission of ionizing radiation resistance to Earth and the extraterrestrial and terrestrial species were genetically compatible, horizontal genetic transfer between species provides a convenient explanation for the lack of vertical inheritance between species. However, it is noted that the scenario outlined is only successful if the characteristic acquired by horizontal gene transfer confers a selective advantage to the recipient, a selective pressure sufficiently strong enough to propagate and maintain the characteristic must be present on Earth, and as discussed that selective pressure could not come from natural sources of ionizing radiation.

6.3.2 Man-made sources of ionizing radiation are the source of extreme ionizing radiation resistant microorganisms

As indicated previously, iterative exposure to high dose ionizing radiation will select for ionizing radiation resistant species [21] and there is evidence that high dose exposure to ionizing radiation in the area surrounding the Chernobyl disaster [48] selects for species with higher tolerance to ionizing radiation, but assessment of available evidence does not currently support the idea that extreme ionizing radiation resistance is the consequence of the use of nuclear power and resulting concentration of waste radioactive materials that have appeared during the past seventy years. While surveys of microbial biodiversity at low-level radioactive waste sites are available, they are for the most part catalogues of the species present [49, 50]. Typically, these environmental characterizations are accomplished through 16S ribosomal DNA catalogs, and do not assess the radioresistance of individual species. Microorganisms exhibiting extreme ionizing radiation resistance have been found associated with nuclear waste sites [51], but such isolations are infrequent, and the radioresistant species found are part of a larger community that is not radioresistant. In the absence of evidence to the contrary, this result suggests that ionizing radiation is not affecting the composition of exposed communities. If recently deposited radioactive materials were creating resistant species among existing flora, or killing off sensitive species, it is expected that radioresistant species would be present in greater numbers at such sites.

6.3.3 Exaptation

Given the very low flux of ionizing radiation present on Earth throughout its history, it seems implausible that naturally-occurring ionizing radiation could act as the selective agent in the evolution of species exhibiting extreme ionizing radiation resistance. As discussed above, there is little advantage to being ionizing radiation resistant on Earth, and so we must assume that extreme ionizing radiation resistance did not evolve to counteract the effects of ionizing radiation *per se*. Instead, it seems clear that extreme ionizing radiation resistance is the byproduct of another cellular process that permits the microbe to tolerate the damage introduced by ionizing radiation; the radioresistant cell may have adapted to exposure to ionizing radiation by co-opting pre-existing functions intended for another purpose [31]. Exaptation describes a specie's use of a cellular function for purposes other than that which it was built for through natural selection [52]. Assuming that extreme ionizing radiation resistance is an exaptation eliminates the need to explain the phenotype as a response to ionizing radiation. If extreme ionizing radiation resistance is an exaptation and the functions necessary to survive exposure to ionizing radiation did not arise in response to ionizing radiation, identifying the selective pressure that is incidentally responsible for the phenotype should help clarify the evolutionary origins of ionizing radiation resistance.

As detailed above, ionizing radiation generates a distinctive pattern of cellular damage. If we assume that ionizing radiation resistance is an exaptation to a stress that causes the same damage, then we would expect this selective pressure to produce reactive oxygen species (ROS) at levels comparable to those produced during high dose gamma irradiation. There is also the possibility that the functions co-opted by ionizing radiation resistant species are not specific to ROS-induced damage. For example, these functions could elevate the cell's capacity to repair any form of DNA damage or ameliorate the effect of a particularly lethal type of DNA damage that occurs in response to ionizing radiation.

Perhaps the most abundant potential source of ROS on Earth is the UV-A and UV-B components of solar radiation [53]. About 8% of the solar radiation reaching Earth is in the ultraviolet (UV) spectrum; infrared and visible light making up the balance [53]. UV-A is about 6.5% of the total and UV-B represents 1.5% of UV exposure. The current dose rate for UV-A on Earth is approximately $1 J/m^2 s$ [54]. This dose rate does not appear to have changed since the Archean era (3.9 to 2.5 Ga). Recent studies estimate that the rate of DNA single-strand breaks and alkali-labile sites appearing in DNA after exposure to $10,000 J/m^2$ UV-A to be similar to the damage caused by 1 Gy gamma radiation [55]. If you assume full exposure (no clouds) to UV-A for 24 h without DNA repair, the total dose is slightly less than that introduced by 9 Gy gamma radiation. The doses of UV-A and UV-B introduced are too small to account for the level of resistance observed in ionizing radiation resistance species. It would require exposure to roughly 1.6 years of uninterrupted sunlight to achieve a dose equivalent of 5000 Gy.

For many years, the isolation of ionizing radiation resistant species was the result of fortuitous accidents, byproducts of attempts to use ionizing radiation to sterilize a variety of materials. Ionizing radiation resistant organisms have been identified in a wide range of environments and at locations throughout the world. Species have been isolated from soil [56], thermal springs [57], hydrothermal vents [58, 59], animal feces [60], processed meats [61, 62], dried foods [63], room dust [64], medical instruments [65], textiles [66, 67], sewage [60, 68], and effluent from paper mills [69]. These studies have revealed two ecological niches in which ionizing radiation resistant species can be reliably identified: dry soils [70, 71] and high temperature aqueous environments [57–59, 72].

In agreement with the observations made empirically during isolation, many ionizing radiation resistant species are also resistant to desiccation and/or elevated temperatures, suggesting that ionizing radiation resistance is the consequence of a species capacity to tolerate desiccation or heat-induced cellular damage. While desiccation and heat do not introduce the same types of damage cause by ionizing radiation-induced ROS, they have the capacity to cause DNA double-strand breaks (DSBs), the lesions that are most commonly associated with ionizing radiation-induced lethality.

Arid soils are rich in species that survive exposure to doses above 5000 Gy [73]. In one large-scale study, representatives of eleven distinct genus-level taxonomic groups, exhibiting elevated resistance to ionizing radiation, were found in soil samples from the Sonoran desert in southern Arizona in the United States. The authors compared this environment to forest soil in southern Louisiana, USA, and concluded that ionizing radiation resistance is a phenotype more typically found associated with a dry environment, suggesting that the capacity to tolerate ionizing radiation enhances the species' ability to survive in the arid environment.

Desiccation resulting from dehydration or vacuum will introduce DSBs into the prokaryotic genome [74–77]. Species displaying high level resistance to ionizing radiation, including members of the genera *Deinococcus* [77] and *Chroococcidiopsis* [78], are also extremely resistant to desiccation. Drying introduces large numbers of DNA DSBs into species within these genera. In one study [77], the number of DSBs after six weeks desiccation appeared equivalent to a 3000 Gy exposure to ionizing radiation.

All of the hyperthermophilic Archaea that have been identified as ionizing radiation resistant were isolated from deep-sea hydrothermal vents [58, 59, 79, 80]. In addition, species of *Deinococcus* [57], *Rubrobacter* [72], and *Trupera* [81] have been identified during surveys of hot spring sites in Portugal, Italy, and the Azores. DNA is inherently unstable, readily undergoing spontaneous deamination and depurination events at low temperatures [82–84]. Depurination, which results following hydrolytic cleavage of the N-glycosyl bond, predisposes the DNA to strand breaks through a β-elimination reaction. At elevated temperature, the rate of depurination and subsequent strand breaks is dramatically increased. For a 40 °C increase in temperature there is an approximately 100-fold increase in depurination events, increasing from 0.5 events per cell at 37 °C to approximately 50 per cell at 80 °C. As described for desiccation-induced DNA damage, heat has the potential to generate DNA DSBs through these single-strand breaks.

6.4 Conclusions

While more exotic explanations are possible, I interpret the evidence outlined above to mean that extreme ionizing radiation resistance is the consequence of an evolutionary process that allowed these microorganisms to survive exposure to something other than ionizing radiation. In other words, ionizing radiation resistant species co-opt mechanisms built by natural selection for another purpose – to survive exposure to high dose ionizing radiation. Given the capacity of ionizing radiation to modify and destroy biological molecules, including DNA and proteins, it seems likely that the co-opted mechanisms increase the cell's tolerance of such damage by either passive protection (i.e., act as an antioxidant) or by heightening the cell's ability to repair the damage. Many of the species that are characterized as extremely resistant to ionizing radiation display resistance to other environmental stresses. Of special interest are those species exhibiting resistance to desiccation and high temperature. These are commonly encountered environmental insults that introduce a pattern of cellular damage that overlaps with that resulting from exposure to ionizing radiation, generating significant numbers of DNA DSBs. Species with the capacity to tolerate dry or high-temperature environments have a selective advantage over species without that capacity in those environments. If the mechanisms used to survive in a dry or hot environment confer better tolerance to ionizing radiation, these same species may display extreme ionizing radiation resistance relative to species lacking those mechanisms.

References

[1] C. D. Jonah, R. B. S. Madhava. Radiation chemistry: Present status and future trends. Amsterdam: Elsevier, 2001.
[2] J. W. T. Spinks, R. J. Woods. Introduction to radiation chemistry. New York: Wiley, 1990.
[3] C. von Sonntag The chemical basis of radiation biology. London: Taylor & Francis Ltd., 1987.
[4] C. Von Sonntag. Free radical-induced DNA damage and its repair. Berlin: Springer, 2006.
[5] C. Von Sonntag. Free radical reactions of carbohydrates as studied by radiation techniques. Adv Carbohydr Chem Biochem 1980, 37, 7–77.
[6] M. A. Shenoy, D. S. Joshi, B. B. Singh, A. R. Gopal-Ayengar. Role of bacterial membranes in radiosensitization. Adv Biol Med Phys 1970, 13, 255–271.
[7] F. Hutchinson. Chemical changes induced in DNA by ionizing radiation. Prog Nucleic Acid Res Mol Biol 1985, 32, 115–154.
[8] J. F. Ward. Molecular mechanisms of radiation-induced damage to nucleic acids. Adv Rad Biol 1975, 5, 181–239.
[9] M. Dizdaroglu, D. Schulte-Frohlinde, C. von Sonntag. Radiation chemistry of DNA, II. Strand breaks and sugar release by gamma-irradiation of DNA in aqueous solution. The effect of oxygen. Z C. Naturforsch 1975, 30, 826–828.
[10] M. Dizdaroglu, C. von Sonntag, D. Schulte-Frohlinde. Letter: Strand breaks and sugar release by gamma-irradiation of DNA in aqueous solution. J Am Chem Soc 1975, 97, 2277–2278.
[11] C. Von Sonntag, D. Schulte-Frohlinde. Radiation-induced degradation of the sugar in model compounds and in DNA. Mol Biol Biochem Biophys 1978, 27, 204–226.
[12] D. Freifelder, B. Trumbo. Matching of single-strand breaks to form double-strand breaks in DNA. Biopolymers 1969, 7, 681–693.
[13] M. J. Daly. A new perspective on radiation resistance based on *Deinococcus radiodurans*. Nat Rev Microbiol 2009, 7, 237–245.

[14] M. J. Daly, E. K. Gaidamakova, V. Y. Matrosova, A. Vasilenko, M. Zhai, R. D. Leapman, B. Lai, B. Ravel, Li SM, K. M. Kemner et al. Protein oxidation implicated as the primary determinant of bacterial radioresistance. PLoS Biol 2007, 5, e92.

[15] M. J. Daly, E. K. Gaidamakova, V. Y. Matrosova, A. Vasilenko, M. Zhai, A. Venkateswaran, M. Hess, M. V. Omelchenko, H. M. Kostandarithes, K. S. Makarova et al. Accumulation of Mn(II) in *Deinococcus radiodurans* facilitates gamma-radiation resistance. Science 2004, 306, 1025–1028.

[16] H. A. Makada, W. M. Garrison. Radiolytic oxidation of peptide derivatives of glycine in aqueous solution. Radiat Res 1972, 50, 48–55.

[17] K. M. Schaich. Free radical initiation in proteins and amino acids by ionizing and ultraviolet radiations and lipid oxidation–Part 22: ultraviolet radiation and photolysis. Critical reviews in food science and nutrition 1980, 13, 131–159.

[18] K. M. Schaich. Free radical initiation in proteins and amino acids by ionizing and ultraviolet radiations and lipid oxidation–part III: free radical transfer from oxidizing lipids. Critical reviews in food science and nutrition 1980, 13, 189–244.

[19] K. M. Schaich. Free radical initiation in proteins and amino acids by ionizing and ultraviolet radiations and lipid oxidation–Part I: ionizing radiation. Critical reviews in food science and nutrition 1980, 13, 89–129.

[20] E. M. Witkin. Inherited differences in sensitivity to radiation in *Escherichia coli*. Proc Natl Acad Sci U S A 1946, 32, 59–68.

[21] D. R. Harris, S. V. Pollock, E. A. Wood, R. J. Goiffon, A. J. Klingele, E. L. Cabot, W. Schackwitz, J. Martin, J. Eggington, T. J. Durfee et al. Directed evolution of ionizing radiation resistance in *Escherichia coli*. J Bacteriol 2009, 191, 5240–5252.

[22] R. Davies, A. J. Sinskey. Radiation-resistant mutants of *Salmonella typhimurium* LT2: development and characterization. J Bacteriol 1973, 113, 133–144.

[23] I. E. Erdman, F. S. Thatcher, K. F. Macqueen. Studies on the irradiation of microorganisms in relation to food preservation. I. The comparative sensitivities of specific bacteria of public health significance. Can J Microbiol 1961, 7, 199–205.

[24] A. Parisi, A. D. Antoine. Increased radiation resistance of vegetative *Bacillus pumilus*. Appl Microbiol 1974, 28, 41–46.

[25] G. J. Silverman, A. J. Sinskey. The Destruction of Microorganisms by Ionizing Radiation. In In: Antiseptic Disinfectants, Fungicides and Chemical and Physical Sterilization. Edited by S. S. Block, C. A. Lawrence. Philadelphia: Lea and Febiger, 1968, 741–761.

[26] A. W. Anderson, H. C. Nordon, R. F. Cain, G. Parrish, D. Duggan. Studies on a radio-resistant micrococcus. I. Isolation, morphology, cultural characteristics, and resistance to gamma radiation. Food Technol 1956, 10, 575–578.

[27] B. W. Brooks, R. G. E. Murray. Nomenclature for "*Micrococcus radiodurans*" and other radiation-resistant cocci: *Deinococcaceae* fam. nov. and *Deinococcus* gen. nov., including five new species. Int J Syst Bacteriol 1981, 31, 353–360

[28] B. E. B. Moseley. Photobiology and radiobiology of *Micrococcus (Deinococcus) radiodurans*. Photochem Photobiol Rev 1983, 7, 223–275.

[29] M. M. Cox, J. R. Battista. *Deinococcus radiodurans* – the consummate survivor. Nat Rev Microbiol 2005, 3, 882–892.

[30] United Nations Scientific Committee on the Effects of Atomic Radiation: Ionizing Radiation: Sources and Biological Effects. Effects of Atomic Radiation Report A/RES/37/87, New York: United Nations 1982.

[31] N. H. Barton, D. E. G. Briggs, J. A. Eisen, D. B. Goldstein, N. H. Patel: Evolution. Cold Spring Harbor: Cold Spring Harbor Laboratory Press, 2007.

[32] P. A. Karam. The evolution of Earth's background radiation field over geologic time. The Ohio State University, 1998.
[33] R. Naudet. Summary report on the Oklo phenomenon. C. E. A. French Report: Bull Infor Scien Tech (Eng Trans) 1974, 193, 7–85.
[34] J. R. De Laeter, K. J. R. Rosman, C. L. Smith. The Oklo natural reactor: Cumulative fission yields and retentivity of the symmetric mass region fission products. Earth Plan Sci Lett 1980, 50, 238–246.
[35] F. Gauthier-Lafaye. 2 billion year old natural analogs for nuclear waste disposal: the natural nuclear fission reactors in Gabon (Africa). C R Phys 2002, 3, 839–849.
[36] Y. V. Petrov, A. I. Nazarov, M. S. Onegin, E. G. Sakhnovsky. Natural nuclear reactor at Oklo and variation of fundamental constants: Computation of neutronics of a fresh core. In: Phys Rev C. vol. c74, 2006: 064610-064611-064610-064617.
[37] M. A. Line. Panspermia in the context of the timing of the origin of life and microbial phylogeny Int J Astrobiol 2007, 36, 249–254.
[38] H. A. Melosh. Exchange of meteorites (and life?) between stellar systems. Astrobiology 2004, 3, 207–215.
[39] H. I. Melosh. Impact ejection, spallation, and the origin of meteorites. Icarus 1984, 59, 234–260.
[40] P. Fajardo-Cavazos, F. Langenhorst, H. J. Melosh, W. L. Nicholson. Bacterial spores in granite survive hypervelocity launch by spallation: implications for lithopanspermia. Astrobiology 2009, 9, 647–657.
[41] G. Horneck, H. Bucker, G. Reitz. Long-term survival of bacterial spores in space. Adv Space Res 1994, 14, 41–45.
[42] G. Horneck, R. Moeller, J. Cadet, T. Douki, R. L. Mancinelli, W. L. Nicholson, C. Panitz, E. Rabbow, P. Rettberg, A. Spry et al. Resistance of bacterial endospores to outer space for planetary protection purposes–experiment PROTECT of the EXPOSE-E mission. Astrobiology 2012, 12, 445–456.
[43] G. Horneck, P. Rettberg, G. Reitz, J. Wehner, U. Eschweiler, K. Strauch, C. Panitz, V. Starke, C. Baumstark-Khan. Protection of bacterial spores in space, a contribution to the discussion on Panspermia. Orig Life Evol Biosph 2001, 31, 527–547.
[44] A. K. Pavlov, V. L. Kalinin, A. N. Konstantinov, V. N. Shelegedin, A. A. Pavlov. Was Earth ever infected by martian biota? Clues from radioresistant bacteria. Astrobiology 2006, 6, 911–918.
[45] S. Onofri, R. de la Torre, J. P. de Vera, S. Ott, L. Zucconi, L. Selbmann, G. Scalzi, K. J. Venkateswaran, E. Rabbow, F. J. Sanchez Inigo et al. Survival of rock-colonizing organisms after 1.5 years in outer space. Astrobiology 2012, 12, 508–516.
[46] P. A. Vaishampayan, E. Rabbow, G. Horneck, K. J. Venkateswaran. Survival of *Bacillus pumilus* spores for a prolonged period of time in real space conditions. Astrobiology 2012, 12, 487–497.
[47] W. L. Nicholson. Ancient micronauts: interplanetary transport of microbes by cosmic impacts. Trends Microbiol 2009, 17, 243–250.
[48] E. Dadachova, R. A. Bryan, X. Huang, T. Moadel, A. D. Schweitzer, P. Aisen, J. D. Nosanchuk, A. Casadevall. Ionizing radiation changes the electronic properties of melanin and enhances the growth of melanized fungi. PLoS One 2007, 2, e457.
[49] E. K. Field, S. D'Imperio, A. R. Miller, M. R. VanEngelen, R. Gerlach, B. D. Lee, W. A. Apel, B. M. Peyton. Application of molecular techniques to elucidate the influence of cellulosic waste on the bacterial community structure at a simulated low-level-radioactive-waste site. Appl Environ Microbiol 2010, 76, 3106–3115.
[50] A. Francis, S. Dobbs, Nine. B. Microbial activity of trench leachates from shallow-land, low-level radioactive waste disposal sites. Appl Environ Microbiol 1980, 40, 108–113.

[51] J. K. Fredrickson, J. M. Zachara, D. L. Balkwill, D. Kennedy, Li SM, H. M. Kostandarithes, M. J. Daly, M. F. Romine, F. J. Brockman. Geomicrobiology of high-level nuclear waste-contaminated vadose sediments at the Hanford site, Washington State. Appl Environ Microbiol 2004, 70, 4230–4241.

[52] S. J. Gould, E. S. Vrba. Exaptation; a missing term in the science of form Paleobiology 1982, 8, 4–15.

[53] J. Moan. Effects of UV radiation and visible light. In: Radiation at home, outdoors and in the workplace. Edited by D. Brune, R. Hellborg, B. R. R. Persson, R. Pääkkönen. Oslo: Scandinavian Science Publisher, 2001, 474–491.

[54] C. S. Cockrell. The Ultraviolet Radiation Environment of Earth and Mars: Past and Present. In: Astrobiology. Edited by G. Horneck, C. Baumstark-Khan, 1st edn. Berlin: Springer Verlag, 2002, 219–232.

[55] A. N. Osipov, N. M. Smetanina, M. V. Pustovalova, E. Arkhangelskaya, D. Klokov. The formation of DNA single-strand breaks and alkali-labile sites in human blood lymphocytes exposed to 365-nm UVA radiation. Free radical biology & medicine 2014, 73C, 34–40.

[56] R. G. E. Murray. The family Deinococcaceae. In: The Prokaryotes. Edited by A. Ballows, H. G. Truper, M. Dworkin, W. Harder, K. H. Scheilefer. New York: Springer-Verlag, 1992, 3732–3744.

[57] A. C. Ferreira, M. F. Nobre, F. A. Rainey, M. T. Silva, R. Wait, J. Burghardt, A. P. Chung, M. S. da Costa. *Deinococcus geothermalis* sp. nov. and *Deinococcus murrayi* sp. nov., two extremely radiation-resistant and slightly thermophilic species from hot springs. Int J Syst Bacteriol 1997, 47, 939–947.

[58] E. Jolivet, E. Corre, S. L'Haridon, P. Forterre, D. Prieur. *Thermococcus marinus* sp. nov. and *Thermococcus radiotolerans* sp. nov., two hyperthermophilic archaea from deep-sea hydrothermal vents that resist ionizing radiation. Extremophiles 2004, 8, 219–227.

[59] E. Jolivet, S. L'Haridon, E. Corre, P. Forterre, Prieur. *Thermococcus gammatolerans* sp. nov., a hyperthermophilic archaeon from a deep-sea hydrothermal vent that resists ionizing radiation. Int J Syst Evol Microbiol 2003, 53, 847–851.

[60] H. Ito, H. Watanabe, M. Takeshia, H. Iizuka. Isolation and identification of radiation-resistant cocci belonging to the genus *Deinococcus* from sewage sludges and animal feeds. Agric Biol Chem 1983, 47, 1239–1247.

[61] N. S. Davis, G. J. Silverman, E. B. Masurovsky. Radiation-Resistant, Pigmented Coccus Isolated from Haddock Tissue. J Bacteriol 1963, 86, 294–298.

[62] I. R. Grant, M. F. Patterson. A novel radiation resistant *Deinobacter* sp. isolated from irradiated pork. Letters Appl Microbiol 1989, 8, 21–24.

[63] N. F. Lewis. Studies on a radio-resistant coccus isolated from Bombay Duck (*Harpodon nehereus*). J Gen Microbiol 1971, 66, 29–35.

[64] E. A. Christensen, H. Kristensen. Radiation-resistance of micro-organisms from air in clean premises. Acta Pathol Microbiol Scand B 1981, 89, 293–301.

[65] E. A. Christensen, H. Kristensen, J. Hoborn, A. Miller. Radiation resistance of microorganisms on unsterilized infusion sets. APMIS 1991, 99, 620–626.

[66] E. A. Christensen, P. Gerner-Smidt, H. Kristensen. Radiation resistance of clinical *Acinetobacter* spp.: a need for concern? J Hosp Infect 1991, 18, 85–92.

[67] H. Kristensen, E. A. Christensen. Radiation-resistant micro-organisms isolated from textiles. Acta Pathol Microbiol Scand B 1981, 89, 303–309.

[68] H. Ito, H. Iizuka. Taxonomic studies on a radio-resistant *Pseudomonas*. XII. Studies on the microorganisms of cereal grain. Agric Biol Chem 1971, 35 1566–1571.

[69] O. M. Vaisanen, A. Weber, A. Bennasar, F. A. Rainey, H. J. Busse, M. S. Salkinoja-Salonen. Microbial communities of printing paper machines. J Appl Microbiol 1998, 84, 1069–1084.

[70] F. A. Rainey, M. Ferreira, M. F. Nobre, K. Ray, D. Bagaley, A. M. Earl, J. R. Battista, B. Gomez-Silva, C. P. McKay, M. S. da Costa. *Deinococcus peraridilitoris* sp. nov., isolated from a coastal desert. Int J Syst Evol Microbiol 2007, 57, 1408–1412.

[71] F. A. Rainey, K. Ray, M. Ferreira, B. Z. Gatz, M. F. Nobre, D. Bagaley, B. A. Rash, M. J. Park, A. M. Earl, N. C. Shank et al. Extensive diversity of ionizing-radiation-resistant bacteria recovered from Sonoran Desert soil and description of nine new species of the genus *Deinococcus* obtained from a single soil sample. Appl Environ Microbiol 2005, 71, 5225–5235.

[72] A. C. Ferreira, M. F. Nobre, E. Moore, F. A. Rainey, J. R. Battista, M. S. da Costa. Characterization and radiation resistance of new isolates of *Rubrobacter radiotolerans* and *Rubrobacter xylanophilus*. Extremophiles 1999, 3, 235–238.

[73] F. A. Rainey, M. F. Nobre, P. Schumann, E. Stackebrandt, M. S. da Costa. Phylogenetic diversity of the deinococci as determined by 16S ribosomal DNA sequence comparison. Int J Syst Bacteriol 1997, 47, 510–514.

[74] A. Bieger-Dose, K. Dose, R. Meffert, M. Mehler, S. Risi. Extreme dryness and DNA-protein cross-links. Adv Space Res 1991, 12, 265–270.

[75] K. Dose, A. Bieger-Dose, B. Ernst, U. Feister, B. Gomez-Silva, A. Klein, S. Risi, C. Stridde. Survival of microorganisms under the extreme conditions of the Atacama Desert. Orig Life Evol Biosph 2001, 31, 287–303.

[76] K. Dose, A. Bieger-Dose, M. Labusch, M. Gill. Survival in extreme dryness and DNA-single strand breaks. Adv Space Res 1992, 12, 221–229.

[77] V. Mattimore, J. R. Battista. Radioresistance of *Deinococcus radiodurans*: functions necessary to survive ionizing radiation are also necessary to survive prolonged desiccation. J Bacteriol 1996, 178, 633–637.

[78] D. Billi, E. I. Friedmann, K. G. Hofer, M. G. Caiola, R. Ocampo-Friedmann. Ionizing-radiation resistance in the desiccation-tolerant cyanobacterium *Chroococcidiopsis*. Appl Environ Microbiol 2000, 66, 1489–1492.

[79] J. DiRuggiero, J. R. Brown, A. P. Bogert, F. T. Robb. DNA repair systems in archaea: mementos from the last universal common ancestor? J Mol Evol 1999, 49, 474–484.

[80] J. DiRuggiero, N. Santangelo, Z. Nackerdien, J. Ravel, F. T. Robb. Repair of extensive ionizing-radiation DNA damage at 95 degrees C in the hyperthermophilic archaeon *Pyrococcus furiosus*. J Bacteriol 1997, 179, 4643–4645.

[81] L. Albuquerque, C. Simoes, M. F. Nobre, N. M. Pino, J. R. Battista, M. T. Silva, F. A. Rainey, M. S. da Costa. *Truepera radiovictrix* gen. nov., sp. nov., a new radiation resistant species and the proposal of *Trueperaceae* fam. nov. FEMS Microbiol Lett 2005, 247, 161–169.

[82] T. Lindahl. Instability and decay of the primary structure of DNA. Nature 1993, 362, 709–715.

[83] T. Lindahl, A. Andersson. Rate of chain breakage at apurinic sites in double-stranded deoxyribonucleic acid. Biochemistry 1972, 11, 3618–3623.

[84] T. Lindahl, B. Nyberg. Rate of depurination of native deoxyribonucleic acid. Biochemistry 1972, 11, 3610–3618.

Jennifer B. Glass, Cecilia Batmalle Kretz, Melissa J. Warren, and Claire S. Ting

7 Current perspectives on microbial strategies for survival under extreme nutrient starvation: evolution and ecophysiology

7.1 Introduction

The biomass of all living organisms consists of approximately 30 out of 92 naturally occurring elements [1, 2]. Nitrogen and carbon are the only two elements that can be directly incorporated into biomass via microbial fixation of gaseous forms; N_2, CO_2, CO, CH_4 and other hydrocarbons can be fixed into biomass. The remaining ~28 elements required for life must be acquired from chemical species dissolved in water or in some cases can be solubilized from rocks and minerals. Nutrients by definition are essential elements or vitamins required for incorporation into cellular biomass, which distinguishes them from energy sources involved in cellular ATP generation. Nutrients may limit cellular growth due to low abundance and/or bioavailability. Extreme environments often possess unique chemical compositions due to extreme temperature or pH, producing nutrient-limited conditions. In some instances, nutrient limitation increases microbial sensitivity to other extreme conditions; for example, *Deinococcus radiodurans* exhibits increased radiation sensitivity when grown under nutrient limitation [3].

Extremophiles have evolved numerous mechanisms to adapt to conditions that are undesirable to other organisms [4]. In this chapter, we review microbial adaptations to nutrient-limited ecosystems where either major or minor elements, or colimitation by both, limit microbial growth. For subsistence in ecosystems where one or more of these elements is limiting, or in habitats where toxic elements of similar size and charge to essential nutrients are present in high abundance, microbes must evolve survival mechanisms to produce biomass out of scarce ingredients. These survival strategies can generally be classified into four categories: scavenging, shifting speciation, storage, and substitution. In the case of scavenging, nutrient-starved organisms will allocate a greater percentage of cellular energy to resource acquisition by up-regulating high-affinity nutrient transport systems. Microbes can also shift chemical speciation of limiting nutrients by altering the pH of their microenvironment or by converting the limiting nutrient to a more bioavailable form using enzymes such as carbonic anhydrase and alkaline phosphatase. Intracellular storage in storage vacuoles or mineral precipitates enables organisms to survive on scavenged resources after they are depleted. Substitution of another element for the one that is limiting is

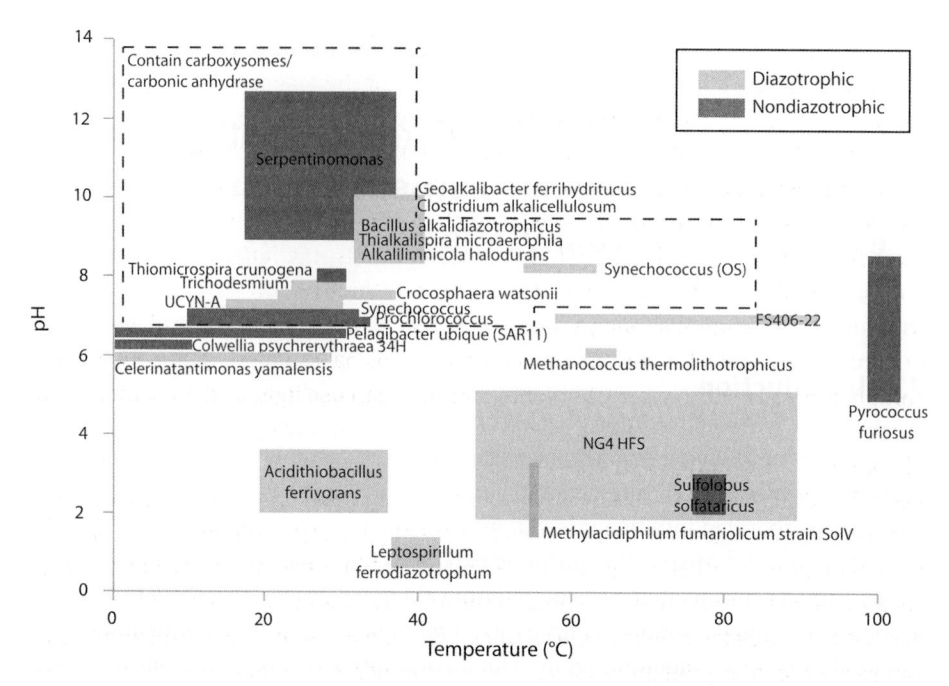

Fig. 7.1. pH-temperature range of nutrient-limited extremophilic Bacteria and Archaea discussed in chapter. Light gray boxes indicate microbes capable of N_2 fixation (diazotrophy). Dark gray boxes indicate non-diazotrophs. Organisms inside of dashed lines contain carboxysomes and carbonic anhydrase enzymes responsible for CO_2 storage and HCO_3^- conversion to CO_2 in neutral to high pH environments that experience CO_2 limitation.

possible for limiting nutrients that are not absolutely essential. Alternatively, a nutrient-starved organism may pass into a dormant stage to survive the duration of nutrient limitation.

In this chapter, we review examples of extremophilic microbial adaptations to nutrient limitation, starting with macronutrients (C, N, and P), and then following with micronutrients (transition metals and rare earth elements). We devote particular attention to nutrient acquisition strategies in marine microbes because the surface ocean is the most expansive nutrient-limited environment with moderate temperature (~ 10–30 °C) and near-neutral pH (~ 8) [5]. Although many extreme eukaryotes exist, we restrict our discussion to Bacteria and Archaea. A pH-temperature diagram highlighting the major nutrient-limited extremophiles is provided in ▶ Fig. 7.1.

7.2 Carbon

Microorganisms inhabiting extreme environments have evolved diverse strategies for obtaining the carbon necessary for their survival. The evolution of metabolic flexibility with regard to carbon sources, energy sources, uptake mechanisms, and carbon fixation pathways could be advantageous in light of the macro- and microscale heterogeneity characteristic of many of these environments. Microorganisms can be classified broadly as heterotrophs (i.e., organisms that utilize organic carbon as their cellular carbon source) or autotrophs (i.e., organisms that use inorganic carbon as their carbon source). While the former category includes chemoorganotrophs, which obtain their energy from the oxidation of organic compounds, the latter category includes most chemolithoautotrophs and photoautotrophs, which obtain their energy from the oxidation of inorganic compounds (e.g., H_2, S, H_2S, NH_3, CO, FeS_2, reduced metal ions) or light, respectively. Notably, many microorganisms have evolved to utilize different carbon and/or energy sources. Photoheterotrophs can use light as their source of energy and organic compounds as their carbon source, and some photoheterotrophs can also grow photoautotrophically. Examples of these organisms include green nonsulfur bacteria, which form microbial mats in hot spring environments and have the ability to use both simple organic compounds and inorganic (H_2, H_2S) compounds as electron donors for photosynthetic reactions [6–8]. Moreover, in some microbial communities, such as thermal environments, chemolithotrophy is a key strategy that has been documented in both heterotrophs and autotrophs [9]. In particular, chemolithoautotrophs can serve as major primary producers in environments in which high temperatures and an absence of sun light inhibit the growth of photoautotrophs [9, 10].

In bacteria, approximately 50% of the dry weight of a cell consists of carbon [11]. Thus, the availability of specific carbon sources and the frequency at which carbon limitation or starvation occurs are expected to drive selection for effective carbon acquisition and/or utilization strategies. Many different organic compounds can serve as potential sources of carbon for heterotrophs, including sugars, amino acids, fatty acids, aromatic compounds, and organic acids. Recent metagenomic studies suggest that a major carbon source for atmospheric microbes may be C1–C4 compounds found in cloud water [12]. *Afipia felis* is an alphaproteobacterium that belongs to the *Bradyrhizobiaceae*; it occurs in aquatic environments, but has also been found to be abundant in the troposphere [12]. The carbon source of this bacterium is dimethyl sulfone ($DMSO_2$) [13, 14], which forms from the oxidation of dimethyl sulfide (DMS), a compound associated with the marine atmosphere [15].

Analyses of the genome sequences of archaea and (hyper)thermophilic bacteria suggest that ATP binding cassette (ABC) transporters have an important role in facilitating the uptake of organic solutes in these microorganisms [16–19]. With these transporters, a substrate first binds to a membrane-associated protein, which typically has a high binding affinity ($K_d < 1\ \mu M$) [16]. Thus, these transporters are efficient in the uptake of solutes present at low concentrations in the environment and they are also characterized by high rates of transport [16]. Substrates of these ABC transporters in extremophiles include glucose, trehalose, maltose, maltotriose, cellobiose, maltodextrin and arabinose [16–18, 20–23]. Studies on the sugar transporters of the hyperthermophilic archaeon, *Pyrococcus furiosus* (▶ Fig. 7.1), suggest that while its maltose/trehalose and maltodextrin transporter system is most similar to ABC transporters of mesophilic bacteria [23, 24], its cellobiose transporter system exhibits greatest similarity to bacterial di/oligopeptide transporters [25]. Similarly, the sugar ABC transporters of the thermoacidophile *Sulfolobus solfataricus* (▸ Fig. 7.1) have also been categorized as carbohydrate transporters and di/oligopeptide transporters [16, 17, 22].

Furthermore, Na^+-coupled secondary transporters have been found in archaea (*Halobacterium halobium* [26, 27], *Haloferax volcanii* [28]), as well as thermophilic and alkaliphilic bacteria (*Clostridium fervidus* [29], *Bacillus* TA2.A1 [30, 31]). Transport of substrates, such as glucose and amino acids, across the cytoplasmic membrane by this mechanism is dependent on an electrochemical gradient of Na^+ ions [16, 26–31]. The phosphoenolpyruvate (PEP)-dependent phosphotransferase system (PTS) is another potential mechanism for the transport of sugars. Compared to other transporters, though, this mechanism does not appear to be as common in extremophiles [16]. Genomic evidence for this mechanism has not been found in several archaea [16] nor in the hyperthermophilic bacterium, *Thermotoga maritima* [32]. However, genes encoding a putative PTS system involved in fructose uptake have been found on a megaplasmid in the radioresistant bacterium, *Deinococcus radiodurans* R1 [33].

In addition to the compounds mentioned above that may be available in the environment, some extremophiles have evolved to utilize cellular organic carbon storage reserves including glycogen, polyhydroxyalkanoates, polyphosphates, triglycerides, and wax esters [34]. These storage compounds are especially prevalent in psychrophiles, which have reduced metabolic activity [34–37]. In thermoacidophilic verrucomicrobial methanotrophs, glycogen is accumulated in cells that are exposed to nitrogen depletion [38]. These glycogen reserves can then be used for maintaining cell viability under conditions of methane starvation [38].

The availability of dissolved inorganic carbon species (DIC, $CO_2 + HCO_3^- + CO_3^{2-}$) in aquatic environments can vary both spatially and temporally, and this creates challenges for autotrophs, which fix inorganic carbon into cellular carbon. A significant decrease in atmospheric and oceanic CO_2 concentrations occurred during the late Proterozoic and present day atmospheric CO_2 concentrations are estimated to be one to three orders of magnitude lower than they were approximately two billion years ago [39–42]. Dissolved inorganic carbon (DIC) species are linked by a pH-dependent equilibrium: $CO_2 + H_2O \leftrightarrow H_2CO_3 \leftrightarrow H^+ + HCO_3^- \leftrightarrow 2H^+ + CO_3^{2-}$, apparent $pK_a[HCO_3^-/CO_2] = 6.3$ [43]. In environments that are slightly alkaline, such as the oceans, the concentration of HCO_3^- is typically higher than the concentration of $CO_2(aq)$, though temperature and salt concentrations impact the solubility of CO_2 and the $CO_2(aq)/HCO_3^-$ equilibrium [43, 44]. Although DIC can be abundant (millimolar concentrations) in surface oceans, its concentrations often exhibit patchiness and regions of depletion can develop due to events such as phytoplankton blooms [45–48]. Furthermore, sea surface $CO_2(aq)$ concentrations are more variable than total DIC concentrations and it has been estimated that while the former can vary by a factor of three, the latter varies by less than 10% [44]. In hydrothermal vent habitats, DIC concentrations (2 to 7 mM) and pH levels (pH 5 to 8) are in constant flux, and CO_2 concentrations between 20 µM to 1 mM have been reported [49–52].

The evolution of mechanisms for the efficient uptake of inorganic carbon and alternative pathways for autotrophic carbon fixation could provide a selective advantage in extreme environments where the availability of inorganic carbon is variable or low. In addition, because of the low affinity of RuBisCO for CO_2(aq) and the oxygenase activity of this enzyme [43, 53, 54], the ability to concentrate CO_2 in the vicinity of RuBisCO would also enhance cellular CO_2 fixation reactions in organisms dependent on the reductive pentose phosphate cycle (Calvin-Benson-Bassham cycle). The carbon concentrating mechanism (CCM) has been studied extensively in cyanobacteria, and provides an effective system for increasing intracellular CO_2/HCO_3^- concentrations [55–58]. In conjunction with the passive movement of CO_2 into the cell and the maintenance of cytoplasmic pH levels, active transporters involved in this mechanism permit the direct uptake and accumulation of inorganic carbon to 20–40 mM [56, 57, 59, 60]. Studies on the chemolithoautotroph *Thiomicrospira crunogena* (▸ Fig. 7.1) suggested that it has also evolved to utilize a CCM [51, 61]. This bacterium was originally isolated from a deep-sea hydrothermal vent, where HCO_3^- is the most abundant species [49]. Laboratory studies suggested that *T. crunogena* can actively transport inorganic carbon and that this ability enables it to generate elevated intracellular DIC levels when it is grown at low CO_2 concentrations [51, 61].

A central element of the CCM is the carboxysome, a microcompartment in which CO_2 is concentrated in the vicinity of RuBisCO, thus enhancing its carboxylase activity [62–64]. These microcompartments are surrounded by a protein shell, which has an important role in regulating metabolite flux [65–67]. In addition, a carbonic anhydrase is associated with the carboxysome and catalyzes the conversion of HCO_3^- and protons to CO_2 and water, thus shifting DIC speciation to favor greater CO_2 availability [68–70]. Carboxysomes have also been studied in detail in cyanobacteria and could potentially contribute to reducing cellular oxidative stress by sequestering intracellular HCO_3^- and CO_2 [71]. Visualization of the near-native architecture of the marine cyanobacterium, *Prochlorococcus*, which typically dominates nutrient-limited open oceans in subtropical and tropical regions, revealed the presence of several carboxysomes in the central cytoplasmic space [72]. Interestingly, strain-specific differences were observed in carboxysome size (90 vs. 130 nm), as well as in the presence of a gene encoding a putative carboxysome shell polypeptide [72]. Studies of marine *Synechococcus* WH8102 indicated that RuBisCO is abundant in carboxysomes, with 232 ± 18 oligomers packed in three to four concentric layers [73].

Carboxysomes have been reported in organisms from other extreme environments. Electron microscopy was used to visualize carboxysomes from the hydrothermal vent chemolithotroph, *T. crunogena* [74], and genes encoding carboxysome-associated proteins are located in a distinct cluster in its genome [75]. Recently, three alkaliphilic betaproteobacterial strains were isolated from the Barnes spring complex in The Cedars of northern California and were classified in the novel genus *Serpenti-*

nomonas [76, ▶ Fig. 7.1]. Spring waters in this area have high pH (~ pH 9–12.5) and Ca^{2+} levels (~ 1 mM), and low concentrations of DIC, which is present mainly as CO_3^{2-} and reacts with calcium to form calcium carbonate ($CaCO_3$) precipitates [76]. Genes encoding putative carboxysome shell polypeptides and a carbonic anhydrase are present in the genomes of all three strains [76]. Notably, $CaCO_3$ is required for the growth of these strains. The authors proposed that cellular aggregation on $CaCO_3$ precipitates could lead to localized decreases in pH through the production of a proton gradient and the conversion of CO_3^- to HCO_3^-; this HCO_3^- could then be transported into cells and assimilated [76].

As a key enzyme associated with carboxysomes, RuBisCO is central to the reductive pentose phosphate cycle (Calvin-Benson-Bassham cycle), a major pathway for autotrophic CO_2 fixation [43]. Although this cycle is present in some thermophiles, it has not been found in hyperthermophiles [43]. It is possible that metabolic intermediates of this pathway are unstable at temperatures above 70 °C and produce toxic forms [43, 77, 78]. At least five other autotrophic CO_2 fixation pathways are presently known, which differ in their energy (ATP) requirements, the active CO_2 species involved (CO_2, HCO_3^-, HCO_2), major enzymes and intermediates, and the identity and number of reductants required [79]. In deep sea hydrothermal vent communities, for instance, at least five of the six autotrophic CO_2 fixation pathways are present [52]. The balance between key biotic (i.e., cellular energy and metabolic status, physiological and biochemical strategies, thermostability) and abiotic (i.e., physicochemical properties of the environment, including presence of oxygen, availability of metals and C1 compounds) factors will determine which pathways might provide a selective advantage in a particular extreme environment.

The reductive citric acid cycle (also known as the reductive tricarboxylic acid cycle, Arnon-Buchanan cycle or rTCA cycle) is another important pathway for autotrophic CO_2 fixation. Although this cycle involves enzymes that are oxygen sensitive, it requires only two ATP equivalents for the synthesis of one pyruvate, compared to the seven required by the reductive pentose phosphate cycle [79]. This pathway is prevalent in anaerobic or microaerobic members of *Aquificae*, *Proteobacteria*, and *Nitrospirae* and also in thermophilic members of the *Aquificae*, which thrive at temperatures of 70 °C or higher [43, 79]. Recent work suggests that an uncultured gammaproteobacterial endosymbiont of the deep-sea tube worm, *Riftia pachyptila*, has both the reductive citric acid and reductive pentose phosphate cycles and can shift between pathways depending on the availability of cellular energy [80].

Additional autotrophic CO_2 fixation pathways include the reductive acetyl-CoA (Wood-Ljungdahl) pathway, the 3-hydroxypropionate/4-hydroxybutyrate (HP/HB) cycle and the dicarboxylate/4-hydroxybutyrate (DC/HB) cycle. Although the reductive acetyl-CoA pathway involves enzymes that are highly oxygen sensitive, it requires approximately one ATP for the synthesis of pyruvate and has been found in both hyperthermophiles, such as *Methanopyrus kandleri* (122 °C) [81] and psychrophiles [43, 79]. The reductive acetyl-CoA pathway, along with the reductive pentose phosphate cycle, are thought to be the primary carbon fixation pathways in microorganisms thriving at low temperatures (< 20° C) [82, 83]. However, the metal and coenzyme requirements of the reductive acetyl-CoA pathway are higher than those of the HP/HB and DC/HB cycles, which were discovered in the *Crenarchaeota*, a group that includes thermoacidophiles and (hyper)thermophiles [43, 84–86]. Recent work demonstrated that although *Thaumarchaeota* (*Nitrosopumilus maritimus*) also utilize the HP/HB cycle, the pathway is more energy efficient in this phylum (five ATP in the *Thaumarchaeota* HP/HB cycle versus nine to ten ATP per pyruvate in the *Crenarchaeota* HP/HB cycle) [87]. Furthermore, most of the enzymes involved are divergent and unrelated to those used in *Crenarchaeota* and these pathways most likely evolved independently in these two phyla [87]. Both the HP/HB and DC/HB pathways assimilate HCO_3^- and might be advantageous in slightly alkaline aquatic environments, such as the oceans [43, 79].

7.3 Nitrogen

Nitrogen limitation, defined as insufficient quantities of fixed nitrogen in either inorganic or organic forms for optimal cellular growth, is widespread in both extreme and temperate ecosystems [88] and likely has been since life's earliest evolution [89]. The most effective, yet energetically expensive, microbial mechanism to combat nitrogen limitation is biological fixation of atmospheric N_2 into NH_3, the most widely bioavailable source of fixed nitrogen for assimilation into cellular biomass. It is commonly believed that the nitrogenase enzyme complex evolved early in Earth's history as a critical adaption to access the vast reservoir of atmospheric N_2 gas [89–93]. Before nitrogenase's evolution, lightning was the main supply of fixed nitrogen input into the biosphere [89, 94, 95]. Geochemical and modeling efforts suggest that a fixed nitrogen crisis occurred ~ 2.2–3.5 billion years (Ga) ago [96, 97]. Isotopic records of sedimentary rocks support an origin of biological nitrogen fixation > 2.5 Ga ago [98, 99]; however the date of this crisis is still debated, with molecular clock estimates giving much younger dates for the evolution of nitrogenase (~ 1.5–2.2 Ga ago) [100–102]. The nitrogen crisis is thought to have provided selective pressure for the widespread occurrence of biological N_2 fixation, which may have evolved before the time of the nitrogen crisis, possibly for detoxification of cyanide [102, 103]. The capability to fix N_2 likely first evolved in a strictly anaerobic methanogenic archaeon [100, 101] and was subsequently horizontally transferred amongst numerous other archaeal and bacterial species, including numerous extremophiles [103]. Inconsistencies between nitrogenase and 16S RNA phylogenies support the importance of horizontal gene transfer after the divergence of Bacteria and Archaea in the distribution of nitrogenase genes in prokaryotes, with the first acquisition in the bacterial domain in the ancestral anaerobic members of the *Firmicutes* [100–104]. There are no known instances of nitrogen fixation in Eukaryota, suggesting that it is limited to the Bacteria and Archaea.

To counter the effects of an oxic environment to the oxygen-sensitive nitrogenase enzyme, cyanobacteria evolved strategies to separate O_2 from N_2 fixation [105, 106]. As reviewed below, nitrogenase genes are widespread in microbes that survive among extreme environmental ranges of temperature and pH [107].

Nitrogen limitation is prevalent in high temperature ecosystems that are typically depleted in oxidized nitrogen, although they can be enriched in reduced ammonia [108]. In 1984, the methanogenic archaeon *Methanococcus thermolithotrophicus* (▶ Fig. 7.1) was discovered to fix N_2 at 64 °C [109]. Subsequent studies discovered nitrogenase genes in fluids from marine hydrothermal vents [110]. The highest temperature measured for nitrogen fixation activity in marine hydrothermal systems is 92 °C by a hyperthermophilic methanogenic archaeon isolated from a deep sea volcanic vent (strain FS406-22; 58–92 °C; ▶ Fig. 7.1) [111]. Other methanogens are capable of diazotrophy at thermophilic temperatures [112]. These (hyper)thermophilic methanogens appear to utilize Mo–Fe nitrogenases for diazotrophy [112, 113], the most common form of nitrogenase in non-extreme environments.

Nitrogen fixation has also been widely studied in terrestrial hydrothermal systems [114]. The highest temperature reported to date for terrestrial N_2 fixation is 89 °C for alphaproteobacteria (clade NG4 HFS; 48–89 °C; ▶ Fig. 7.1) enrichments from hot springs at Yellowstone National Park [115]. Unicellular cyanobacteria belonging to *Synechococcus* (Octopus Spring = OS) ecotypes fix N_2 in Yellowstone microbial mats at 54–63 °C [116, ▶ Fig. 7.1]. Adaptations allowing for ultra-high temperature diazotrophy may include extra *nif* genes of unknown function (i.e. *nifN*) [107]. It is likely that many more thermophilic diazotrophs remain to be discovered; one report of thermophilic *Streptomyces thermoautotrophicus* isolated from a burning charcoal pile with optimal growth temperature of 65 °C suggested a link between diazotrophy and superoxide metabolism [117]; this pathway remains to be found in another organism.

Nitrogen limitation is also widespread in ultra-cold ecosystems. Psychrophilic nitrogen fixation occurs in extreme ecosystems such as the dry valleys of Antarctica [118], deep sea marine sediments [119–122], and permafrost [123] (*Celerinatantimonas yamalensis*; ▶ Fig. 7.1). The marine psychrophile *Colwellia psychrerythraea* 34H (▶ Fig. 7.1) possesses genes involved in degradation of cyanophycin, a protein-like polymer that may serve as a nitrogen reserve [124]. The McMurdo Dry Valley in Antarctica, once thought to be lifeless due to extreme limitations on water, nutrients, and energy supplies as well as the extremely cold temperatures, is home to active microbial consortia in highly arid soils during the summer months [118]. The most transcriptionally active diazotrophs at 0.7 °C along the McMurdo Ice Shelf are heterocystous cyanobacteria of the genus *Nostoc* [125]. Recent proteomic analyses of psychrophilic diazotrophs suggest the importance of additional genes (i.e. *nifU* and copper sensitivity suppressor genes) for low temperature diazotrophy [126]. Warmer arid regions have diazotrophic communities in biological soil crusts similarly dominated by heterocystous cyanobacteria [127–129] as do hypersaline microbial communities, particularly in dry conditions [88, 130].

Several species of Fe(II)-oxidizing acidophilic bacteria, including *Leptospirillum ferrodiazotrophum* and *Acidithiobacillus ferrivorans*, fix N_2 at extremely low pH (0.7–1.3 for *L. ferrodiazotrophum*; 2.1–3.5 for *A. ferrivorans*) under low-O_2 conditions at mesophilic temperatures [131–134, ▶ Fig. 7.1]. The methanotrophic verrumicrobium *Methylacidiphilum fumariolicum* strain SolV fixes N_2 at pH 2 and 55 °C (▶ Fig. 7.1) at microaerophilic conditions (0.5% O_2) and was not inhibited by NH_4^+ up to 94 mM [135]. Previously mentioned diazotrophic thermoacidophilic alphaproteobacteria of clade NG4 HFS fix N_2 at pH 1.9–5 in Yellowstone hot springs (▶ Fig. 7.1) [115]. These studies suggest that acidophilic diazotrophy is likely widespread, but that these aerobic diazotrophs must minimize O_2 requirements in order to protect nitrogenase from O_2 inactivation. Thus, they do not seem to have the same O_2/N_2 fixation separation strategies as diazotrophic cyanobacteria.

Early studies of haloalkaliphilic diazotrophy performed in the littoral zone of Mono Lake, California, pH ~9.8 revealed the importance of non-heterocystous cyanobacteria [136, 137]. Subsequently, other haloalkaliphilic diazotrophic bacteria have been isolated from soda lake sediments, including the aerobic gammaproteobacteria *Thialkalispira microaerophila* and *Alkalilimnicola halodurans* [138], as well as obligate anaerobes *Clostridium alkalicellulosum* [139] and *Geoalkalibacter ferrihydrituscus* [140] (▶ Fig. 7.1). Recently, a novel diazotroph from RNA group 6, *Bacillus alkalidiazotrophicus* (▶ Fig. 7.1), was isolated from a Mongolian soda soil at pH 10, the first time that diazotrophy has been discovered in that bacterial clade [141, 142].

In marine environments, diazotrophic cyanobacteria are the main input of new nitrogen, contributing significant global biological N_2 fixation [143, 144]. These cyanobacteria include unicellular as well as both heterocystous and non-heterocystous filamentous species [145]. Oligotrophic, tropical regions tend to be dominated by the non-heterocystous filamentous cyanobacterium *Trichodesmium*, which forms large surface blooms [146, 147, ▶ Fig. 7.1]. More recently, the global importance of unicellular marine cyanobacteria including *Crocosphaera watsonii* and the uncultured UCYN clade has become apparent [145, 148, 149, ▶ Fig. 7.1]. UCYN-A is unique in that it lacks PSII and thus cannot perform oxygenic photosynthesis; instead it utilizes a photofermentative metabolism [150, 151].

Non-diazotrophic microbes must rely on fixed sources of nitrogen for growth. These commonly include NH_4^+ and NO_3^-, as well as NO_2^- and organic N. *Prochlorococcus*, the most abundant photosynthetic organism in the oligotrophic ocean, has numerous ecotypes specialized for uptake of the most abundant N species in their econiche. While *Prochlorococcus* strains cultured to date cannot generally assimilate oxidized N species, uncultured strains have more diverse N utilization abilities [152]. Moreover, the abundance of N assimilation genes in small, streamlined genomes of *Prochlorococcus* is higher in N-limited regions, suggesting that these N-limited ecotypes evolved to optimize nitrogen uptake strategies [153, 154]. Similarly, the ubiquitous heterotrophic alphaproteobacterium *Pelagibacter ubique* of the SAR11 clade (▶ Fig. 7.1) possesses numerous transporters for ammonia, urea, and amino acids in its small, streamlined genome [155] and devotes a large percentage of expressed proteins to nutrient transport [156, 157]. Combined with streamlined genomes and small cell size, a newly discovered mechanism for marine microbial survival in oligotrophic regions is reduction of the N content of amino acids, resulting in reduction in total cellular N budget by 3–10% [158].

7.4 Phosphorus

Phosphorus (P) plays a central role in life, accounting for ~3% of the dry weight of all organisms due to its presence in nucleic acids, ATP, phospholipids, and other biomolecules. The absence of an atmospheric reservoir of P makes it the ultimate limiting nutrient in many ecosystems. Its inorganic form, PO_4^{3-}, is the preferred source of phosphorus for microbial growth. Microbes likely evolved mechanisms to scavenge and store P early in Earth's history [159]. One mechanism for P storage involves the intracellular production of P storage organelles, acidic calcium and P-rich membrane-bound structures called acidocalcisomes, in which P can be present as short- and long-chain polyphosphate (polyP), inorganic phosphate, or pyrophosphate [160, 161]. Under conditions of P limitation, diverse extremophiles are capable of acidocalcisome dissolution to use polyP for organic compound synthesis [162].

Around two-thirds of ocean surface waters are P-limited [163] and PO_4^{3-} is particularly scarce (low nM) in subtropical gyres such as the Sargasso Sea in the north Atlantic [163]. Scavenging, storage, and substitution mechanisms for combatting P starvation are all utilized by marine microbes. Open ocean *Prochlorococcus* and *Synechococcus* possess numerous P transporters and utilize a wide variety of phosphate sources [154, 164–167], in contrast to coastal strains that contain fewer P acquisition genes [153, 168]. Thus, as for nitrogen, there is an apparent evolutionary link between gene content for nutrient acquisition and survival in P-limited ecosystems. Moreover, in *Prochlorococcus*, most nutrient acquisition genes are found in genomic islands that are thought to have arisen by horizontal gene transfer and represent local and recent adaptation [154, 169]. *Trichodesmium* possesses alkaline phosphatases for stripping PO_4^{3-} groups off of a variety of molecules [170, 171] and can also use organic P sources of phosphonate [172]. In the Sargasso Sea, cyanobacteria reduce cellular P demand by replacing phospholipids in their cellular membranes with nitrogen- and sulfur-containing lipids [173]. Using this mechanism, severely P-limited cyanobacteria are capable of lowering their P demands for lipid synthesis from ~17 to 1%. Heterotrophic marine bacteria were not found to have this ability. This evolutionary capability may contribute to the dominance of cyanobacteria worldwide in oligotrophic marine regions [173]. However, heterotrophic marine bacteria of the SAR11 clade from the Sargasso Sea have alternative mechanisms for scavenging organic P, including the ability to acquire P from methylphosphonate, producing methane as a byproduct [174]. In contrast to the canonical view that polyP is only utilized as a luxury storage molecule, microbial polyP accumulation and rapid recycling was recently observed in the Sargasso Sea in parallel with other indicators of P stress, including alkaline phosphatase activity and substitution of sulfolipids for phospholipids [175, 176].

Terrestrial P-limited extremophilic communities have also evolved strategies for coping with P limitation. When placed under extreme P limitation, the thermophilic diazotrophic cyanobacterium *Synechococcus* strain OS-B', isolated from microbial mats at Octopus Spring, Yellowstone National Park, can utilize organic phosphonates in place of inorganic P [177, 178]. In P-limited Lake Matano in Indonesia, betaine and glycolipids replace phospholipids [179].

While alternative lipids and sources of P can be utilized by some extremophiles, microbes have an absolute requirement for P for nucleic acids, and will go to great lengths to scavenge P when it is scarce or when toxic analogues, such as arsenic, are more abundant. Recently, it was proposed that *Halomonas* strain GFAJ-1, isolated from hypersaline and alkaline Mono Lake, could substitute AsO_4^{3-} in place of PO_4^{3-} in nucleic acids to sustain growth [180]. Arsenate (AsO_4^{3-}) and phosphate (PO_4^{3-}) have many chemical and physical similarities, with nearly identical pK_a values, similarly charged oxygen atoms, and thermochemical radii within 4% [181]. Subsequent studies found no detectable arsenic in GFAJ-1's nucleic acids and an inability of the strain to grow in media containing < 0.3 μM PO_4^{3-} and 40 mM AsO_4^{3-} [182, 183]. A unique PO_4^{3-} binding protein (PBP) located on the periplasmic portion of the ABC phosphate transport system, which can discriminate between PO_4^{3-} and AsO_4^{3-} over a 4,500-fold concentration difference, has been proposed to be the molecular mechanism for highly specific PO_4^{3-} selection in high AsO_4^{3-} environments [181]. Phosphate is apparently irreplaceable in nucleic acid biochemistry; no other molecule appears to be capable of filling its singular role [184].

7.5 Iron

Life likely evolved in an anoxic environment with abundant soluble Fe(II), and microbial metalloenzyme inventories show that Fe is the most important redox-active trace metal in biology [1, 185]. Indeed, Fe-binding protein fold families predate most others [186–188]. There is only one known organism that completely lacks Fe requirements, the pathogenic spirochete involved in Lyme disease, *Borrelia burgdorferi*, which substitutes manganese for functions requiring iron in other organisms [189]. Today, numerous extreme environments – most notably the oligotrophic open ocean – are Fe-limited due to the low solubility of Fe(III) in oxic conditions with circumneutral pH. About 40% of the world's ocean is Fe limited, including so-called "high nitrate low chlorophyll" (HNLC) regions in the equatorial and north Pacific as well as the Southern Ocean [190]. Marine microbes have evolved a wide range of mechanisms to cope with Fe limitation, including but not limited to: (i) production of iron-binding siderophores, (ii) expression of iron-specific transporters, (iii) reduction of insoluble

Fe(III) to more soluble Fe(II), (iv) uptake of organic Fe sources such as heme, (v) Fe storage, and (vi) replacement of Fe-containing proteins for those containing alternative metals or no metals at all [191–194]. In addition, some marine bacteria have evolved to be "cheaters" that acquire low level Fe by utilizing siderophores made by other organisms [195].

Iron acquisition strategies for *Trichodesmium* have been particularly well studied due to its extraordinary Fe requirements, which are ~5x higher when fixing N_2 than when assimilating NH_4^+ [196, 197]. Under Fe limitation, *Trichodesmium erythraeum* IMS101 selectively sacrifices N_2 fixation to conserve Fe for photosynthesis and electron transport and initiates a complex stress response involving up-regulation of flavodoxin (*isiB*) genes that replace Fe-containing ferredoxin as well as Fe(III) (*idiA*) and Fe(II) (*feoB*) transporters [198–200]. Additionally, the presence of *dps* and *tonB* genes, respectively, suggests that *Trichodesmium* may have the ability to store Fe in ferritin and to actively transport siderophores [191, 199, 200]. Another unique adaptation of *Trichodesmium* was recently discovered in samples from the Red Sea that actively acquire and utilize Fe from dust by forming ball-shaped "puff" colonies which are effective at dissolving and trapping Fe from dust [201]. Iron conservation in the unicellular marine diazotroph, *Crocosphaera watsonii*, functions differently than in *Trichodesmium* [202, 203]. *C. watsonii* minimizes its iron quota by shifting Fe from photosynthesis in the day to N_2 fixation at night, thus requiring half the Fe quota of *Trichodesmium*, which fixes N_2 during the day [202, 204].

7.6 Other micronutrients

In addition to Fe, additional metal micronutrients – most commonly Mn, Co, Cu, Zn and Mo – are required as cofactors in microbial metalloenzymes. In some organisms, Ni and W may also be required for specific metalloenzymes that tend to be more common in anaerobes [205]. Evidence for limitation of microbial growth by these trace metals is far less conclusive than for Fe in the open ocean, but laboratory culture experiments have clearly shown that cobalt availability limits growth of marine cyanobacteria *Prochlorococcus* and *Synechococcus*, likely due to the presence of cobalt in carbonic anhydrase [185, 206]. Thus, cobalt and carbon requirements are coupled in marine cyanobacteria through the need for cobalt-requiring carbon anhydrase activity to convert HCO_3^- to CO_2 for carbon fixation. The importance of cobalt for marine cyanobacteria is substantiated by the discovery of production of cobalt-binding ligands by *Synechococcus* in the open ocean [207]. *Prochlorococcus* uses calcium in place of zinc in its alkaline phosphatase enzyme, likely due to the ultra-low zinc concentrations in P-depleted ocean regions [208]. *Synechococcus* and *Trichodesmium* have also been shown to be capable of utilizing nickel for urease and superoxide dismutase (SOD) enzyme activity [209–211]. In the case of Ni-SOD, this is possibly an evolutionary advantage to minimize Fe demand by reducing usage of Fe-SOD [212].

Substitution of limiting trace elements for more bioavailable ones has also been reported for microbes at both temperature extremes. In hyperthermophilic enzymes, tungsten may be used in place of molybdenum due to greater thermal stability and lower redox potential of W complexes, enabling them to operate at very low potentials in ultra reducing hydrothermal settings [1]. The slower kinetics of tungstoenzymes coupled with their increased oxygen sensitivity is thought to limit their distribution to only the highest temperature anoxic ecosystems, such as volcanoes and hot springs. Indeed, tungsten was possibly an important micronutrient for the earliest microbes [213] and only hyperthermophilic archaea are known to be obligately tungsten dependent [214]. However, it has recently been suggested that uncultivated psychrophilic archaea from marine sediments utilize tungstoenzymes for key carbon metabolisms [215, 216], suggesting a broader role for tungsten in marine microbes.

The trace element plasticity of extremophiles is exemplified by *Methylacidiphilum fumariolicum* strain SolV, the thermoacidophilic methanotroph previously highlighted for its abilities to store carbon and fix N_2, and the most acidophilic bacterium capable of methane oxidation [217, 218, ▶ Fig. 7.1]. The grown of *M. fumariolicum* is strictly dependent on the presence of rare earth elements in the growth media [219]. Such rare earth elements, including lanthanum, cerium, praseodymium, and neodymium, are elevated in geothermal environments compared to soil and aquatic ecosystems. *M. fumariolicum* has been shown to substitute lanthanides in place of Ca^{2+} in the key enzyme methanol dehydrogenase. Lanthanide has also been observed to induce methanol dehydrogenase protein expression in the radiation resistant bacterium *Methylobacterium radiotolerans* [220]. It is likely that new technologies will continue to reveal novel metalloenzymes in extremophiles [221].

7.7 Conclusions

Microbial evolution has resulted in numerous adaptations to allow survival in nutrient starved ecosystems, which can broadly be grouped into the four "Ss": scavenging, speciation shifting, storage, and substitution. While diverse bacteria and archaea possess one or more of these strategies, certain phyla – such as cyanobacteria – are particularly well adapted for low nutrient ecosystems [88]. This is due to their widespread capabilities for CO_2 accumulation via carbonic anhydrase and carboxysomes, the common occurrence of N_2 fixation in cyanobacterial clades, and their diverse micronutrient acquisition strategies. These mechanisms enable cyanobacteria to proliferate in extremely nutrient-limited systems such as the oligotrophic open ocean, where *Prochlorococcus*, *Synechococcus*, and *Trichodesmium* dominate by way of niche adaptation. In most cases, the presence of extra nutrient acquisition genes, while not energetically favorable, may confer an ecological advantage in low nutrient regions. The impacts of global climate change and other consequences of rising CO_2, such as ocean acidification, on nutrient availability are a current topic of active research. New studies show that ocean acidification may exacerbate nutrient limitation [222], suggesting that microbes may evolve novel strategies to survive the unique challenges posed by the Anthropocene.

References

[1] J. J. R. F. da Silva, R. J. P. Williams. The Biological Chemistry of the Elements: The Inorganic Chemistry of Life. Oxford: Clarendon Press, 1991.
[2] R. W. Sterner, J. J. Elser. Ecological Stoichiometry: The Biology of Elements from Molecules to the Biosphere. Princeton, NJ: Princeton University Press, 2002.
[3] A. Venkateswaran, S. C. McFarlan, D. Ghosal, et al. Physiologic determinants of radiation resistance in *Deinococcus radiodurans*. Appl Environ Microbiol 2000, 66, 2620–2626.
[4] L. J. Rothschild, R. L. Mancinelli. Life in extreme environments. Nature 2001, 409, 1092–1101.
[5] C. Moore, M. Mills, K. Arrigo, et al. Processes and patterns of oceanic nutrient limitation. Nat Geosci 2013, 6, 701–710.
[6] B. K. Pierson, R. W. Castenholz. A phototrophic gliding filamentous bacterium of hot springs, *Chloroflexus aurantiacus*, gen. and sp. nov. Arch Microbiol 1974, 100, 5–24.
[7] B. K. Pierson, R. W. Castenholz. Studies of pigments and growth in *Chloroflexus aurantiacus*, a phototrophic filamentous bacterium. Arch Microbiol 1974, 100, 283–305.
[8] M. T. Madigan, S. R. Petersen, T. D. Brock. Nutritional studies on *Chloroflexus*, a filamentous photosynthetic, gliding bacterium. Arch Microbiol 1974, 100, 97–103.
[9] E. Burgess, I. Wagner, J. Wiegel. Thermal environments and their biodiversity. In: C. Gerday, N. Glansdorff, eds. Physiology and Biochemistry of Extremophiles. ASM Press, 2007, 13–29.
[10] C. E. Blank, S. L. Cady, N. R. Pace. Microbial composition of near-boiling silica-depositing thermal springs throughout Yellowstone National Park. Appl Environ Microbiol 2002, 68, 5123–5135.
[11] K. M. Fagerbakke, M. Heldal, S. Norland. Content of carbon, nitrogen, oxygen, sulfur and phosphorus in native aquatic and cultured bacteria. Aquat Microb Ecol 1996, 10, 15–27.
[12] N. DeLeon-Rodriguez, T. L. Lathem, L. M. Rodriguez-R, et al. Microbiome of the upper troposphere, species composition and prevalence, effects of tropical storms, and atmospheric implications. Proc Natl Acad Sci 2013, 110, 2575–2580.
[13] B. Scola, L. Barrassi, D. Raoult. Isolation of new fastidious α Proteobacteria and *Afipia felis* from hospital water supplies by direct plating and amoebal co-culture procedures. FEMS Microbiol Ecol 2000, 34, 129–137.
[14] S. A. Moosvi, C. C. Pacheco, I. R. McDonald, et al. Isolation and properties of methanesulfonate-degrading *Afipia felis* from Antarctica and comparison with other strains of *A. felis*. Environ Microbiol 2005, 7, 22–33.
[15] D. Davis, G. Chen, P. Kasibhatla, et al. DMS oxidation in the Antarctic marine boundary layer, Comparison of model simulations and held observations of DMS, DMSO, $DMSO_2$, H_2SO_4 (g), MSA (g), and MSA (p). J Geophys Res Atmos 1998, 103, 1657–1678.
[16] S.-V. Albers, J. L. Vossenberg, A. J. Driessen, W. N. Konings. Bioenergetics and solute uptake under extreme conditions. Extremophiles 2001, 5, 285–294.
[17] W. N. Konings, S.-V. Albers, S. Koning, A. J. Driessen. The cell membrane plays a crucial role in survival of bacteria and archaea in extreme environments. Antonie Leeuwen 2002, 81, 61–72.
[18] S. M. Koning, S.-V. Albers, W. N. Konings, A. J. Driessen. Sugar transport in (hyper) thermophilic archaea. Res Microbiol 2002, 153, 61–67.

[19] S. Lalithambika, L. Peterson, K. Dana, P. Blum. Carbohydrate hydrolysis and transport in the extreme thermoacidophile *Sulfolobus solfataricus*. Appl Environ Microbiol 2012, 78, 7931–7938.

[20] K. B. Xavier, L. O. Martins, R. Peist, M. Kossmann, W. Boos, H. Santos. High-affinity maltose/trehalose transport system in the hyperthermophilic archaeon *Thermococcus litoralis*. J Bacteriol 1996, 178, 4773–4777.

[21] R. Horlacher, K. B. Xavier, H. Santos, J. DiRuggiero, M. Kossmann, W. Boos. Archaeal binding protein-dependent ABC transporter, molecular and biochemical analysis of the trehalose/maltose transport system of the hyperthermophilic archaeon *Thermococcus litoralis*. J Bacteriol 1998, 180, 680–689.

[22] M. G. Elferink, S. V. Albers, W. N. Konings, A. J. Driessen. Sugar transport in *Sulfolobus solfataricus* is mediated by two families of binding protein-dependent ABC transporters. Mol Microbiol 2001, 39, 1494–1503.

[23] A. G. Evdokimov, D. E. Anderson, K. M. Routzahn, D. S. Waugh. Structural basis for oligosaccharide recognition by *Pyrococcus furiosus* maltodextrin-binding protein. J Mol Biol 2001, 305, 891–904.

[24] S. M. Koning, W. N. Konings, A. J. Driessen. Biochemical evidence for the presence of two α-glucoside ABC-transport systems in the hyperthermophilic archaeon *Pyrococcus furiosus*. Archaea 2002, 1, 19–25.

[25] S. M. Koning, M. G. Elferink, W. N. Konings, A. J. Driessen. Cellobiose uptake in the hyperthermophilic archaeon *Pyrococcus furiosus* is mediated by an inducible, high-affinity ABC transporter. J Bacteriol 2001, 183, 4979–4984.

[26] R. V. Greene, R. E. MacDonald. Partial purification and reconstitution of the aspartate transport system from *Halobacterium halobium*. Arch Biochem Biophys 1984, 229, 576–584.

[27] N. Kamo, Y. Wakamatsu, K. Kohno, Y. Kobatake. On the glutamate transport through cell envelope vesicles of *Halobacterium halobium*. Biochem Biophys Res Comm 1988, 152, 1090–1096.

[28] E. Tawara, N. Kamo. Glucose transport of *Haloferax volcanii* requires the Na^+-electrochemical potential gradient and inhibitors for the mammalian glucose transporter inhibit the transport. Biochim Biophys Acta 1991, 1070, 293–299.

[29] G. Speelmans, W. De Vrij, W. Konings. Characterization of amino acid transport in membrane vesicles from the thermophilic fermentative bacterium *Clostridium fervidus*. J Bacteriol 1989, 171, 3788–3795.

[30] C. J. Peddie, G. M. Cook, H. W. Morgan. Sucrose transport by the alkaliphilic, thermophilic *Bacillus* sp. strain TA2.A1 is dependent on a sodium gradient. Extremophiles 2000, 4, 291–296.

[31] C. J. Peddie, G. M. Cook, H. W. Morgan. Sodium-dependent glutamate uptake by an alkaliphilic, thermophilic *Bacillus* strain, TA2.A1. J Bacteriol 1999, 181, 3172–3177.

[32] K. E. Nelson, R. A. Clayton, S. R. Gill, et al. Evidence for lateral gene transfer between Archaea and Bacteria from genome sequence of *Thermotoga maritima*. Nature 1999, 399, 323–329.

[33] O. White, J. A. Eisen, J. F. Heidelberg, et al. Genome sequence of the radioresistant bacterium *Deinococcus radiodurans* R1. Science 1999, 286, 1571–1577.

[34] J. P. Bowman. Genomic analysis of psychrophilic prokaryotes. In: C. Gerday, N. Glansdorff, eds. Psychrophiles: From Biodiversity to Biotechnology. Springer, 2008, 265–284.

[35] A. Danchin. An interplay between metabolic and physicochemical constrains: lessons from the psychrophilic prokaryote genomes. Physiology and Biochemistry of Extremophiles 2007, 208–220.

[36] D. Kadouri, E. Jurkevitch, Y. Okon, S. Castro-Sowinski. Ecological and agricultural significance of bacterial polyhydroxyalkanoates. Crit Rev Microbiol 2005, 31, 55–67.

[37] Ng C, M. Z. DeMaere, T. J. Williams, et al. Metaproteogenomic analysis of a dominant green sulfur bacterium from Ace Lake, Antarctica. ISME J 2010, 4, 1002–1019.

[38] A. F. Khadem, M. C. van Teeseling, L. van Niftrik, M. S. Jetten, H. J. Op Den Camp, A. Pol. Genomic and physiological analysis of carbon storage in the verrucomicrobial methanotroph "*Ca.* Methylacidiphilum fumariolicum" SolV. Front Microbiol 2012, 3, 345.

[39] J. A. Raven. Implications of inorganic carbon utilization, ecology, evolution, and geochemistry. Can J Bot 1991, 69, 908–924.

[40] R. Rye, P. H. Kuo, H. D. Holland. Atmospheric carbon dioxide concentrations before 2.2 billion years ago. Nature 1995, 378, 603–605.

[41] P. D. Tortell. Evolutionary and ecological perspectives on carbon acquisition in phytoplankton. Limnol Oceangr 2000, 45, 744–750.

[42] A. J. Kaufman, S. Xiao. High CO_2 levels in the Proterozoic atmosphere estimated from analyses of individual microfossils. Nature 2003, 425, 279–282.

[43] I. A. Berg. Ecological aspects of the distribution of different autotrophic CO_2 fixation pathways. Appl Environ Microbiol 2011, 77, 1925–1936.

[44] J. R. Reinfelder. Carbon concentrating mechanisms in eukaryotic marine phytoplankton. Ann Rev Mar Sci 2011, 3, 291–315.

[45] L. Codispoti, G. Friederich, R. Iverson, D. Hood. Temporal changes in the inorganic carbon system of the southeastern Bering Sea during spring 1980. Nature 1982, 296, 242–245.

[46] D. M. Karl, B. D. Tilbrook, G. Tien. Seasonal coupling of organic matter production and particle flux in the western Bransfield Strait, Antarctica. Deep Sea Res A 1991, 38, 1097–1126.

[47] A. Murata, Y. Kumamoto, C. Saito, et al. Impact of a spring phytoplankton bloom on the CO_2 system in the mixed layer of the northwestern North Pacific. Deep Sea Res II 2002, 49, 5531–5555.

[48] F. J. Millero. Chemical Oceanography. CRC Press, 2013.

[49] H. W. Jannasch, C. O. Wirsen, D. C. Nelson, L. A. Robertson. *Thiomicrospira crunogena* sp. nov., a colorless, sulfur-oxidizing bacterium from a deep-sea hydrothermal vent. Int J Syst Bact 1985, 35, 422–424.

[50] K. S. Johnson, J. J. Childress, C. L. Beehler. Short-term temperature variability in the Rose Garden hydrothermal vent field: an unstable deep-sea environment. Deep Sea Res A 1988, 35, 1711–1721.

[51] K. P. Dobrinski, D. L. Longo, K. M. Scott. The carbon-concentrating mechanism of the hydrothermal vent chemolithoautotroph *Thiomicrospira crunogena*. J Bacteriol 2005, 187, 5761–5766.

[52] Z. Minic, P. D. Thongbam. The biological deep sea hydrothermal vent as a model to study carbon dioxide capturing enzymes. Mar Drug 2011, 9, 719–738.

[53] G. Bowes. Photosynthesis and photo-respiration. Aquat Bot 1989, 34, 1–299.

[54] M. R. Badger, T. J. Andrews, S. Whitney, et al. The diversity and coevolution of Rubisco, plastids, pyrenoids, and chloroplast-based CO_2-concentrating mechanisms in algae. Can J Bot 1998, 76, 1052–1071.

[55] M. R. Badger, G. D. Price. CO_2 concentrating mechanisms in cyanobacteria, molecular components, their diversity and evolution. J Exp Bot 2003, 54, 609–622.

[56] M. R. Badger, G. D. Price, B. M. Long, F. J. Woodger. The environmental plasticity and ecological genomics of the cyanobacterial CO_2 concentrating mechanism. J Exp Bot 2006, 57, 249–265.

[57] G. D. Price, M. R. Badger, F. J. Woodger, B. M. Long. Advances in understanding the cyanobacterial CO_2-concentrating-mechanism (CCM), functional components, Ci transporters, diversity, genetic regulation and prospects for engineering into plants. J Exp Bot 2008, 59, 1441–1461.

[58] E. V. Kupriyanova, M. A. Sinetova, S. M. Cho, Y.-I. Park, D. A. Los, N. A. Pronina. CO_2-concentrating mechanism in cyanobacterial photosynthesis, organization, physiological role, and evolutionary origin. Photosyn Res 2013, 117, 133–146.

[59] D. Sültemeyer, G. D. Price, Yu J-W, M. R. Badger. Characterisation of carbon dioxide and bicarbonate transport during steady-state photosynthesis in the marine cyanobacterium *Synechococcus* strain PCC7002. Planta 1995, 197, 597–607.

[60] A. Kaplan, L. Reinhold. CO_2 concentrating mechanisms in photosynthetic microorganisms. Ann Rev Plant Biol 1999, 50, 539–570.

[61] K. P. Dobrinski, A. J. Boller, K. M. Scott. Expression and function of four carbonic anhydrase homologs in the deep-sea chemolithoautotroph *Thiomicrospira crunogena*. Appl Environ Microbiol 2010, 76, 3561–3567.

[62] J. Shively, F. L. Ball, B. W. Kline. Electron microscopy of the carboxysomes (polyhedral bodies) of *Thiobacillus neapolitanus*. J Bacteriol 1973, 116, 1405–1411.

[63] G. C. Cannon, C. E. Bradburne, H. C. Aldrich, S. H. Baker, S. Heinhorst, J. M. Shively. Microcompartments in prokaryotes, carboxysomes and related polyhedra. Appl Environ Microbiol 2001, 67, 5351–5361.

[64] T. O. Yeates, C. A. Kerfeld, S. Heinhorst, G. C. Cannon, J. M. Shively. Protein-based organelles in bacteria, carboxysomes and related microcompartments. Nat Rev Microbiol 2008, 6, 681–691.

[65] C. A. Kerfeld, M. R. Sawaya, S. Tanaka, et al. Protein structures forming the shell of primitive bacterial organelles. Science 2005, 309, 936–938.

[66] M. G. Klein, P. Zwart, S. C. Bagby, et al. Identification and structural analysis of a novel carboxysome shell protein with implications for metabolite transport. J Mol Biol 2009, 392, 319–333.

[67] J. N. Kinney, S. D. Axen, C. A. Kerfeld. Comparative analysis of carboxysome shell proteins. Photosyn Res 2011, 109, 21–32.

[68] S. H. Baker, D. S. Williams, H. C. Aldrich, A. C. Gambrell, J. M. Shively. Identification and localization of the carboxysome peptide Csos3 and its corresponding gene in *Thiobacillus neapolitanus*. Arch Microbiol 2000, 173, 278–283.

[69] So AK-C, G. S. Espie, E. B. Williams, J. M. Shively, S. Heinhorst, G. C. Cannon. A novel evolutionary lineage of carbonic anhydrase (ε class) is a component of the carboxysome shell. J Bacteriol 2004, 186, 623–630.

[70] M. R. Sawaya, G. C. Cannon, S. Heinhorst, et al. The structure of β-carbonic anhydrase from the carboxysomal shell reveals a distinct subclass with one active site for the price of two. J Biol Chem 2006, 281, 7546–7555.

[71] C. S. Ting. The architecture of cyanobacteria, archetypes of microbial innovation. In: M. Homann-Marriott, ed. The Structural Basis of Biological Energy Generation. Springer, 2014, 249–275.

[72] C. S. Ting, C. Hsieh, S. Sundararaman, C. Mannella, M. Marko. Cryo-electron tomography reveals the comparative three-dimensional architecture of *Prochlorococcus*, a globally important marine cyanobacterium. J Bacteriol 2007, 189, 4485–4493.

[73] C. V. Iancu, H. J. Ding, D. M. Morris, et al. The structure of isolated *Synechococcus* strain WH8102 carboxysomes as revealed by electron cryotomography. J Mol Biol 2007, 372, 764–773.

[74] K. Scott, M. Bright, C. Fisher. The burden of independence, Inorganic carbon utilization strategies of the sulphur chemoautotrophic hydrothermal vent isolate *Thiomicrospira crunogena* and the symbionts of hydrothermal vent and cold seep vestimentiferans. Cah Biol Mar 1998, 39, 379–381.

[75] K. M. Scott, S. M. Sievert, F. N. Abril, et al. The genome of deep-sea vent chemolithoautotroph *Thiomicrospira crunogena* XCL-2. PLoS Biology 2006, 4, e383.
[76] S. Suzuki, J. G. Kuenen, K. Schipper, et al. Physiological and genomic features of highly alkaliphilic hydrogen-utilizing Betaproteobacteria from a continental serpentinizing site. Nature Comm 2014, 5.
[77] S. A. Phillips, P. J. Thornalley. The formation of methylglyoxal from triose phosphates. Eur J Biochem 1993, 212, 101–105.
[78] H. Imanaka, T. Fukui, H. Atomi, T. Imanaka. Gene cloning and characterization of fructose-1, 6-bisphosphate aldolase from the hyperthermophilic archaeon *Thermococcus kodakaraensis* KOD1. J Biosci Bioeng 2002, 94, 237–243.
[79] I. A. Berg, D. Kockelkorn, W. H. Ramos-Vera, et al. Autotrophic carbon fixation in archaea. Nat Rev Microbiol 2010, 8, 447–460.
[80] S. Markert, C. Arndt, H. Felbeck, et al. Physiological proteomics of the uncultured endosymbiont of *Riftia pachyptila*. Science 2007, 315, 247–250.
[81] K. Takai, K. Nakamura, T. Toki, et al. Cell proliferation at 122 °C and isotopically heavy CH_4 production by a hyperthermophilic methanogen under high-pressure cultivation. Proc Natl Acad Sci 2008, 105, 10949–10954.
[82] M. Hügler, S. M. Sievert. Beyond the Calvin cycle, autotrophic carbon fixation in the ocean. Mar Sci 2011, 3.
[83] L. Montoya, L. B. Celis, E. Razo-Flores, Á. G. Alpuche-Solís. Distribution of CO_2 fixation and acetate mineralization pathways in microorganisms from extremophilic anaerobic biotopes. Extremophiles 2012, 16, 805–817.
[84] I. A. Berg, D. Kockelkorn, W. Buckel, G. Fuchs. A 3-hydroxypropionate/4-hydroxybutyrate autotrophic carbon dioxide assimilation pathway in Archaea. Science 2007, 318, 1782–1786.
[85] K. S. Auernik, C. R. Cooper, R. M. Kelly. Life in hot acid, pathway analyses in extremely thermoacidophilic archaea. Curr Opin Biotechnol 2008, 19, 445–453.
[86] H. Huber, M. Gallenberger, U. Jahn, et al. A dicarboxylate/4-hydroxybutyrate autotrophic carbon assimilation cycle in the hyperthermophilic Archaeum *Ignicoccus hospitalis*. Proc Natl Acad Sci 2008, 105, 7851–7856.
[87] M. Könneke, D. M. Schubert, P. C. Brown, et al. Ammonia-oxidizing archaea use the most energy-efficient aerobic pathway for CO_2 fixation. Proc Natl Acad Sci 2014, 111, 8239–8244.
[88] H. W. Paerl, J. L. Pinckney, T. F. Steppe. Cyanobacterial–bacterial mat consortia, examining the functional unit of microbial survival and growth in extreme environments. Environ Microbiol 2000, 2, 11–26.
[89] D. E. Canfield, A. N. Glazer, P. G. Falkowski. The evolution and future of Earth's nitrogen cycle. Science 2010, 330, 192–196.
[90] J. Raymond. The evolution of biological carbon and nitrogen cycling – a genomic perspective. Rev Mineral Geochem 2005, 59, 211–231.
[91] P. G. Falkowski. Evolution of the nitrogen cycle and its influence on the biological sequestration of CO_2 in the ocean. Nature 1997, 387, 272–275.
[92] R. Mancinelli, C. McKay. The evolution of nitrogen cycling. Origins of Life and Evolution of the Biosphere 1988, 18, 311–325.
[93] L. V. Godfrey, J. B. Glass. The geochemical record of the ancient nitrogen cycle, nitrogen isotopes, and metal cofactors. Methods Enzymol 2011, 486, 483.
[94] L. V. Godfrey, P. G. Falkowski. The cycling and redox state of nitrogen in the Archaean ocean. Nat Geosci 2009, 2, 725–729.
[95] Y. L. Yung, M. B. McElroy. Fixation of nitrogen in the prebiotic atmosphere. Science 1979, 203, 1002–1004.

[96] K. Fennel, M. Follows, P. G. Falkowski. The co-evolution of the nitrogen, carbon and oxygen cycles in the Proterozoic ocean. Am J Sci 2005, 305, 526.
[97] R. Navarro-Gonzalez, C. P. McKay, D. N. Mvondo. A possible nitrogen crisis for Archaean life due to reduced nitrogen fixation by lightning. Nature 2001, 412, 61–64.
[98] X. Zhang, D. M. Sigman, F. M. Morel, A. M. Kraepiel. Nitrogen isotope fractionation by alternative nitrogenases and past ocean anoxia. Proc Natl Acad Sci 2014, 111, 4782–4787.
[99] V. Beaumont, F. Robert. Nitrogen isotope ratios of kerogens in Precambrian cherts: a record of the evolution of atmosphere chemistry? Precambrian Res 1999, 96, 63–82.
[100] E. Boyd, A. Anbar, S. Miller, T. Hamilton, M. Lavin, J. Peters. A late methanogen origin for molybdenum-dependent nitrogenase. Geobiology 2011, 9, 221–232.
[101] E. S. Boyd, T. L. Hamilton, J. W. Peters. An alternative path for the evolution of biological nitrogen fixation. Front Microbiol 2011, 2, 205.
[102] E. S. Boyd, J. W. Peters. New insights into the evolutionary history of biological nitrogen fixation. Front Microbiol 2013, 4, 201.
[103] J. Raymond, J. L. Siefert, C. R. Staples, R. E. Blankenship. The natural history of nitrogen fixation. Mol Biol Evol 2004, 21, 541–554.
[104] K. J. Kechris, J. C. Lin, P. J. Bickel, A. N. Glazer. Quantitative exploration of the occurrence of lateral gene transfer by using nitrogen fixation genes as a case study. Proc Natl Acad Sci 2006, 103, 9584–9589.
[105] I. Berman-Frank, P. Lundgren, P. Falkowski. Nitrogen fixation and photosynthetic oxygen evolution in cyanobacteria. Res Microbiol 2003, 154, 157–164.
[106] P. Fay. Oxygen relations of nitrogen fixation in cyanobacteria. Microbio Mol Biol Rev 1992, 56, 340–373.
[107] P. C. Dos Santos, Z. Fang, S. W. Mason, J. C. Setubal, R. Dixon. Distribution of nitrogen fixation and nitrogenase-like sequences amongst microbial genomes. BMC Genomics 2012, 13, 162.
[108] S. E. Humphris, R. A. Zierenberg, L. S. Mullineaux, R. E. Thomson. Seafloor hydrothermal systems: physical, chemical, biological, and geological interactions. American Geophysical Union, 1995.
[109] N. Belay, R. Sparling, L. Daniels. Dinitrogen fixation by a thermophilic methanogenic bacterium. Nature 1984, 312, 286–288.
[110] M. P. Mehta, D. A. Butterfield, J. A. Baross. Phylogenetic diversity of nitrogenase (*nifH*) genes in deep-sea and hydrothermal vent environments of the Juan de Fuca Ridge. Appl Environ Microbiol 2003, 69, 960–970.
[111] M. P. Mehta, J. A. Baross. Nitrogen fixation at 92°C by a hydrothermal vent archaeon. Science 2006, 314, 1783–1786.
[112] M. Nishizawa, J. Miyazaki, A. Makabe, K. Koba, K. Takai. Physiological and isotopic characteristics of nitrogen fixation by hyperthermophilic methanogens: Key insights into nitrogen anabolism of the microbial communities in Archean hydrothermal systems. Geochem Cosmochim Acta 2014, 138, 117–135.
[113] S. E. McGlynn, E. S. Boyd, J. W. Peters, V. J. Orphan. Classifying the metal dependence of uncharacterized nitrogenases. Front Microbiol 2013, 3, 419.
[114] S. T. Loiacono, D. A. R. Meyer-Dombard, J. R. Havig, A. T Poret-Peterson, H. E. Hartnett, E. L. Shock. Evidence for high-temperature in situ *nifH* transcription in an alkaline hot spring of Lower Geyser Basin, Yellowstone National Park. Environ Microbiol 2012, 14, 1272–1283.
[115] T. L. Hamilton, R. K. Lange, E. S. Boyd, J. W. Peters. Biological nitrogen fixation in acidic high-temperature geothermal springs in Yellowstone National Park, Wyoming. Environ Microbiol 2011, 13, 2204–2215.

[116] A. S. Steunou, D. Bhaya, M. M. Bateson, et al. In situ analysis of nitrogen fixation and metabolic switching in unicellular thermophilic cyanobacteria inhabiting hot spring microbial mats. Proc Natl Acad Sci 2006, 103, 2398–2403.

[117] M. Ribbe, D. Gadkari, O. Meyer. N_2 fixation by *Streptomyces thermoautotrophicus* involves a molybdenum dinitrogenase and a manganese-superoxide oxidoreductase that couple N_2 reduction to the oxidation of superoxide produced from O_2 by a molybdenum-CO dehydrogenase. J Biol Chem 1997, 272, 26627–26633.

[118] S. C. Cary, I. R. McDonald, J. E. Barrett, D. A. Cowan. On the rocks, the microbiology of Antarctic Dry Valley soils. Nat Rev Microbiol 2010, 8, 129–138.

[119] J. Miyazaki, R. Higa, T. Toki, et al. Molecular characterization of potential nitrogen fixation by anaerobic methane-oxidizing archaea in the methane seep sediments at the Number 8 Kumano Knoll in the Kumano Basin, offshore of Japan. Appl Environ Microbiol 2009, 75, 7153–7162.

[120] A. Dekas, R. S. Poretsky, V. J. Orphan. Deep-sea archaea fix and share nitrogen in methane-consuming microbial consortia. Science 2009, 326, 422–426.

[121] A. E. Dekas, V. J. Orphan. Identification of diazotrophic microorganisms in marine sediment via fluorescence in situ hybridization coupled to nanoscale secondary ion mass spectrometry (FISH-NanoSIMS). Methods Enzymol 2011, 486, 281–305.

[122] A. Pernthaler, A. E. Dekas, C. T. Brown, S. K. Goffredi, T. Embaye, V. J. Orphan. Diverse syntrophic partnerships from deep-sea methane vents revealed by direct cell capture and metagenomics. Proc Natl Acad Sci 2008, 105, 7052–7057.

[123] V. Shcherbakova, N. Chuvilskaya, E. Rivkina, et al. *Celerinatantimonas yamalensis* sp. nov., a cold-adapted diazotrophic bacterium from a cold permafrost brine. Int J Syst Evol Microbiol 2013, 63, 4421–4427.

[124] B. A. Methé, K. E. Nelson, J. W. Deming, et al. The psychrophilic lifestyle as revealed by the genome sequence of *Colwellia psychrerythraea* 34H through genomic and proteomic analyses. Proc Natl Acad Sci 2005, 102, 10913–10918.

[125] A. D. Jungblut, B. A. Neilan. *nifH* gene diversity and expression in a microbial mat community on the McMurdo Ice Shelf, Antarctica. Antarctic Sci 2010, 22, 117–122.

[126] D. C. Suyal, A. Yadav, Y. Shouche, R. Goel. Differential proteomics in response to low temperature diazotrophy of Himalayan psychrophilic nitrogen fixing *Pseudomonas migulae* S10724 strain. Curr Microbiol 2013, 68, 1–8.

[127] C. M. Yeager, J. L. Kornosky, R. E. Morgan, et al. Three distinct clades of cultured heterocystous cyanobacteria constitute the dominant N_2-fixing members of biological soil crusts of the Colorado Plateau, USA. FEMS Microbiol Ecol 2007, 60, 85–97.

[128] R. M. Abed, Al S. Kharusi, A. Schramm, M. D. Robinson. Bacterial diversity, pigments and nitrogen fixation of biological desert crusts from the Sultanate of Oman. FEMS Microbiol Ecol 2010, 72, 418–428.

[129] S. L. Strauss, T. A. Day, F. Garcia-Pichel. Nitrogen cycling in desert biological soil crusts across biogeographic regions in the Southwestern United States. Biogeochemistry 2012, 108, 171–182.

[130] A. C. Yannarell, T. F. Steppe, H. W. Paerl. Genetic variance in the composition of two functional groups (diazotrophs and cyanobacteria) from a hypersaline microbial mat. Appl Environ Microbiol 2006, 72, 1207–1217.

[131] G. W. Tyson, Lo I, B. J. Baker, E. E. Allen, P. Hugenholtz, J. F. Banfield. Genome-directed isolation of the key nitrogen fixer *Leptospirillum ferrodiazotrophum* sp. nov. from an acidophilic microbial community. Appl Environ Microbiol 2005, 71, 6319–6324.

[132] P. R. Norris, J. C. Murrell, D. Hinson. The potential for diazotrophy in iron-and sulfur-oxidizing acidophilic bacteria. Arch Microbiol 1995, 164, 294–300.

[133] Parro Vc, M. Moreno-Paz. Nitrogen fixation in acidophile iron-oxidizing bacteria: The *nif* regulon of *Leptospirillum ferrooxidans*. Res Microbiol 2004, 155, 703–709.
[134] K. B. Hallberg, E. González-Toril, D. B. Johnson. *Acidithiobacillus ferrivorans*, sp. nov.; facultatively anaerobic, psychrotolerant iron-, and sulfur-oxidizing acidophiles isolated from metal mine-impacted environments. Extremophiles 2010, 14, 9–19.
[135] A. F. Khadem, A. Pol, M. S. Jetten, H. J. O. den Camp. Nitrogen fixation by the verrucomicrobial methanotroph '*Methylacidiphilum fumariolicum*' SolV. Microbiol 2010, 156, 1052–1059.
[136] R. S. Oremland. Nitrogen fixation dynamics of two diazotrophic communities in Mono Lake, California. Appl Environ Microbiol 1990, 56, 614–622.
[137] D. B. Herbst. Potential salinity limitations on nitrogen fixation in sediments from Mono Lake, California. Int J Salt Lake Res 1998, 7, 261–274.
[138] T. P. Tourova, E. M. Spiridonova, I. A. Berg, N. V. Slobodova, E. S. Boulygina, D. Y. Sorokin. Phylogeny and evolution of the family Ectothiorhodospiraceae based on comparison of 16S rRNA, *cbbL* and *nifH* gene sequences. Int J Syst Evol Microbiol 2007, 57, 2387–2398.
[139] T. Zhilina, V. Kevbrin, T. Tourova, A. Lysenko, N. Kostrikina, G. Zavarzin. *Clostridium alkalicellum* sp. nov., an obligately alkaliphilic cellulolytic bacterium from a soda lake in the Baikal region. Microbiol 2005, 74, 557–566.
[140] D. Zavarzina, T. Kolganova, E. Boulygina, N. Kostrikina, T. Tourova, G. Zavarzin. *Geoalkalibacter ferrihydriticus* gen. nov. sp. nov., the first alkaliphilic representative of the family Geobacteraceae, isolated from a soda lake. Microbiol 2006, 75, 673–682.
[141] I. Sorokin, E. Zadorina, I. Kravchenko, E. Boulygina, T. Tourova, D. Sorokin. *Natronobacillus azotifigens* gen. nov., sp. nov., an anaerobic diazotrophic haloalkaliphile from soda-rich habitats. Extremophiles 2008, 12, 819–827.
[142] I. D. Sorokin, I. K. Kravchenko, T. P. Tourova, T. V. Kolganova, E. S. Boulygina, D. Y. Sorokin. *Bacillus alkalidiazotrophicus* sp. nov., a diazotrophic, low salt-tolerant alkaliphile isolated from Mongolian soda soil. Int J Syst Evol Microbiol 2008, 58, 2459–2464.
[143] J. A. Sohm, E. A. Webb, D. G. Capone. Emerging patterns of marine nitrogen fixation. Nat Rev Microbiol 2011, 9, 499–508.
[144] N. Gruber, J. N. Galloway. An Earth-system perspective of the global nitrogen cycle. Nature 2008, 451, 293–296.
[145] J. P. Zehr. Nitrogen fixation by marine cyanobacteria. Trends Microbiol 2011, 19, 162–173.
[146] D. G. Capone, J. P. Zehr, H. W. Paerl, B. Bergman, E. J. Carpenter. *Trichodesmium*, a globally significant marine cyanobacterium. Science 1997, 276, 1221–1229.
[147] B. Bergman, G. Sandh, S. Lin, J. Larsson, E. J. Carpenter. *Trichodesmium*–a widespread marine cyanobacterium with unusual nitrogen fixation properties. FEMS Microbiol Rev 2013, 37, 286–302.
[148] J. P. Zehr, J. B. Waterbury, P. J. Turner, et al. Unicellular cyanobacteria fix N_2 in the subtropical North Pacific Ocean. Nature 2001, 412, 635–638.
[149] J. P. Montoya, C. M. Holl, J. P. Zehr, A. Hansen, T. A. Villareal, D. G. Capone. High rates of N_2 fixation by unicellular diazotrophs in the oligotrophic Pacific Ocean. Nature 2004, 430, 1027–1032.
[150] J. P. Zehr, S. R. Bench, B. J. Carter, et al. Globally distributed uncultivated oceanic N_2-fixing cyanobacteria lack oxygenic photosystem II. Science 2008, 322, 1110–1112.
[151] H. J. Tripp, S. R. Bench, K. A. Turk, et al. Metabolic streamlining in an open-ocean nitrogen-fixing cyanobacterium. Nature 2010, 464, 90–94.
[152] A. C. Martiny, S. Kathuria, P. M. Berube. Widespread metabolic potential for nitrite and nitrate assimilation among *Prochlorococcus* ecotypes. Proc Natl Acad Sci 2009, 106, 10787–10792.

[153] C. S. Batmalle, H.-I. Chiang, K. Zhang, M. W. Lomas, A. C. Martiny. Development and bias assessment of a method for targeted metagenomic sequencing of marine cyanobacteria. Appl Environ Microbiol 2014, 80, 1116–1125.

[154] A. C. Martiny, M. L. Coleman, S. W. Chisholm. Phosphate acquisition genes in *Prochlorococcus* ecotypes, evidence for genome-wide adaptation. Proc Natl Acad Sci 2006, 103, 12552–12557.

[155] S. J. Giovannoni, H. J. Tripp, S. Givan, et al. Genome streamlining in a cosmopolitan oceanic bacterium. Science 2005, 309, 1242–1245.

[156] S. M. Sowell, L. J. Wilhelm, A. D. Norbeck, et al. Transport functions dominate the SAR11 metaproteome at low-nutrient extremes in the Sargasso Sea. ISME J 2009, 3, 93–105.

[157] D. P. Smith, J. C. Thrash, C. D. Nicora, et al. Proteomic and transcriptomic analyses of "*Candidatus* Pelagibacter ubique" describe the first PII-independent response to nitrogen limitation in a free-living alphaproteobacterium. mBio 2013, 4, e00133–00112.

[158] J. J. Grzymski, A. M. Dussaq. The significance of nitrogen cost minimization in proteomes of marine microorganisms. ISME J 2011, 6, 71–80.

[159] M. A. Pasek, J. P. Harnmeijer, R. Buick, M. Gull, Z. Atlas. Evidence for reactive reduced phosphorus species in the early Archean ocean. Proc Natl Acad Sci 2013, 110, 10089–10094.

[160] R. Docampo, W. de Souza, K. Miranda, P. Rohloff, S. N. Moreno. Acidocalcisomes, conserved from bacteria to man. Nat Rev Microbiol 2005, 3, 251–261.

[161] M. J. Seufferheld, H. M. Alvarez, M. E. Farias. Role of polyphosphates in microbial adaptation to extreme environments. Appl Environ Microbiol 2008, 74, 5867–5874.

[162] A. Orell, C. A. Navarro, M. Rivero, J. S. Aguilar, C. A. Jerez. Inorganic polyphosphates in extremophiles and their possible functions. Extremophiles 2012, 16, 573–583.

[163] D. M. Karl. Microbially mediated transformations of phosphorus in the sea: new views of an old cycle. Ann Rev Mar Sci 2014, 6, 279–337.

[164] L. Moore, M. Ostrowski, D. Scanlan, K. Feren, T. Sweetsir. Ecotypic variation in phosphorus-acquisition mechanisms within marine picocyanobacteria. Aquat Microb Ecol 2005, 39, 257–269.

[165] V. Tai, B. Palenik. Temporal variation of *Synechococcus* clades at a coastal Pacific Ocean monitoring site. ISME J 2009, 3, 903–915.

[166] R. Feingersch, A. Philosof, T. Mejuch, et al. Potential for phosphite and phosphonate utilization by *Prochlorococcus*. ISME J 2011, 6, 827–834.

[167] A. Martínez, M. S. Osburne, A. K. Sharma, E. F. DeLong, S. W. Chisholm. Phosphite utilization by the marine picocyanobacterium *Prochlorococcus* MIT9301. Environ Microbiol 2012, 14, 1363–1377.

[168] B. Palenik, Q. Ren, C. L. Dupont, et al. Genome sequence of *Synechococcus* CC9311, insights into adaptation to a coastal environment. Proc Natl Acad Sci 2006, 103, 13555–13559.

[169] G. C. Kettler, A. C. Martiny, K. Huang, et al. Patterns and implications of gene gain and loss in the evolution of *Prochlorococcus*. PLoS Genetics 2007, 3, e231.

[170] E. D. Orchard, E. A. Webb, S. T. Dyhrman. Molecular analysis of the phosphorus starvation response in *Trichodesmium* spp. Environ Microbiol 2009, 11, 2400–2411.

[171] B. A. Van Mooy, L. R. Hmelo, L. E. Sofen, et al. Quorum sensing control of phosphorus acquisition in *Trichodesmium* consortia. ISME J 2012, 6, 422–429.

[172] S. Dyhrman, P. Chappell, S. Haley, et al. Phosphonate utilization by the globally important marine diazotroph *Trichodesmium*. Nature 2006, 439, 68–71.

[173] B. A. Van Mooy, H. F. Fredricks, B. E. Pedler, et al. Phytoplankton in the ocean use non-phosphorus lipids in response to phosphorus scarcity. Nature 2009, 458, 69–72.

[174] P. Carini, A. E. White, E. O. Campbell, S. J. Giovannoni. Methane production by phosphate-starved SAR11 chemoheterotrophic marine bacteria. Nature Comm 2014, 5.

[175] E. D. Orchard, C. R. Benitez-Nelson, P. J. Pellechia, M. W. Lomas, S. T. Dyhrman. Polyphosphate in *Trichodesmium* from the low-phosphorus Sargasso Sea. Limnol Oceangr 2010, 55, 2161–2169.
[176] P. Martin, S. T. Dyhrman, M. W. Lomas, N. J. Poulton, B. A. Van Mooy. Accumulation and enhanced cycling of polyphosphate by Sargasso Sea plankton in response to low phosphorus. Proc Natl Acad Sci 2014, 111, 8089–8094.
[177] M. M. Adams, M. R. Gómez-García, A. R. Grossman, D. Bhaya. Phosphorus deprivation responses and phosphonate utilization in a thermophilic *Synechococcus* sp. from microbial mats. J Bacteriol 2008, 190, 8171–8184.
[178] M. R. Gomez-Garcia, M. Davison, M. Blain-Hartnung, A. R. Grossman, D. Bhaya. Alternative pathways for phosphonate metabolism in thermophilic cyanobacteria from microbial mats. ISME J 2011, 5, 141–149.
[179] C. Jones, S. Crowe, B. Viehweger, et al. Entire community of microbes lacks phospholipids. Goldschmidt Conference. Florence, Italy2013.
[180] F. Wolfe-Simon, J. S. Blum, T. R. Kulp, et al. A bacterium that can grow by using arsenic instead of phosphorus. Science 2011, 332, 1163–1166.
[181] M. Elias, A. Wellner, K. Goldin-Azulay, et al. The molecular basis of phosphate discrimination in arsenate-rich environments. Nature 2012, 491, 134–137.
[182] T. J. Erb, P. Kiefer, B. Hattendorf, D. Günther, J. A. Vorholt. GFAJ-1 is an arsenate-resistant, phosphate-dependent organism. Science 2012, 337, 467–470.
[183] M. L. Reaves, S. Sinha, J. D. Rabinowitz, L. Kruglyak, R. J. Redfield. Absence of detectable arsenate in DNA from arsenate-grown GFAJ-1 cells. Science 2012, 337, 470–473.
[184] F. H. Westheimer. Why nature chose phosphates. Science 1987, 235, 1173–1178.
[185] M. A. Saito, D. M. Sigman, F. M. M. Morel. The bioinorganic chemistry of the ancient ocean, the co-evolution of cyanobacterial metal requirements and biogeochemical cycles at the Archean-Proterozoic boundary? Inorg Chim Acta 2003, 356, 308–318.
[186] C. L. Dupont, A. Butcher, R. E. Ruben, P. E. Bourne, G. Caetano-Anolles. History of biological metal utilization inferred through phylogenomic analysis of protein structures. Proc Natl Acad Sci 2010, 107, 10567–10572.
[187] C. L. Dupont, S. Yang, B. Palenik, P. E. Bourne. Modern proteomes contain putative imprints of ancient shifts in trace metal geochemistry. Proc Natl Acad Sci 2006, 103, 17822.
[188] A. Harel, Y. Bromberg, P. G. Falkowski, D. Bhattacharya. Evolutionary history of redox metal-binding domains across the tree of life. Proc Natl Acad Sci 2014, 111, 7042–7047.
[189] J. D. Aguirre, H. M. Clark, M. McIlvin, et al. A manganese-rich environment supports superoxide dismutase activity in a Lyme disease pathogen, *Borrelia burgdorferi*. J Biol Chem 2013, 288, 8468–8478.
[190] P. W. Boyd, T. Jickells, C. Law, et al. Mesoscale iron enrichment experiments 1993–2005, Synthesis and future directions. Science 2007, 315, 612–617.
[191] M. Castruita, M. Saito, P. C. Schottel, et al. Overexpression and characterization of an iron storage and DNA-binding Dps protein from *Trichodesmium erythraeum*. Appl Environ Microbiol 2006, 72, 2918–2924.
[192] J. Morrissey, C. Bowler. Iron utilization in marine cyanobacteria and eukaryotic algae. Front Microbiol 2012, 3, 43.
[193] B. M. Hopkinson, K. A. Barbeau. Iron transporters in marine prokaryotic genomes and metagenomes. Environ Microbiol 2012, 14, 114–128.
[194] M. Sandy, A. Butler. Microbial iron acquisition: marine and terrestrial siderophores. Chemical reviews 2009, 109, 4580–4595.

[195] O. X. Cordero, L.-A. Ventouras, E. F. DeLong, M. F. Polz. Public good dynamics drive evolution of iron acquisition strategies in natural bacterioplankton populations. Proc Natl Acad Sci 2012, 109, 20059–20064.
[196] I. Berman-Frank, J. T. Cullen, Y. Shaked, R. M. Sherrell, P. G. Falkowski. Iron availability, cellular iron quotas, and nitrogen fixation in *Trichodesmium*. Limnol Oceangr 2001, 46, 1249–1260.
[197] A. B. Kustka, S. A. Sanudo-Wilhelmy, E. J. Carpenter, D. Capone, J. Burns, W. G. Sunda. Iron requirements for dinitrogen- and ammonium-supported growth in cultures of *Trichodesmium* (IMS 101): Comparison with nitrogen fixation rates and iron: carbon ratios of field populations. Limnol Oceangr 2003, 48, 1869–1884.
[198] T. Shi, Y. Sun, P. G. Falkowski. Effects of iron limitation on the expression of metabolic genes In the marine cyanobacterium *Trichodesmium erythraeum* IMS101. Environ Microbiol 2007, 9, 2945–2956.
[199] P. D. Chappell, J. W. Moffett, A. M. Hynes, E. A. Webb. Molecular evidence of iron limitation and availability in the global diazotroph *Trichodesmium*. ISME J 2012, 6, 1728–1739.
[200] P. D. Chappell, E. A. Webb. A molecular assessment of the iron stress response in the two phylogenetic clades of *Trichodesmium*. Environ Microbiol 2010, 12, 13–27.
[201] M. Rubin, I. Berman-Frank, Y. Shaked. Dust-and mineral-iron utilization by the marine dinitrogen-fixer *Trichodesmium*. Nat Geosci 2011, 4, 529–534.
[202] C. Tuit, J. Waterbury, G. Ravizza. Diel variation of molybdenum and iron in marine diazotrophic cyanobacteria. Limnol Oceangr 2004, 49, 978–990.
[203] I. Hewson, R. S. Poretsky, R. A. Beinart, et al. *In situ* transcriptomic analysis of the globally important keystone N_2-fixing taxon *Crocosphaera watsonii*. ISME J 2009, 3, 618–631.
[204] M. A. Saito, E. M. Bertrand, S. Dutkiewicz, et al. Iron conservation by reduction of metalloenzyme inventories in the marine diazotroph *Crocosphaera watsonii*. Proc Natl Acad Sci 2011, 108, 2184–2189.
[205] A. L. Zerkle, C. H. House, S. L. Brantley. Biogeochemical signatures through time as inferred from whole microbial genomes. Am J Sci 2005, 305, 467.
[206] M. A. Saito, J. W. Moppett, S. W. Chisholm, J. B. Waterbury. Cobalt limitation and uptake in Prochlorococcus. Limnol Oceanogr 2002, 47, 1629–1636.
[207] M. A. Saito, G. Rocap, J. W. Moffett. Production of cobalt binding ligands in a *Synechococcus* feature at the Costa Rica upwelling dome. Limnol Oceanogr 2005, 50, 279–290.
[208] S. Kathuria, A. C. Martiny. Prevalence of a calcium-based alkaline phosphatase associated with the marine cyanobacterium *Prochlorococcus* and other ocean bacteria. Environ Microbiol 2011, 13, 74–83.
[209] C. L. Dupont, K. Barbeau, B. Palenik. Ni uptake and limitation in marine *Synechococcus*. Appl Environ Microbiol 2007, 74, 23–31.
[210] Ho T-Y. Nickel limitation of nitrogen fixation in *Trichodesmium*. Limnol Oceanogr 2013, 58, 112–120.
[211] J. Nuester, S. Vogt, M. Newville, A. B. Kustka, B. S. Twining. The unique biochemical signature of the marine diazotroph *Trichodesmium*. Front Microbiol 2012, 3, 1–15.
[212] C. Dupont, K. Neupane, J. Shearer, B. Palenik. Diversity, function and evolution of genes coding for putative Ni-containing superoxide dismutases. Environ Microbiol 2008, 10, 1831–1843.
[213] B. Schoepp-Cothenet, R. van Lis, P. Philippot, A. Magalon, M. J. Russell, W. Nitschke. The ineluctable requirement for the trans-iron elements molybdenum and/or tungsten in the origin of life. Sci Rep 2012, 2, 263.
[214] A. Kletzin, M. W. Adams. Tungsten in biological systems. FEMS Microbiol Rev 1996, 18, 5–63.

[215] J. B. Glass, Yu H, J. A. Steele, et al. Geochemical, metagenomic and metaproteomic insights into trace metal utilization by methane-oxidizing microbial consortia in sulphidic marine sediments. Environ Microbiol 2014, 16, 1592–1611.

[216] K. G. Lloyd, L. Schreiber, D. G. Petersen, et al. Predominant archaea in marine sediments degrade detrital proteins. Nature 2013, 496, 215–218.

[217] P. F. Dunfield, A. Yuryev, P. Senin, et al. Methane oxidation by an extremely acidophilic bacterium of the phylum Verrucomicrobia. Nature 2007, 450, 879–882.

[218] A. Pol, K. Heijmans, H. R. Harhangi, D. Tedesco, M. S. Jetten, H. J. O. den Camp. Methanotrophy below pH 1 by a new Verrucomicrobia species. Nature 2007, 450, 874–878.

[219] A. Pol, T. R. Barends, A. Dietl, et al. Rare earth metals are essential for methanotrophic life in volcanic mudpots. Environ Microbiol 2014, 16, 255–264.

[220] Y. Hibi, K. Asai, H. Arafuka, M. Hamajima, T. Iwama, K. Kawai. Molecular structure of La3+-induced methanol dehydrogenase-like protein in *Methylobacterium radiotolerans*. J Biosci Bioeng 2011, 111, 547–549.

[221] A. Cvetkovic, A. L. Menon, M. P. Thorgersen, et al. Microbial metalloproteomes are largely uncharacterized. Nature 2010, 466, 779–782.

[222] D. Spungin, I. Berman-Frank, O. Levitan. *Trichodesmium's* strategies to alleviate phosphorus limitation in the future acidified oceans. Environ Microbiol 2014, doi:10.1111/462-2920.12424.

Joseph Seckbach and Pabulo Henrique Rampelotto
8 Polyextremophiles

8.1 Introduction

Over the last decades, scientists have been intrigued by the fascinating organisms that inhabit extreme environments. Such organisms, known as extremophiles, thrive in habitats which for other terrestrial life-forms are intolerably hostile or even lethal [1–3]. They may be found thriving from the frigid environments of the Antarctic, to the superheated waters of the hydrothermal vents, from the bottom of 11-km deep ocean trenches to the high altitudes of the atmosphere, from acidic to alkaline places [4]; some may grow in toxic waste, organic solvents, heavy metals, or in several other habitats that were previously considered inhospitable for life [5]. For every extreme environmental condition investigated, a variety of organisms have shown that they not only can tolerate these conditions, but that they also often require those conditions for survival [6]. Such discoveries have fueled research aimed to understand the unique survival strategies evolved by these extremophilic forms of life [7]. Examples of extremophiles include, but are not limited to: thermophiles (high temperature), psychrophiles (low temperature), acidophiles (low pH), alkaliphiles (high pH), piezophiles (high pressure, formerly called barophiles), halophiles (high salt concentration), osmophiles (high concentration of organic solutes), oligotrophs (low concentration of solutes and/or nutrients), and xerophiles (very dry environment). The term 'extremophile' also includes microorganisms able to grow in the presence of high metal concentrations or high doses of radiation [8].

Whereas extremophiles are usually defined by one extreme, many natural environments pose two or more extremes [9]. To cite a few examples, many hot springs are acid or alkaline at the same time, and usually rich in metal content; several hypersaline lakes are very alkaline; the deep ocean is generally cold, oligotrophic (very low nutrient content), and exposed to high pressure. Interestingly, an increasing number of species and strains isolated from these environments have been found to tolerate multiple extremes. Such organisms are known as polyextremophiles and may be found in the three domains of life [3]. For example, hydrothermal chimney vents harbor communities of prokaryotes and multicellular organisms (e.g. tubeworms, clams, and a variety of grazers) at high pressure, temperature, and acidic pH [10].

Although the impacts of individual extremes on microorganisms have been widely researched, attempts to define their collective influences on life are scarce and we are just beginning to understand the mechanisms of evolution underlying adaptation to multiple extremes.

In this chapter, we review the knowledge gained over the last decades on the main specimens of polyextremophiles. Examples from the three domains of life are provided. These specimens give a glimpse of how intriguing and fascinating these living beings are.

8.2 Bacteria

8.2.1 *Deinococcus radiodurans*: Conan the bacterium

Deinococcus radiodurans is a polyextremophile, known for its ability to withstand a multitude of extreme environments including severe cold, dehydration, vacuum, acid, and intense radiation. For these reasons, the bacterium has been given the nickname, "Conan the Bacterium" and currently holds the Guinness World Record for the world's toughest bacterium [11].

D. radiodurans was discovered in 1956 by Arthur Anderson at the Oregon Agricultural Experiment Station [12]. The accidental discovery of this polyextremophile occurred while trying to sterilize canned ham. After exposing the canned ham to extreme radiation it still spoiled. Anderson cultured the spoiled ham and found the only remaining bacteria to be *D. radiodurans* R1.

D. radiodurans is a red-pigmented spherical bacterium that is usually found in a tetrad grouping of four cells. Although the cell walls resemble a gram-negative bacterium, the cells stain gram positive [13]. *D. radiodurans* is an obligate aerobe and uses organic materials as its source of energy and carbon. Despite its ubiquitous presence in nature, its populations are minor compared with other bacteria occupying the same ecological niches. It is easily cultured on nutrient agar with 1% glucose at 30 °C.

The complete DNA sequence of *D. radiodurans* was published in 1999 [14], and an analysis of the genome appeared in 2001 [15]. The bacterium is a polyploid which contains four copies of the genome in its non-replicating state and ten genome equivalents during replication. This contributes to its efficient homologous recombination based repair mechanism for double-strand breaks but does not entirely explain the bacterium's high resistance to ionizing and UV radiation.

D. radiodurans' ability to survive in a multitude of extreme environments stems from its ability to repair intense damage to its DNA [16]. When the DNA is damaged, chromosome fragments are reconnected by a process called single-strand annealing [17]. After this, double-strand breaks are repaired through homologous recombination, utilizing the multiple copies of the genome in order to generate at least one complete copy of the genome [18]. The repair to the DNA is usually accomplished in 12 to 24 hours, and does not introduce any more errors in the genome than a normal DNA replication cycle [19]. These capabilities make *D. radiodurans* able to withstand up to 5,000 Gy of ionizing radiation without effect on viability (500 times the fatal human dose), and can survive at levels of 15,000 Gy with a somewhat lower 37% viability [20].

In recent years, novel studies have shown that high intracellular levels of manganese(II) in *D. radiodurans* protect proteins from being oxidized by radiation [21, 22]. Such studies suggest the hypothesis that protein, rather than DNA, is the principal target of the biological action of ionizing radiation in sensitive bacteria, and extreme resistance in Mn-accumulating bacteria is based on protein protection [23, 24]. According to this new paradigm, the extreme radiation resistance presented by *D. radiodurans* is due to a highly efficient protection against proteome, instead of genome, damage [25].

These unique characteristics have exciting applications. Genetically engineered strains of *D. radiodurans* are ideal candidates for cleaning up toxic waste. For example, researchers created a strain of *D. radiodurans* that can be used to detoxify ionic mercury in radioactive waste by incorporating the gene for mercuric reductase in the genome [26]. Another engineered strain can break down toluene, an organic chemical found in radioactive waste sites [27, 28]. New strains can even be used to efficiently precipitate uranium over a wide range of input U concentrations [29, 30]. *D. radiodurans* can also be used as a means of long term information storage. In 2003, scientists translated the song "It's a Small World" into a series of base pairs that was subsequently inserted into the genome of the bacterium [31]; 100 generations later, the complete message was still intact. In the future, if we ever wish to make extremely robust storage devices, that might survive even a nuclear catastrophe, we could probably learn to store data in this bacterium.

The origin and evolution of such extreme radiation resistance has been the topic of much research and speculation, in particular, because no radiation sources on Earth are known that could produce doses comparable to those of *D. radiodurans* resistance [32]. Normal levels of background radiation on Earth are around 0.4 mGy per year, with the highest known levels being 260 mGy near Ramsar, Iran. This raises the question of why a bacterium would evolve to withstand radiation doses of 10,000 Gy when exposure to such high levels is very unlikely [33]. Some scientists pose a theory that *D. radiodurans* may have originated in outer space where cosmic radiation is very significant [34]. The bacterium could then have been brought to earth by way of a meteorite. But this is an unlikely possibility, considering the similarities of *D. radiodurans* to other prokaryotes, which suggests a terrestrial origin. Studies have indicated that the effects of dehydration on DNA are very similar to the effects of radiation [35]. Although normal strains of *D. radiodurans* are resistant to both desiccation and radiation, mutated strains have become susceptible to damage from both conditions [20, 36]. This has led some scientists to conclude that *D. radiodurans* originally evolved to withstand long periods without water, which causes very similar DNA damage to ionizing radiation [37, 38]. The radiation resistance that has come to define this bacterium may actually have been a byproduct of the mechanism for surviving dehydration.

8.2.2 Chroococcidiopsis

Among Bacteria, the best adapted group to various extreme conditions is the *Cyanobacteria*. They often form microbial mats with other bacteria from Antarctic ice to continental hot springs. Cyanobacteria can also develop in hypersaline and alkaline lakes, support high metal concentrations, tolerate xerophilic conditions (i.e., low availability of water), and form endolithic communities in desertic regions [2]. An interesting group is the genus *Chroococcidiopsis*, considered one of the most primitive cyanobacteria [39]. This photosynthetic, coccoidal bacterium reproduces internally by binary fission, forming small cells (baeocytes) contained within an enlarged mother cell. It occurs throughout a wide geographic range and different habitat types, encompassing different life strategies, from extreme cold (Dry Valleys in Antarctica) to extreme hot (e.g. Atacama Desert in Chile) deserts. *Chroococcidiopsis* species grow as litho-, endo-, chasmoendo-, crypto-, and hypo-liths in and on rocks; in fresh, brackish or salt waters, also as photobionts of rock and soil inhabiting lichens of the order Lichinales; and rarely as free-living cells in soil [40–45]. In its natural environment, to escape the harsh outside climate, *Chroococcidiopsis* occupies the last refuges for life inside porous rocks or at the stone-soil interfaces, where it survives in a dry, ametabolic state for prolonged periods.

Chroococcidiopsis is known for its ability to survive extreme environmental conditions, including both high and low temperatures, ionizing radiation, and high salinity. Strains of *Chroococcidiopsis* are able of withstanding years of desiccation [46], ionizing radiation up to 15 kGy [47], and UVC doses as high as 13 kJ/m^2 [48]. With the main prerequisites for survival in space and under Martian conditions, *Chroococcidiopsis* has been used in ground-based simulations and space exposure [49]. A monolayer survived 10 min of simulated unattenuated Martian UV-flux [50] and a multilayer overlain by grounded sandstone survived space and Martian simulations [51], while cells augmented to an epilithic community tolerated 584 days in real space [52]. Because of these abilities to survive in extreme environments, *Chroococcidiopsis* has been of interest to astrobiology [53, 54] and has been proposed as a suitable organism for the terraforming of Mars [55, 56].

Despite the great interest in this genus, many aspects of its biology remain poorly understood. Its biogeography, for example, is still debated. Fewer et al. (2002) confirmed the non-existence of biogeographical patterns suggesting close relationships between strains from very distant geographical origins [57]. However, a recent study indicates that *Chroococcidiopsis* variants from different hot and cold deserts around the world are specific to their habitat, as a result of the ancient legacy due to a very early separation of these lineages [58]. These contrasting results may be due to differences in the genetic marker used. There are studies indicating that the choice of genetic marker influences the results in the detection of biogeographic relationships.

In terms of origin and evolution, the earliest morphologically based reports of *Chroococcidiopsis* are derived from 400 million years old lichenized fossil from the

Early Devonian Rhynie Chert [59, 60]. A relaxed-clock phylogenetic analysis however indicates an age between 3.1–1.9 billion years for free-living variants of *Chroococcidiopsis* [58]. These studies indicate that the genus is one of the evolutionary oldest cyanobacteria. Nevertheless, the exact phylogenetic position of the genus *Chroococcidiopsis* is still unclear. Historically, *Chroococcidiopsis* has been classified within the order *Pleurocapsales* [61, 62], based on a unique reproduction modus by baeocytes. On the other hand, phylogenetic analyses suggest a closer relationship of this genus to the order *Nostocales* [63]. Further phylogenetic studies based on saltwater and freshwater phenotypes provide evidence that the genus is polyphyletic and should be reclassified to improve clarity in the literature [64].

So, it is clear that future research should continue with the search for mechanisms that help to understand the patterns of ecology and evolution of the genus *Chroococcidiopsis*.

8.3 Archaea

Archaea is the main group to thrive in extreme environments. Although members of this group are generally less versatile than bacteria and eukaryotes, they are quite skilled in adapting to different extreme conditions, frequently holding extremophily records [65]. Within the domain Archaea, one of the most interesting polyextremophiles is the *Halobacterium salinarum* NRC-1, as described in the following section.

8.3.1 *Halobacterium salinarum* NRC-1: a model organism

Halobacterium salinarum NRC-1 is an extreme halophilic archaeon that grows optimally at 4.3 M NaCl and is capable of growth between 2.6 and 5.1 M (29.8%) NaCl [66]. Interestingly, this microorganism is metabolically versatile. In addition to its aerobic metabolic capacity, it possesses facultative growth capabilities through anaerobic respiration, utilizing dimethyl sulfoxide (DMSO) and trimethylamine N-oxide (TMAO), and via arginine fermentation [67]. *H. salinarum* NRC-1 may be found in hypersaline environments all over the world, including salt production facilities, brine inclusions, salt mines, as well as natural lakes and ponds [68].

In addition to its ability to survive in hypersaline conditions (it is one of the few species known that can live in saturated salt solutions), *H. salinarum* NRC-1 is adapted to desiccation [69], ionizing radiation [70], transition metals [71], different temperatures [72], high pressure [73], oxygen limitation [74], and different regimes of UV radiation [75]. Furthermore, the long term survival of *H. salinarum* has been evidenced with the isolation of a strain from a single fluid inclusion in a 97,000-year-old halite crystal from Death Valley [76], and experiments conducted by Stan-Lotter et al. (2003) showed that *H. salinarum* NRC-1 is able to survive a simulated Martian atmosphere (6 mbar pressure, 98% carbon dioxide, and an average temperature of −60 °C) for 6 hours of exposure [77]. This incredible robustness and ability to survive and adapt to different and extreme environmental stresses makes these archaea of great interest in the search for extraterrestrial life.

H. salinarum NRC-1 is easy to culture and manipulate in the laboratory, which has made it an excellent model organism in the domain Archaea for the study of genetics, functional genomics and mechanisms of adaptive evolution [78, 79]. The genome of *H. salinarum* NRC-1 was the first halobacterial genome to be completely sequenced [80].

8.4 Eukaryota

When we think of extremophiles, Bacteria and Archaea first come to mind. However, eukaryotic life is also found actively growing in almost any extreme condition where there is a source of energy to sustain it, with the only exception of high temperature (> 70 °C) and possibly the deep subsurface biosphere. Recent studies based on molecular ecology have demonstrated that eukaryotic organisms are as adaptable as Bacteria and Archaea, although most habitats have not been sufficiently sampled to date. The majority of eukaryotes found living in extreme environments are microbial but many multicellular organisms are also known to thrive or tolerate such extreme conditions. In the following, we describe interesting examples of eukaryotic polyextremophiles.

8.4.1 *Cyanidiophyceae*

The class *Cyanidiophyceae* is comprised of asexual, unicellular red algae that are known to grow in acidic (pH 0.2–4.0) and salty (up to 10% solutions) environments and at moderately high temperatures (up to 56 °C) [81, 82]. They are also known to proliferate even when exposed to a stream of 100% CO_2 or 1N H_2SO_4 solution [81] as well as tolerate heavy metals [83, 84]. For example, the microalga *Galdieria sulphuraria* tolerates desiccation [85] and high concentrations of toxic metal ions such as cadmium, mercury, aluminum, and nickel, suggesting potential application in bioremediation [83]. *G. sulphuraria* may represent up to 90% of the biomass in extreme habitats, such as hot sulfur springs with pH 0–4.

The prime habitats for *Cyanidiophyceae* are thermo-acidic environments such as warm-hot acidic springs and pools that emanate from geothermal sources [86, 87]. Among photosynthetic organisms, the environmental niche for *Cyanidiophyceae* is unique since no other phototrophs are expected to occur in this combination of low pH and elevated temperature. Furthermore, they were also observed in cryptoendolithic layers of rocks [88, 89] and caves [90]. Heavy metals are common in the *Cyanidiophyceae* habitats, although their concentrations can vary considerably [84].

The taxonomy and phylogeny of *Cyanidiophyceae* has been quite confusing until the application of genetic approaches [91–93]. Molecular phylogenetic studies demonstrated that this group of algae is phylogenetically distinct from the main line of descent in the red algae and split quite early in geological time, i.e. ~ 1.3 Ga [94, 95]. Based on these studies, a new classification of the *Rhodophyta* (red algae) was proposed [96], with the establishment of the new subphylum *Cyanidiophytina* that included the single class *Cyanidiophyceae* (former order *Cyanidiales*).

With the total genome sequencing of *Cyanidioschyzon merolae* [97] and *G. sulphuraria* [98], studies involving *Cyanidiophyceae* species has drawn significant interest from the scientific community. Due to the simplicity of their cells and genomes, members of this ancient group may be used as biological models for providing key information on the basic and essential genes involved in the lives of photosynthetic eukaryotes.

In terms of origins and evolution, the class *Cyanidiophyceae* constitutes a very early and distinct branch of the *Rhodophyta* and may be considered pioneering eukaryotic cells on early Earth. The consensus is that the ancestor of all red algae presumably acquired a cyanobacterium as an endosymbiont [99]. According to Castenholz & McDermott (2010), the ancient eukaryotic host incorporated a thermophilic cyanobacterium as its endosymbiont, which maintained an internal pH close to neutrality, creating a comfortable climate for the cyanobacterium, which turned, in time, into chloroplasts [84].

8.4.2 Lichens

Lichens are examples of symbiotic associations between, at least, two microorganisms: a fungus (the 'mycobiont') and an associated photosynthetic partner (the 'photobiont'). The photobiont can be either a green alga or a cyanobacterium, and some lichens contain both [100]. The physiological interactions that occur between the partners result in the formation of a distinctive lichen body called a thallus, with no division into root, stems, or leaves [101]. In this body, the photobionts are housed to the best advantage for photosynthesis.

The lichen lifestyle is remarkably adaptable and they often have different photobionts at different stages of the life cycle. At certain times, it may be advantageous for the lichen to contain cyanobacteria because they are able to fix atmospheric nitrogen to form amino acids and other complex organic molecules [102]. The lichen fungus is able to then break down these compounds. One of the great advantages of this ability is that lichens are extremely effective colonizers and grow in almost all natural environments and in a variety of different climatic conditions [103]. It has been estimated that they comprise the dominant components of vegetation on 8% of the Earth's terrestrial surface. They play a very important ecological role as carbon fixers and recyclers of the elements and nutrients nitrogen, phosphorous, and sulphur [104].

A key characteristic of lichens is that they have a remarkable ability to tolerate extreme environmental conditions and many lichen species may be considered polyextremophiles due to their high tolerance to temperature, radiation, and desiccation [105]. For example, in high mountains and polar regions, lichens are well adapted to long-term desiccation, temperatures between −40 °C and 60 °C, and high doses of radiation, including UV [106].

The basic physiology of lichens is what determines their efficiency at growing in extreme environments [104]. Because they lack roots, they are able to grow on rock surfaces and bark, relying on absorbing substances throughout their upper surface, which lacks a waxy protective outer cuticle. Minute pores are often present and the upper layer contains polysaccharides (carbohydrates), which attract water. The algae rest just below the surface, which also contains light-screening compounds to protect them under high levels of ultraviolet radiation. Green algae produce the sugar alcohol ribitol and cyanobacteria produce glucose (sugar), which are passed over to the fungus and rapidly converted to the sugar alcohol mannitol. This process ensures that lichens have the extraordinary ability to maintain themselves during very long dry periods. Furthermore, the huge chemical arsenal lichens possess helps ensure their survival under extreme conditions [107]. Some species contain up to 30 percent dry weight of organic compounds, which act as 'stress metabolites' [108].

Due to these interesting features, lichens have been successfully used in experiments under simulated space conditions [109–111], and they have been launched into space in the experiment Lichens [112] and Lithopanspermia [113] in order to better understand the effects of extreme space conditions on multicellular eukaryotic organisms. Such studies demonstrated that lichens are among the only organisms able to survive exposure to the complex matrix of all parameters of space including solar extraterrestrial UV radiation [114]. The thalli from this symbiotic system have certain morphologies and include adaptations, such as a thick and pigmented cortex covering the algal layer, which provide efficient shielding against the hostile parameters of outer space [115]. Many experiments performed after space exposure (as well as space and Mars simulation experiments) have shown their capacity to maintain physiological and photosynthetic activity, and their capacity to germinate and grow after being exposed to space parameters [116].

The evolution of lichens is complex and not well understood. Lichens were a component of the early terrestrial ecosystems, and the estimated age of the oldest terrestrial lichen fossil is 600 Ma [117]. In 1995, Gargas and colleagues proposed that there were at least five independent origins of lichenization [118]. However, Lutzoni et al. (2000) suggested that there may have been only a single origin of lichenization in the *Ascomycota*, but that there have been several reversals [119]. Some non-lichen-forming fungi may have secondarily lost the ability to form a lichen association. For this reason, lichenization has been viewed as a highly successful nutritional strategy [120].

The presence of a 600 million year-old fossil lichen associated with cyanobacteria [117] is consistent with an ancient origin for a lichenized state in fungi and is consistent with the "Protolichenes hypothesis" [121], which suggested that fungi existed as lichens in terrestrial paleoecosystems prior to the diversification of land plants. A conservative interpretation of recent data indicates that lichenization evolved multiple times in *Ascomycota* [122].

8.4.3 Tardigrades: nature's toughest animal

Tardigrades (also known as waterbears or moss piglets) are microscopic invertebrates (approx. 0.1–1.2 mm) with a well-developed organization including brain and sensory organs, muscles, a complex feeding apparatus and alimentary tract, and reproductive and osmoregulatory organs [123]. They form the phylum Tardigrada with over 1,000 species identified [124]. It is an ancient group, with fossils dating from 530 million years ago, in the Cambrian period [125]. Their first observations were reported over two centuries ago by the German pastor Johann A. E. Goeze [126]. Tardigrades have representatives distributed across all continents, and in most terrestrial, freshwater and marine habitats: from the top of the Himalayas to the ocean floor [127–129].

When subjected to environmental stress, tardigrades are one of the few groups that are able to enter a state of suspended animation called cryptobiosis [130]. Depending on the environment, they may enter this state via anhydrobiosis (dehydration), cryobiosis (extremely low temperatures), osmobiosis (high salt concentration), or anoxybiosis (lack of oxygen), with desiccation induced anhydrobiosis and freezing induced cryobiosis being the most extensively studied states. While in this state (referred to as a tun), their metabolism lowers to less than 0.01% of normal and their water content can drop to 1% of normal [131].

The genetic and biochemical mechanisms behind desiccation tolerance of anhydrobiotic organisms are not fully understood yet, but several biochemical components are likely to be involved, including sugars (e.g. trehalose), stress proteins (e.g. heat shock proteins), and an efficient DNA repair system [132–136]. Such molecular mechanisms are responsible for maintaining the structural integrity of the cell components and membranes during the processes of desiccation and rehydration, as well as during the dry state.

As a likely by-product of the adaptations for desiccation and freezing, tardigrades in the dormant state show a very high tolerance to multiple extreme conditions. Interestingly, it has been reported that tardigrades in the active state also exhibit high tolerance to environmental stresses [137, 138]. Consequently, these intriguing organisms are classical examples of multicellular polyextremophiles. They are able to survive extreme temperatures from −272 °C (just 1 degree above absolute zero!) to 151 °C for a few minutes [6, 139], high hydrostatic pressure of 7.5 GPa (which corresponds to the pressure at the depth of about 180 km below the surface of the Earth!) [140], high radiation doses of 5,000 Gy (of gamma rays) and 6,200 Gy (of heavy ions) [141], and remain viable after a long period in the anhydrobiotic state (nearly 10 years) [142]. They can also tolerate exposure to CO_2, H_2S, and the biocide methyl bromide, and also immersion in ethanol, butanol, and hexanol [143–146]. In addition, they can even survive in space conditions for a limited time [147–150]. Thus, tardigrades are interesting multicellular model organisms for space and astrobiological studies [146, 151].

Tardigrades are also enigmatic in terms of evolution. The phylogenetic position of this phylum in the tree of life is still uncertain. Although it is known that Tardigrada belong to the recently recognized clade of molting animals (*Ecdysozoa*), it is still debated if they are more closely related to nematodes and nematomorphs or to arthropods and onychophorans [152–154]. MicroRNA and phylogenomic studies show that the grouping of tardigrades with nematodes found in a number of molecular studies is a long branch attraction artifact, and provide evidence that *Onychophora* is the sister group of *Arthropoda* [155]. Nevertheless, recent analyses of the organization of the nervous system in three tardigrade species speak against the proposed sister group relationship of tardigrades with nematodes [156]. They suggest instead that *Tardigrada* and *Arthropoda* are sister groups (to the exclusion of *Onychophora*).

Tardigrade genomes are generally small [157, 158], probably due to extensive gene loss, and most gene sequences show high substitution rates. This causes problems with phylogenetic reconstructions using molecular datasets due to the so-called Long Branch Attraction artifact. This means that taxa with a high substitution rate artificially cluster together. Most morphological data seem to be equally problematic, as key characteristics might have been lost due to the miniaturization of these animals and the remaining characteristics might be highly derived. Consequently, it is difficult to interpret their origin and evolution.

8.5 Conclusion

The study of recently discovered polyextremophiles have extended the known physical and chemical boundaries for life and have provided excellent models for the study and characterization of novel physiologies as well as biochemical and evolutionary pathways. The above discussion gives an overview of the potential adaptive and metabolic capacities of some specimens of these fascinating forms of life. The sequencing of their genomes is providing new insights into the evolution of interesting adaptive strategies, with *D. radiodurans* as one of the best examples; the radiation resistance that has come to define this bacterium may actually have been a byproduct of the mechanism for surviving dehydration. The authors hope that this chapter will encourage other scientists to investigate and characterize novel polyextremophiles and further expand the current limited knowledge of their ecology, biochemistry, and mechanism of adaptive evolution.

References

[1]   K. Horikoshi, G. Antranikian, A. T. Bull, F. T. Robb, K. O. Stetter. Extremophiles Handbook. Dordrecht, The Netherlands, Springer, 2011.
[2]   J. Seckbach. Algae and Cyanobacteria in Extreme Environments. Dordrecht, The Netherlands, Springer, 2007.
[3]   J. Seckbach, A. Oren, H. Stan-Lotter. Polyextremophiles: Life Under Multiple Forms of Stress. Dordrecht, Netherlands, Springer, 2013.
[4]   H. Stan-Lotter, S. Fendrihan. Adaption of Microbial Life to Environmental Extremes. New York, USA, Springer, 2012.
[5]   L. J. Rothschild, R. L. Mancinelli. Life in extreme environments. Nature 2001, 409, 1092–1101.
[6]   P. H. Rampelotto. Resistance of microorganisms to extreme environmental conditions and its contribution to astrobiology. Sustainability 2010, 2, 1602–1623.
[7]   P. H. Rampelotto. Extremophiles and extreme environments. Life 2013, 3, 482–485.
[8]   C. Gerday, N. Glansdorff. Physiology and Biochemistry of Extremophiles. Washington, USA, ASM Press, 2007.
[9]   J. P. Harrison, N. Gheeraert, D. Tsigelnitskiy, C. S. Cockell. The limits for life under multiple extremes. Trends Microbiol 2013, 21, 204–212.
[10]   M. R. Islam, D. Schulze-Makuch. Adaptations to environmental extremes by multicellular organisms. Int J Astrobiol 2007, 6, 199–215.
[11]   Huyghe, P. Conan the Bacterium. The Sciences 1998, 38, 16–19.
[12]   A. W. Anderson, H. C. Nordon, R. F. Cain, G. Parrish, D. Duggan. Studies on a radio-resistant micrococcus. I. Isolation, morphology, cultural characteristics, and resistance to γ radiation. Food Technol 1956, 10, 575–578.
[13]   J. R. Battista. Against all odds: the survival strategies of *Deinococcus radiodurans*. Annu Rev Microbiol 1997, 51, 203–224.
[14]   O. White, J. A. Eisen, J. F. Heidelberg. Genome sequence of the radioresistant bacterium *Deinococcus radiodurans* R1. Science 1999, 286, 1571–1517.
[15]   K. S. Makarova, L. Aravind, Y. I. Wolf, R. L. Tatusov, K. W. Minton, E. V. Koonin, M. J. Daly. Genome of the extremely radiation-resistant bacterium Deinococcus radiodurans viewed from the perspective of comparative genomics. Microbiology and molecular biology reviews: Microbiol Mol Biol Rev 2001, 65, 44–79.
[16]   M. M. Cox, J. L. Keck, J. R. Battista. Rising from the ashes: DNA repair in *Deinococcus radiodurans*. PLoS Genet 2010, 6(1), e1000815.
[17]   C. Pasternak, B. Ton-Hoang, G. Coste, et al. Irradiation-induced D. radiodurans genome fragmentation triggers transposition of a single resident insertion sequence. PLoS Genet 2009, 6: e1000799.
[18]   K. Zahradka, D. Slade, A. Bailone, et al. Reassembly of shattered chromosomes in *Deinococcus radiodurans*. Nature 2006, 443, 569–573.
[19]   D. Slade, A. B. Lindner, G. Paul, M. Radman. Recombination and replication in DNA repair of heavily irradiated *Deinococcus radiodurans*. Cell 2009, 136: 1044–1055.
[20]   Cox, M.M; Battista, J.R. *Deinococcus radiodurans* – the consummate survivor. Nat Rev Microbiol 2005, 3 11, 882 892.
[21]   M. J. Daly, E. K. Gaidamakova, V. Y. Matrosova, et al. Protein oxidation implicated as the primary determinant of bacterial radioresistance. PLoS Biol 2007, 5, e92.
[22]   M. J. Daly, E. K. Gaidamakova, V. Y. Matrosova, et al. Small-molecule antioxidant proteomeshields in *Deinococcus radiodurans*. PLoS ONE 2010, 5, e12570.
[23]   M. J. Daly. A new perspective on radiation resistance based on *Deinococcus radiodurans*. Nat Rev Microbiol 2009, 7, 237–245.

[24] D. Slade, M. Radman. Oxidative stress resistance in *Deinococcus radiodurans*. Microbiol Mol Biol Rev 2011, 75, 133–191.
[25] A. Krisko, M. Radman. Biology of extreme radiation resistance: the way of *Deinococcus radiodurans*. Cold Spring Harb Perspect Biol 2013, 5, pii: a012765.
[26] H. Brim, S. C. McFarlan, J. K. Fredrickson, et al. Engineering *Deinococcus radiodurans* for metal remediation in radioactive mixed waste environments. Nat Biotechnol 2000, 18, 85–90.
[27] C. C. Lange, L. P. Wackett, K. W. Minton, M. J. Daly. Engineering a recombinant *Deinococcus radiodurans* for organopollutant degradation in radioactive mixed waste environments. Nat Biotechnol 1998, 16, 929–933.
[28] H. Brim, J. P. Osborne, H. M. Kostandarithes, J. K. Fredrickson, L. P. Wackett, M. J. Daly. *Deinococcus radiodurans* engineered for complete toluene degradation facilitates Cr(VI) reduction. Microbiology 2006, 152, 2469–2477.
[29] D. Appukuttan, A. S. Rao, S. K. Apte. Engineering of *Deinococcus radiodurans* R1 for bioprecipitation of uranium from dilute nuclear waste. Appl Environ Microbiol 2006, 72, 7873–7878.
[30] S. Kulkarni, A. Ballal, S. K. Apte. Bioprecipitation of uranium from alkaline waste solutions using recombinant *Deinococcus radiodurans*. J Hazard Mater 2013, 262, 853–861.
[31] N. McDowell. Data stored in multiplying bacteria. New Scientist, 2003-01-08.
[32] J. R. Battista, A. M. Earl, M. J. Park. Why is *Deinococcus radiodurans* so resistant to ionizing radiation? Trends Microbiol 1999, 7, 362–365.
[33] S. Levin-Zaidman, J. Englander, E. Shimoni, A. K. Sharma, K. W. Minton, A. Minsky. Ringlike structure of the *Deinococcus radiodurans* genome: a key to radioresistance? Science 2003, 299, 254–256.
[34] A. K. Pavlov, V. L. Kalinin, A. N. Konstaninov, V. N. Shelegedin, A. A. Pavlov. Was Earth ever infected by Martian Biota? Clues from radioresistant bacteria. Astrobiology 2006, 6, 911–918.
[35] D. Billi, M. Potts. Life and death of dried prokaryotes. Res Microbiol 2002, 153, 7–12.
[36] M. Blasius, U. Hübscher, S. Sommer. *Deinococcus radiodurans*: what belongs to the survival kit? Crit Rev Biochem Mol Biol 2008, 43, 221–238.
[37] V. Mattimore, J. R. Battista.. Radioresistance of *Deinococcus radiodurans*: Functions necessary to survive ionizing radiation are also necessary to survive prolonged desiccation. J Bacteriol 1996, 178, 633–637.
[38] M. Toueille, S. Sommer. Life in extreme conditions: *Deinococcus radiodurans*, an organism able to survive prolonged desiccation and high doses of ionizing radiation. In: M. Gargaud, López-Garcìa P, H. Martin, eds. Origins and Evolution of Life: An Astrobiological Perspective. Cambridge, UK, Cambridge University Press, 2011, 347–358.
[39] E. I. Friedmann, R. Ocampo-Friedmann. A primitive cyanobacterium as pioneer microorganism for terraforming Mars. Adv Space Res 1995, 15, 3, 243–246.
[40] W. H. Schlesinger, J. S. Pippen, M. D. Wallenstein, K. S. Hofmockel, D. M. Klepeis, B. E. Mahall. Community composition and photosynthesis by photoautotrophs under quartz pebbles, southern Mojave Desert. Ecology 2003, 84, 3222–3231.
[41] G. Boison, A. Mergel, H. Jolkver, H. Bothe. Bacterial life and dinitrogen fixation at a gypsum rock. Appl Environ Microbiol 2004, 70, 7070–7077.
[42] B. Budel, B. Weber, M. Kuhl, H. Pfanz, D. Sultemeyer, D. C. J. Wessels. Reshaping of sandstone surfaces by cryptoendolithic cyanobacteria: bioalkalization causes chemical weathering in arid landscapes. Geobiology 2004, 2, 261–268.
[43] K. Warren-Rhodes, K. Rhodes, S. Pointing, et al. Hypolithic cyanobacteria, dry limit of photosynthesis, and microbial ecology in the hyperarid Atacama Desert. Microbiol Ecol 2006, 52, 389–398.

[44] Y. Chan, D. C. Lacap, M. C. Y. Lau, et al. Hypolithic microbial communities: between a rock and a hard place. Environ Microbiol 2012, 14, 2272–2282.
[45] K. Olsson-Francis, A. E. Simpson, D. Wolff-Boenisch, C. S. Cockell. The effect of rock composition on cyanobacterial weathering of crystalline basalt and rhyolite. Geobiology 2012, 10, 434–444.
[46] D. Billi. Subcellular integrities in *Chroococcidiopsis* sp. CCMEE 029 survivors after prolonged desiccation revealed by molecular probes and genome stability assays. Extremophiles 2009, 13, 49–57.
[47] D. Billi, E. I. Friedmann, K. G. Hofer, M. Grilli Caiola, R. Ocampo-Friedmann (2000) Ionizing-radiation resistance in the desiccation-tolerant cyanobacterium *Chroococcidiopsis*. Appl Environ Microbiol 66, 1489–1492.
[48] M. Baqué, E. Viaggiu, G. Scalzi, D. Billi. Endurance of the endolithic desert cyanobacterium *Chroococcidiopsis* under UVC radiation. Extremophiles 2013, 17, 161–169.
[49] M. Baqué, J. P. de Vera, P. Rettberg, D. Billi. The BOSS and BIOMEX space experiments on the EXPOSE-R2 mission: endurance of the desert cyanobacterium *Chroococcidiopsis* under simulated space vacuum, Martian atmosphere, UVC radiation and temperature extremes. Acta Astronaut 2013b, 91, 180–186.
[50] C. S. Cockell, A. C. Schuerger, D. Billi, E. I. Friedmann, C. Panitz. Effects of a simulated martian UV flux on the cyanobacterium, *Chroococcidiopsis* sp. 029. Astrobiology 2005, 5, 127–140.
[51] D. Billi, E. Viaggiu, C. S. Cockell, E. Rabbow, G. Horneck, S. Onofri. Damage escape and repair in dried *Chroococcidiopsis* spp. from hot and cold deserts exposed to simulated space and Martian conditions. Astrobiology 2011, 11, 65–73.
[52] C. S. Cockell, P. Rettberg, E. Rabbow, K. Olsson-Francis. Exposure of phototrophs to 548 days in low Earth orbit: microbial selection pressures in outer space and on early earth. ISME J, 2011, 10, 1671–1682.
[53] D. Billi. Anhydrobiotic rock-inhabiting cyanobacteria: potential for astrobiology and biotechnology. In: H. Stan-Lotter, F. Fendrihan, eds. Adaptation of Microbial Life Organisms in Extreme Environments: Research and Application. New York, USA, Springer, 2012, 119–132.
[54] D. Billi, M. Baqué, H. D. Smith, C. P. McKay. Cyanobacteria from extreme deserts to space. Adv Microbiol 2013, 3, 80–86.
[55] M. Baqué, G. Scalzi, E. Rabbow, P. Rettberg, D. Billi. Biofilm and planktonic lifestyles differently support the resistance of the desert cyanobacterium *Chroococcidiopsis* under space and Martian simulations. Orig Life Evol Biosph 2013a, 43, 377–389.
[56] H. D. Smith, M. Baqué, A. G. Duncan, C. R. Lloyd, C. P. McKay, D. Billi. Comparative analysis of cyanobacteria inhabiting rocks with different light transmittance in the Mojave Desert: a Mars terrestrial analogue. Int J Astrobiol 2014, doi:10.1017/S1473550414000056.
[57] D. Fewer, T. Friedl, B. Budel. *Chroococcidiopsis* and heterocyst-differentiating cyanobacteria are each other's closest living relatives. Mol Phylogenet Evol 2002, 23, 82–90.
[58] J. Bahl, M. C. Y. Lau, G. J. D. Smith, et al. Ancient origins determine global biogeography of hot and cold desert cyanobacteria. Nat Commun 2011, 2, 163.
[59] T. N. Taylor, H. Hass, W. Remy, H. Kerp. The oldest fossil lichen. Nature 1995, 378, 244.
[60] T. N. Taylor, H. Hass, H. Kerp. A cyanolichen from the Lower Devonian Rhyniechert. Am J Bot 1997, 84, 992.
[61] T. Ishida, M. M. Watanabe, J. Sugiyama, A. Yokota. Evidence for polyphyletic origin of the members of the orders of Oscillatoriales and Pleurocapsales as determined by 16S rRNA analysis. FEMS Microbiol Lett 2001, 201, 79–82.
[62] F. Kauff, B. Budel. Phylogeny of cyanobacteria: An overview. Prog Bot 2011, 72, 209–224.
[63] P. S. Seo, A. Yokota. The phylogenetic relationships of cyanobacteria inferred from 16S rRNA, gyrB, rpoC1 and rpoD1 gene sequences. J Gen Appl Microbiol 2003, 49, 191–203.

[64] J. Cumbers, L. J. Rothschild. Salt tolerance and polyphyly in the cyanobacterium *Chroococcidiopsis* (Pleurocapsales). J Phycol 2014, doi: 10.1111/jpy.12169
[65] R. A. Garrett, H. P. Klenk. Archaea: Evolution, Physiology, and Molecular Biology. Hoboken, USA, Wiley-Blackwell, 2007.
[66] S. Leuko, M. J. Raftery, B. P. Burns, M. R. Walter, B. A. Neilan. Global protein-level responses of *Halobacterium salinarum* NRC-1 to prolonged changes in external sodium chloride concentrations. J Proteome Res 2009, 8, 2218–2225.
[67] J. A. Müller, S. DasSarma. Genomic analysis of anaerobic respiration in the archaeon *Halobacterium* sp. strain NRC-1: dimethyl sulfoxide and trimethylamine N-oxide as terminal electron acceptors. J Bacteriol 2005, 187, 1659–1667.
[68] B. K. Baxter, C. D. Litchfield, K. Sowers, J. D. Griffith, P. A. DasSarma, S. DasSarma. Microbial diversity of Great Salt Lake. In: N. Gunde-Cimerman, A. Oren, A. Plemenitas, eds. Adaptation to Life at High Salt Concentrations In Archaea, Bacteria, and Eukarya. Dordrecht, Netherlands, Springer, 2005, 9–26.
[69] M. Kottemann, A. Kish, C. Iloanusi, S. Bjork, J. DiRuggiero. Physiological responses of the halophilic archaeon *Halobacterium* sp. strain NCR-1 to desiccation and gamma irradiation. Extremophiles 2005, 9, 219–227.
[70] L. C. de Veaux, J. A. Müller, J. Smith, J. Petrisko, D. P. Wells, S. DasSarma. Extremely radiation-resistant mutants of a halophilic archaeon with increased single-stranded DNA-binding protein (RPA) gene expression. Radiat Res 2007, 168, 507–514.
[71] A. Kaur, M. Pan, M. Meislin, M. T. Facciotti, R. El-Gewely, N. S. Baliga. A systems view of haloarchaeal strategies to withstand stress from transition metals. Genome Res 2006, 16, 841–854.
[72] J. A. Coker, P. DasSarma, J. Kumar, J. A. Müller, S. DasSarma. Transcriptional profiling of the model archaeon *Halobacterium sp.* NRC-1: responses to changes in salinity and temperature. Saline Systems 2007, 3, 6.
[73] A. Kish, P. L. Griffin, K. L. Rogers, M. L. Fogel, R. J. Hemley, A. Steele. High-pressure tolerance in *Halobacterium salinarum* NRC-1 and other non-piezophilic prokaryotes. Extremophiles 2012, 16, 355–361.
[74] P. DasSarma, R. C. Zamora, J. A. Müller, S. DasSarma. Genome-wide responses of the model archaeon *Halobacterium* sp. strain NRC-1 to oxygen limitations. J Bacteriol 2012, 194, 5530–5537.
[75] N. S. Baliga, S. J. Bjork, R. Bonneau, M. Pan, C. Iloanusi, M. C. H. Kottemann, L. Hood, J. DiRugggiero. Systems level insights into the stress response to UV radiation in the halophilic archaeon *Halobacterium* NRC-1. Genome Res 2004, 14, 1025–1035.
[76] M. R. Mormile, M. A. Biesen, M. C. Gutierrez, et al. Isolation of *Halobacterium salinarum* retrieved directly from halite brine inclusions. Environ Microbiol 2003, 5, 1094–1102.
[77] H. Stan-Lotter, C. Radax, C. Gruber, et al. Astrobiology with haloarchaea from Permo-Triassic rock salt. Int J Astrobiol 2003, 1, 271–284.
[78] S. DasSarma, B. R. Berquist, J. A. Coker, P. DasSarma, J. A. Müller. Post-genomics of the model haloarchaeon *Halobacterium sp.* NRC-1. Saline Systems 2006, 16, 2:3.
[79] F. Pfeiffer, S. C. Schuster, A. Broicher, et al. Evolution in the laboratory: the genome of *Halobacterium salinarum* strain R1 compared to that of strain NRC-1. Genomics 2008, 91, 335–346.
[80] Ng WV, S. P. Kennedy, G. G. Mahairas, et al. Genome sequence of *Halobacterium* species NRC-1. Proc Natl Acad Sci USA 2000, 97, 12176–12181.
[81] J. Seckbach. Evolutionary Pathways and Enigmatic Algae: *Cyanidium caldarium* (Rhodophyta) and Related Cells. Dordrecht, The Netherlands, Kluwer Academic Publishers, 1994.

[82] G. Pinto. Cyanidiophyceae: looking back–looking forward. In: J. Seckbach, ed. Algae and Cyanobacteria in Extreme Environments. Dordrecht, The Netherlands, Springer, 2007, 387–397.

[83] A. P. M. Weber, G. G. Barbier, R. P. Shrestha, R. J. Horst, A. Minoda, C. Oesterhelt. A genomics approach to understanding the biology of thermo-acidophilic red algae. In: J. Seckbach, ed. Algae and Cyanobacteria in Extreme Environments. Dordrecht, The Netherlands, Springer, 2007, 503–518.

[84] Castenholz, RW, T. R. McDermott. The Cyanidiales: ecology, biodiversity, and biogeography. In: J. Seckbach, D. J. Chapman, eds. Red Algae in the Genomic Age. Dordrecht, The Netherlands, Springer, 2009, 355–369.

[85] K. M. Müller, M. D. J. Lynch, R. G. Sheath. Bangiophycidae no more: from one class to six: where do we go from here? Moving the Bangiophytes into the genomic age. In: J. Seckbach, D. J. Chapman, eds. Red Algae in the Genomic Age. Dordrecht, The Netherlands, Springer, 2010, 239–257.

[86] W. Gross, I. Heilmann, D. Lenze, C. Schnarrenberger. Biogeography of the Cyanidiaceae (Rhodophyta) based on 18S ribosomal RNA sequence data. Eur J Phycol 2001,36, 275–280.

[87] I. Enami, H. Adachi, J. R. Shen. Mechanisms of acido-tolerance and characteristics of photosystems in an acido- and thermophilic red alga Cyanidium caldarium. In: J. Seckbach, D. J. Chapman, eds. Red Algae in the Genomic Age. Dordrecht, The Netherlands, Springer, 2009, 371–387.

[88] H. S. Yoon, C. Ciniglia, Wu M, et al. Establishment of endolithic populations of extremeophilic Cyanidiales (Rhodophyta). BMC Evol Biol 2006, 6, 78.

[89] V. Reed, D. Bhattacharya. The thermo-acidophilic Cyanidiaceae (Cyanidiales). In: J. Seckbach, D. J. Chapman, eds. Red Algae in the Genomic Age. Dordrecht, The Netherlands, Springer, 2010, 418–426.

[90] A. Azua-Bustos. Chilean Cave Cyanidium. In: J. Seckbach, D. J. Chapman, eds. Red Algae in the Genomic Age. Dordrecht, The Netherlands, Springer, 2010, 425–438.

[91] P. Albertano, C. Ciniglia, G. Pinto, A. Pollio. The taxonomic position of Cyanidium, Cyanidioschyzon and Galdieria: an update. Hydrobiologia 2000, 433, 137–133.

[92] C. Ciniglia, H. S. Yoon, A. Pollio, G. Pinto, D. Bhattacharya. Hidden biodiversity of the extremophilic Cyanidiales red algae. Mol Ecol 2004, 13, 1827–1838.

[93] J. M. Lopez-Bautista. Red algal genomics: a synopsis. In: J. Seckbach, D. J. Chapman, eds. Red Algae in the Genomic Age. Dordrecht, The Netherlands, Springer, 2010, 225–238.

[94] S. Y. Yoon, J. D. Hackett, G. Pinto, D. Bhattacharya. The single, ancient origin of chromist plastids. Proc Natl Acad Sci USA 2002, 99, 15507–15512.

[95] S. Y. Yoon, J. D. Hackett, C. Ciniglia, G. Pinto, D. Bhattacharya. A molecular timeline for the origin of photosynthetic eukaryotes. Mol Biol Evol 2004, 21, 809–818.

[96] H. S. Yoon, K. M. Muller, R. G. Sheath, F. D. Ott, D. Bhattacharya. Defining the major lineages of red algae (Rhodophyta). J. Phycol 2006, 42, 482–492.

[97] M. Matsuzaki, O. Misumi, I. T. Shin, et al. Genome sequence of the ultrasmall unicellular red alga *Cyanidioschyzon merolae* 10d. Nature 2004, 428, 653–657.

[98] G. Barbier, C. Oesterhelt, M. D. Larson, et al. Comparative genomics of two closely related unicellular thermo-acidophilic red algae, *Galdieria sulphuraria* and *Cyanidioschyzon merolae*, reveals the molecular basis of the metabolic flexibility of *Galdieria sulphuraria* and significant differences in carbohydrate metabolism of both algae. Plant Physiol 2005, 137, 460–474.

[99] J. Seckbach. Overview on cyanidian biology. In: J. Seckbach, D. J. Chapman, eds. Red Algae in the Genomic Age. Dordrecht, The Netherlands, Springer, 2010, 418–426.

[100] F. A. Seymour, P. D. Crittenden, P. S. Dyer. Sex in the extremes: lichen-forming fungi. Mycologist 2005, 19, 51–58.

[101] L. Muggia, L. Vancurova, P. Škaloud, O. Peksa, M. Wedin, M. Grube. The symbiotic playground of lichen thalli–a highly flexible photobiont association in rock-inhabiting lichens, FEMS Microbiol Ecol 2013, 85, 313–323.
[102] V. Ahmadjian. *The Lichen Symbiosis*. New York, USA, John Willey & Sons, 1993.
[103] W. Purvis. Lichens. London, UK, Natural History Museum, 2000.
[104] T. H. Nash. Lichen Biology. 2nd edn. Cambridge, UK, Cambridge University Press, 2008.
[105] J. P. de Vera, P. Rettberg, S. Ott. Life at the limits: capacities of isolated and cultured lichen symbionts to resist extreme environmental stresses. Orig Life Evol Biosph 2008, 38: 457–468.
[106] D. O. Øvstedal, R. I. Lewis-Smith. Lichens of Antarctica and South Georgia. A Guide to Their Identification and Ecology. Cambridge, UK, Cambridge University Press, 2001.
[107] E. Stocker-Wörgötter. Metabolic diversity of lichen-forming ascomycetous fungi: culturing, polyketide and shikimate metabolite production, and PKS genes. Nat Prod Rep 2008, 25, 188–200.
[108] J. Boustie, M. Grube. Lichens-a promising source of bioactive secondary metabolites. Plant Gen Res 2005, 3, 273–287.
[109] J. P. de Vera, G. Horneck, P. Rettberg, S. Ott. The potential of the lichen symbiosis to cope with the extreme conditions of outer space. I. Influence of UV radiation and space vacuum on the vitality of lichen symbiosis and germination capacity. Int J Astrobiol 2003, 1, 285–293.
[110] J. P. de Vera, G. Horneck, P. Rettberg, S. Ott. The potential of the lichen symbiosis to cope with the extreme conditions of outer space. II. Germination capacity of lichen ascospores in response to simulated space conditions. Adv Space Res 2004, 33, 1236–1243.
[111] R. de la Torre, L. G. Sancho, A. Pintado, et al. Biopan experiment Lichens on the Foton M2 mission pre-flight verification tests of the Rhizocarpon geographicum granite ecosystem. Adv Space Res 2007, 40, 1665–1671.
[112] L. G. Sancho, R. De La Torre, G. Horneck, C. Ascaso, A. De Los Ríos, A. Pintado, J. Wierzchos, M. Schuster. Lichens survive in space: results from the 2005 LICHENS experiment. Astrobiology 2007, 7, 443–454.
[113] R. de la Torre, L. G. Sancho, G. Horneck, et al. Survival of lichens and bacteria exposed to outer space conditions. Results of the Lithopanspermia experiments. Icarus 2010, 208, 735–748.
[114] L. G. Sancho, R. De La Torre, A. Pintado. Lichens, new and promising material from experiments in exobiology. Fungal Biol Rev 2008, 22, 103–109.
[115] J. Raggio, A. Pintado, C. Ascaso, et al. Whole lichen thalli survive exposure to space conditions: results of Lithopanspermia experiment with *Aspicilia fruticulosa*. Astrobiology 2011, 11, 281–292.
[116] J. P. de Vera. Lichens as survivors in space and on Mars. Fungal Ecol 2012, 5, 472–479.
[117] X. Yuan, S. Xiao, T. N. Taylor. Lichen-like symbiosis 600 million years ago. Science 2005, 308, 1017–1020.
[118] A. Gargas, P. T. de Priest, M. Grube, et al. Multiple origins of lichen symbioses in fungi suggested by SSU rDNA phylogeny. Science 1995, 268, 1492–1495.
[119] F. Lutzoni, M. Pagel, V. Reeb. Major fungal lineages are derived from lichen symbiotic ancestors. Nature 2001, 411, 937–940.
[120] M. Wedin, H. Döring, G. Gilenstam. Saprotrophy and lichenization as options for the same fungi species on different substrata: environmental plasticity and fungal lifestyles in the Strictis-Conotremacomplex. New Phytol 2004, 16, 4459–4465.
[121] O. E. Eriksson. Ascomyceternas ursprung och evolution Protolichenes-hypotesen. Sven Mykol Tidskr 2005, 26, 22–29.

[122] C. L. Schoch, G. H. Sung, F. López-Giráldez, et al. The Ascomycota tree of life: a phylum-wide phylogeny clarifies the origin and evolution of fundamental reproductive and ecological traits. Syst Biol 2009, 58, 224–239.

[123] D. Schulze-Makuch, J. Seckbach. Tardigrades: An example of multicellular extremophiles. In: J. Seckbach, A. Oren, H. Stan-Lotter, eds. Polyextremophiles: Life Under Multiple Forms of Stress. Dordrecht, Netherlands, Springer, 2013.

[124] P. Degma, R. Bertolani, R. Guidetti. 2014. Actual checklist of Tardigrada species. (accessed 31-05-2014, at http://www.tardigrada.modena.unimo.it/miscellanea/Actual{%}20checklist{%}20of{%}20Tardigrada.pdf).

[125] A. Maas, D. Waloszek. Cambrian derivatives of the early arthropod stem lineage, pentastomids, tardigrades and lobopodians – An Orsten perspective. Zool Anz 2001, 240, 451–459.

[126] J. A. E. Goeze. Herrn Karl Bonnets Abhandlungen aus der Insektologie aus d. Franz. übers. u mit einigen Zusätzen hrsg. v. Joh. August Ephraim Goeze Gebauer, Halle, 414. 1773.

[127] D. R. Nelson. Current status of the Tardigrada: evolution and ecology. Integr Comp Biol 2002, 42, 652–659.

[128] S. J. McInnes, P. J. A. Pugh. An attempt to revisit the global biogeography of limno-terrestrial Tardigrada. J Limnol 2007, 66, 90–96.

[129] J. R. Garey, S. J. McInnes, P. B. Nichols. Global diversity of tardigrades (Tardigrada) in freshwater. Hydrobiologia 2008, 595, 101–106.

[130] R. Guidetti, T. Altiero, L. Rebecchi. On dormancy strategies in tardigrades. J Insect Physiol 2011, 57, 567–576.

[131] Wełnicz W, M. A. Grohme, L. Kaczmarek, R. O. Schill, M. Frohme. Anhydrobiosis in tardigrades-the last decade. J Insect Physiol 2011, 57, 577–583.

[132] S. Hengherr, A. G. Heyer, H. R. Köhler, R. O. Schill. Trehalose and anhydrobiosis in tardigrades-evidence for divergence in responses to dehydration. FEBS J 2008, 275, 281–288.

[133] S. Neumann, A. Reuner, F. Brümmer, Schill, RO. DNA damage in storage cells of anhydrobiotic tardigrades. Comp Biochem Physiol 2009, 153, 425–429.

[134] A. M. Rizzo, M. Negroni, T. Altiero, G. Montorfano, P. Corsetto, P. Berselli, B. Berra, R. Guidetti, L. Rebecchi. Antioxidant defences in hydrated and desiccated states of the tardigrade Paramacrobiotus richtersi. Comp Biochem Physiol B 2010, 156, 115–121.

[135] E. Schokraie, A. Hotz-Wagenblatt, U. Warnken, B. Mali, M. Frohme, F. Förster, T. Dandekar, S. Hengherr, R. O. Schill, M. Schnölzer. Proteomic analysis of Tardigrades: towards a better understanding of molecular mechanisms by anhydrobiotic organisms. PLOS ONE 2010, 5, e9502.

[136] D. D. Horikawa, J. Cumbers, I. Sakakibara, et al. Analysis of DNA repair and protection in the Tardigrade *Ramazzottius varieornatus* and *Hypsibius dujardini* after exposure to UVC radiation. PLoS One 2013, 8, e64793.

[137] K. A. Halberg, D. Persson, H. Ramlov, P. Westh, R. M. Kristensen, N. Mobjerg. Cyclomorphosis in Tardigrada: Adaptation to environmental constraints. J Exp Biol 2009, 212, 2803–2811.

[138] N. Møbjerg, K. A. Halberg, A. Jørgensen, et al. Survival in extreme environments – on the current knowledge of adaptations in tardigrades. Acta Physiol 2011, 202, 409–420.

[139] K. I. Jönsson. Tardigrades as a potential model organism in space research. Astrobiology 2007, 7, 757–766.

[140] F. Ono, M. Saigusa, T. Uozumi, Y. Matsushima, H. Ikeda, N. L. Saini, M. Yamashita. Effect of high hydrostatic pressure on to life of the tiny animal tardigrade. J. Phys Chem, Solids 2008, 69, 2297–2230.

[141] D. D. Horikawa, T. Sakashita, Katagiri, C, at al. Radiation tolerance in the tardigrade. Int J Radiat Biol 2006, 82, 12, 843–848.

[142] R. Guidetti, K. I. Jönsson. Long-term anhydrobiotic survival in semi-terrestrial micrometazoans. J Zool 2002, 257, 181–187.

[143] H. Ramløv, P. Westh. Cryptobiosis in the Eutardigrade Adorybiotus (Richter sius) coronifer: tolerance to alcohols, temperature and de novo protein synthesis. Zoolog Anz 2001, 240, 517–523

[144] K. I. Jönsson, R. Bertolani. Facts and fiction about long-term survival in tardigrades. J Zool 2001, 255, 121–123.

[145] R. Bertolani, R. Guidetti, K. I. Jönsson, T. Altiero, D. Boschini, L. Rebecchi. Experiences with dormancy in tardigrades. J Limnol 2004, 63, 16–25.

[146] R. O. Schill, G. B. Fritz. Desiccation tolerance in embryonic stages of the tardigrade. J Zool 2008, 276, 103–107.

[147] K. I. Jönsson, E. Rabbow, R. O. Schill, M. Harms-Ringdahl, P. Rettberg. Tardigrades survive exposure to space in low Earth orbit. Curr Biol 2008, 18, R729–731.

[148] L. Rebecchi, T. Altiero, R. Guidetti, M. Cesari, R. Bertolani, M. Negroni, A. M. Rizzo. Tardigrade resistance to space effects: first results of experiments on the LIFE-TARSE mission on FOTON-M3 (September 2007). Astrobiology 2009, 9, 581–591.

[149] L. Rebecchi, T. Altiero, M. Cesari, R. Bertolani, A. M. Rizzo, P. A. Corsetto, R. Guidetti. Resistance of the anhydrobiotic eutardigrade Paramacrobiotus richtersi to space flight (LIFE–TARSE mission on FOTON-M3). J. Zoolog Syst Evol Res 2011, 49, 98–103.

[150] D. Persson, K. A. Halberg, A. Jørgensen, C. Ricci, N. Møbjerg, R. M. Kristensen. Extreme stress tolerance in tardigrades: surviving space conditions in low earth orbit. J Zoolog Syst Evol Res 2011, 49, 90–97.

[151] R. Guidetti, A. M. Rizzo, T. Altiero, L. Rebecchi. What can we learn from the toughest animals of the Earth? Water bears (Tardigrades) as multicellular model organisms in order to perform scientific preparations for lunar exploration. Planet Space Sci 2012, 74, 97–102.

[152] C. J. Sands, S. J. McInnes, N. J. Marley, W. P. Goodall-Copestake, P. Convey, K. Linse. Phylum Tardigrada: an "individual" approach. Cladistics 2008, 24, 861–871.

[153] A. Jørgensen, S. Faurby, J. G. Hansen, N. Møbjerg, R. M. Kristensen. Molecular phylogeny of Arthrotardigrada (Tardigrada). Mol Phylogenet Evol 2010, 54, 1006–1015.

[154] N. Guil, G. Giribet. A comprehensive molecular phylogeny of tardigrades – adding genes and taxa to a poorly resolved phylum-level phylogeny. Cladistics 2012, 28, 21–49.

[155] L. I. Campbell, O. Rota-Stabelli, G. D. Edgecombe, et al. MicroRNAs and phylogenomics resolve the relationships of Tardigrada and suggest that velvet worms are the sister group of Arthropoda. Proc Natl Acad Sci U S A 2011, 108, 15920–15924.

[156] G. Mayer, C. Martin, J. Rüdiger, et al. Selective neuronal staining in tardigrades and onychophorans provides insights into the evolution of segmental ganglia in panarthropods. BMC Evol Biol 2013, 13, 230.

[157] Garagna, S, L. Rebecchi, A. Guidi. Genome size variation in Tardigrada. Zool J Linn Soc 996, 116, 115–121.

[158] J. A. Toplin, T. B. Norris, C. R. Lehr, T. R. McDermott, R. W. Castenholz. Biogeographic and phylogenetic diversity of thermoacidophilic Cyanidiales in Yellowstone National Park, Japan, and New Zealand. Appl Environ Microbiol 2008, 74, 2822–2833.

William F. Martin, Sinje Neukirchen, and Filipa L. Sousa
9 Early life

Life is a chemical reaction. All organisms great and small have some form of energy releasing chemical reaction at the core of their living process. Such bioenergetic reactions have a myriad of substrates and products in nature [1, 2] but they all go forward according to the second law of thermodynamics: the conversion of substrates to products, with cell mass as a biological byproduct, releases energy [3]. Some of the energy released can be conserved in chemical form that allows the cell to perform a bit of metabolic work. Life's main energy currency today is adenosine triphosphate (ATP), but there are other "energy-rich" compounds that play an important role in biological energy conservation [4] among them thioesters like acetyl-CoA or reduced ferredoxin [5]. Bioenergetic reactions constitute the main flux of matter and energy through the cell. For an adult human, the core bioenergetic reaction is the oxygen-dependent burning of fats and sugars in our mitochondria. This reaction generates about 1 kg of CO_2 and 0.4 kg of H_2O as end products and about 60–100 kg (roughly a body weight) of ATP per day, per person. For the bacterium *Escherichia coli*, a cell division requires about 60 billion molecules of ATP, corresponding to about 50 body weights of ATP synthesized per cell division [5]. The numbers for an archaeon [6] can be readily calculated from Y_{ATP}, the yield in grams of cell mass per mole ATP consumed, and are very similar to *E. coli* values.

For all forms of life, from archaea and bacteria to humans, when the core bioenergetic reaction stops, so does life. Bioenergetic reactions have been running in a sequence of uninterrupted continuity since the first prokaryotes arose on Earth more than 3.5 billion years ago [7]. During that time, evolution has tinkered mightily with the proteins and cofactors that cells use to harness energy, but not even a force as powerful as evolution can tinker with the second law of thermodynamics. Rather, evolution is the byproduct – the innately creative byproduct – of one long and continuous bioenergetic reaction. When we talk about early life, the title of this chapter, we are basically asking: how did that reaction get started and how did the first cells make a living?

The geochemical record harbors only few and faint traces of the very ancient microbial past that might help to answer such questions or that might bear upon our understanding of the nature of earliest life. Geochemical evidence indicates that methanogenesis goes back about 3.5 billion years [8], approaching the 3.8 billion year old age of the oldest rocks harboring evidence for life [9]. We can be relatively sure that there was no molecular oxygen around at life's origin, because oxygen is a biological product [10]. Beyond that, geological evidence generally tells us much more about the rocks-and-water setting for life's origin than it does about the first kind of microbes that emerged there [7, 11]. But geochemistry is not the only window we have into the ancient past. The nature of the bioenergetic reactions that the very first microbes used belongs to the subject matter of evolutionary microbiology, so a comparative approach can unearth some insights.

If we want to address the ancestral state of microbial physiology [12], we have to start with a process of elimination, because there are so many possibilities. There are many different kinds of organisms known that use hundreds of different main redox reactions to harness environmentally available energy [1], also among thermophilic habitats [2]. Using common sense and a few simple pruning rules, we can narrow the possibilities.

Biologists have always known that anaerobes are ancient and that anaerobic environments should harbor primitive kinds of bioenergetic reactions [13, 14]. Since life arose in a world without molecular oxygen, the first cells had to be anaerobes. Because eukaryotes arose from a symbiosis of prokaryotes [15–18], the first cells were prokaryotes, not eukaryotes. There are also good reasons to think that biochemistry, and microbial life, started off from CO_2, rather than from some kind of preformed organic soup [19, 20]. Following this reasoning, we can infer that the first organisms were anaerobic, prokaryotic autotrophs. There is nothing new about this approach to the problem of early life, it has a long and robust tradition in biology [13, 14, 21].

We can narrow down the possibilities further. If we accept the reasonable proposition that the deepest branch in the prokaryotic tree of life separates the Bacteria from the Archaea [22–24], the foregoing three criteria place anaerobic autotrophs at the root of the prokaryotic tree. Accordingly, the founder lineages at the base of both the archaeal and the bacterial domains should have the same pathway of CO_2 fixation, namely the one used in by the last universal common ancestor (LUCA). The criterion of CO_2 fixation narrows down the possibilities to a more specific set of candidate ancient lineages for early life. This is because there are at present only six pathways of core CO_2 fixation known [25, 26]: the Calvin cycle, the reductive citric acid cycle, the 3-hydroxypropionate/4-hydroxybutyrate cycle, the dicarboxylate/4-hydroxybutyrate cycle, the 3-hydroxypropionate bi-cycle, and the acetyl-CoA pathway. Yet of those six CO_2 fixation pathways, only the acetyl-CoA – also called the Wood-Ljungdahl (WL) – pathway occurs in both Bacteria and Archaea [25]. For this reason, and for other reasons based on bioenergetic considerations, the simplicity of its chemistry [27, 28], and the prevalence of transition metal catalysis in its main reactions [29], the acetyl-CoA pathway is considered to be the most ancient of the CO_2 fixation pathways [25].

Among the lineages of bacterial and archaeal anaerobes that use the acetyl-CoA pathway, the acetogens and methanogens stand out in particular. This is because in acetogens (Bacteria) and methanogens (Archaea), the acetyl-CoA pathway is not only the main route of CO_2 fixation, it is also integral to energy conservation in that acetogens and methanogens generate their ATP with the help of ion gradients that are generated in the process of reducing CO_2 with electrons from H_2 [30]. The other five pathways of CO_2 fixation do not generate ATP, they consume ATP that is generated via an independent energy metabolism involving cytochromes and quinones (or quinone analogues). This sets methanogens and acetogens apart from all other autotrophs, as does the circumstance that both among acetogens [31] and among methanogens [6], lineages occur that lack both quinones and cytochromes.

Today, when environmental microbiologists probe ancient anaerobic niches deep in the Earth's crust, they find acetogens (Bacteria) [32–34] and methanogens (Archaea) [35] in addition to sulfate reducers [36], the very same groups that biologists always thought were ancient [13]. Although such anaerobic environments present steep bioenergetic challenges because they harbor so little energy to harness [30, 37, 38], they also are home to the organisms that arguably possess the simplest, and what might be the most primitive, forms of energy metabolism: acetogens and methanogens that lack cytochromes [39].

Given an environmental supply of nitrogen and trace elements, microbial physiology is, at the most basic level, a matter of carbon and energy metabolism. Carbon metabolism points to acetogens and methanogens as ancient and possibly ancestral. What about energy metabolism? Energy metabolism concerns the way(s) in which cells harness environmentally available sources of energy and convert them into chemical forms that are accessible to metabolism. With or without oxygen, life as we know it uses only two basic mechanisms to tap environmentally available energy and harness it as ATP: i) substrate-level phosphorylation and ii) chemiosmotic coupling. Here, critics will be quick to interject "But what about life as we *don't* know it – at the origin of life maybe energy was harnessed differently!" That could be, but for evolutionary biologists the issues are i) how known microbial life forms arose (not ones that we can imagine) and ii) what is the origin of biological energy conversion as it occurs in *life*, the thing that origin of life research is supposed to be explaining. So what are the two mechanisms that cells use to conserve energy?

The first mechanism, substrate level phosphorylation (SLP) is simple: metabolism generates highly reactive phosphate-containing compounds that phosphorylate ADP to make ATP [37, 41]. The energy that is conserved in ATP is then released in a subsequent reaction that can do some chemical work for the cell or allow more sluggish reactions to go forward. There are only a handful of six or eight highly reactive phosphate-containing compounds that are widely used for SLP [37]; they include acyl phosphates like acetyl phosphate, which contains a highly labile ("high energy") mixed anhydride bond. Importantly, the high energy bonds in these highly reactive compounds that fuel SLP are not mined from deposits in the environment. They are generated during conversions of carbon compounds. Their synthesis is driven by environmental sources of chemical energy such as H_2 plus CO_2 (or sugar plus O_2) that are harnessed during synthesis of thermodynamically more favorable end products, like methane and acetate (or water and CO_2).

The second mechanism that cells use to harness energy involves ion gradients and is called chemiosmotic coupling, a mechanism discovered by Peter Mitchell [42]. Here, an exergonic reaction is coupled to the pumping of ions across a membrane from inside the cell to the outside. The most common ions used for this purpose are protons, rendering the inside of a cell alkaline relative to the outside, but sodium ions are often used in organisms from low energy environments [37]. The energy stored in the ion gradient is then harnessed by a rotor-stator type ATPase to phosphorylate ADP. Even the anaerobic energy misers, methanogens and acetogens, are obligately chemiosmotic. They use an ATPase, but diverge in the mechanism by which they generate their ion gradient [25, 30, 40]. During microbial growth, chemiosmosis and SLP always harness redox energy, the natural tendency of electrons to flow from donors (such as H_2) to acceptors (such as CO_2).

Since Oparin and Haldane [43], scientists long thought that the first free-living cells were pure fermenters, organisms that we know today live from SLP alone. Though all cells use both SLP and chemiosmotic coupling in their overall metabolism (fermenters spend energy gleaned through SLP to make ion gradients for nutrient import), pure fermenters that use SLP as their sole source of energy harnessing are always derived, in the phylogenetic sense, from chemiosmotic forms [44], and fermenters always live from compounds produced by autotrophs, which are always chemiosmotic. That means that whatever the nature of the first free-living cells, they were chemiosmotic – they harnessed ion gradients [45].

But in modern cells, harnessing chemiosmotic potential always requires proteins, posing a chicken-and-egg kind of problem: what kind of energy harnessing allowed genes and proteins to arise? A possibility is that the first prebiotic energy-harnessing reactions involved SLP [41, 46, 47] but that the first free-living cells were chemiosmotic. In that view, life arose where a constant source of reactive carbon compounds was available, but the ability to harness geochemically generated chemiosmotic gradients with the help of proteins was pivotal for the emergence of free-living cells [40]. Where would geochemically generated ion gradients come from? Mike Russell and coworkers suggested that hydrothermal vents could create pH gradients [48–50]. Martin and Russell [50] suggested that those natural pH gradients could have been used by the precursors of the first free-living cells. That is, the ATPase arose in the common ancestor of all prokaryotes, which in turn arose in and inhabited an environment where ion gradients were generated by geochemical processes. That would help explain why the rotor-stator type ATPase is as universal among cells as the ribosome and the genetic code [40, 50], whereas respiratory chains [51, 52] and the many other mechanisms that cells use to generate ion gradients across membranes are as varied as the hundreds of environmental redox couples that modern life forms harness [2].

All this points to two kinds of energy at life's origin: high energy compounds and chemiosmosis. In terms of environments, what on Earth could provide both chemical reactivity and chemiosmotic gradients? Since their discovery, submarine hydrothermal vents have attracted intense interest in this context because they harbor geological manifestations of both kinds of energy that are used by life: chemically reactive compounds [53, 54] and natural proton gradients [48, 49]. In addition, hydrothermal vents reside in the crust and are thus chock full of catalytic transition metals [49, 53].

They furthermore generate vast networks of inorganic microcompartments that i) provide a natural mechanism to concentrate any organic compounds that might have been formed early on, rendering the steep hurdles en route to chemical complexity more readily surmountable [50] and ii) provide a system of naturally formed inorganic territories within which the first replicating systems, once they arose, could have existed and competed in the form of the organic contents of those territories [23].

Indeed, the closer we look at hydrothermal vents, the stronger their similarities with biological energy conversions become [55]. Both chemical reactivity and ion gradients within hydrothermal vents come from the process of serpentinization [56–59]. During serpentinization, seawater circulating through hydrothermal systems reacts with Fe^{2+} in the submarine crust; Fe^{2+} reduces water to H_2, generating up to 50 mM H_2 in vent effluents (and Fe^{3+} in the crust). At the same time, CO_2 is reduced to methane and formate, which occur at 1 mM and 0.1 mM concentrations in the effluent of Lost City (14), a low-temperature hydrothermal vent discovered in 2000 by Deborah Kelly and her colleagues at the University of Washington [60–62]. Serpentinization and its accompanying CO_2 reduction are energy releasing geochemical reactions [58, 59, 63]. And chemiosmosis? The process of serpentinization not only generates a strongly reducing environment at Lost City, it also makes the effluent alkaline [56]. Lost City effluent has a pH of about 10 [58–60, 63], far more alkaline on the inside than ocean water, either now or 4 billion years ago, making these vents naturally chemiosmotic [48, 49]. The natural proton gradients at Lost City are the same in their magnitude and orientation as those in modern autotrophic cells.

A number of ideas about energy-releasing reactions at life's origins have been proposed that have nothing in common with how life actually works. Among such suggestions are pyrite synthesis [64], UV radiation [43], lightning [65], or NiS-based hydrogen generation [66], but none of those processes actually operates in the energy metabolism of modern cells. By contrast, the geochemical synthesis of methane at Lost City [63] and other serpentinizing systems [67], represents a spontaneously occurring geochemical reaction that appears to be homologous to a biological mechanism of energy metabolism: the reduction of CO_2 to methane [55]. That is a strong reason in favour of alkaline hydrothermal vents like Lost City as particularly interesting as models for the origin of early life.

This is especially true when we look closer at the acetyl-CoA pathway in an evolutionary context. The acetyl-CoA pathway consists of two segments: methyl synthesis from H_2 and CO_2 and acetyl synthesis from the methyl moiety and CO [25]. Curiously, the acetyl synthesis segment is highly conserved across the acetogen-methanogen (bacterial-archaeal) divide and entails exclusively transition metals as catalysts. While the methyl synthesis segment is conserved within acetogens and within methanogens, but across the bacterial-archaeal divide the enzymes of the methyl segment are not related, having arisen independently in the common ancestors of the two groups [47]. This circumstance suggests that the acetyl-CoA pathway, while being the most ancient of known CO_2 assimilation pathways, reflects two phases in early evolution: an ancient phase in a geochemically confined and non-free-living universal common ancestor, in which acetyl thioester synthesis proceeded spontaneously with the help of geochemically supplied methyl groups, and a later phase that reflects the primordial divergence of the bacterial and archaeal stem groups, which independently invented genetically-encoded means to synthesize methyl groups via enzymatic reactions. Sustained spontaneous synthesis of thioesters would be a ready source for acyl phosphates and SLP [41, 46] and it is, at least in principle, possible that processes of that type, as sketched in ▶ Fig. 9.1 A, gave rise to genes and proteins [68].

The simplest interpretation of the circumstance that methyl synthesis in acetogens and methanogens differs is that the two ur-lineages of prokaryotes diverged before methyl synthesis had been invented and the pathways arose independently prior to the origin of free-living cells. But the ATPase of bacteria (F-type) and archaea (A-type) is conserved [37], indicating that it was present in their common ancestor as a means of harnessing naturally existing ion gradients and converting their geochemically generated energy into biochemical currency as ATP (▶ Fig. 9.1 B). The function of the ATPase required the thickness of a biological lipid bilayer membrane, but those earliest hydrophobic layers need not have been the products of genetically encoded enzymatic pathways, because Fischer-Tropsch type synthesis at hydrothermal vents can generate hydrophobic compounds [58]. Also, membranes are more porous to protons than they are to sodium ions, especially in environments where short chain organic acids abound. Thus, an early event would have been the conversion of the geochemical proton gradient into a sodium gradient, with a simple antiporter [40], as sketched in ▶ Fig. 9.1 C. The first ATPases were either sodium-utilizing [69], or more likely promiscuous for protons and sodium, like modern sodium-utilizing ATPases [70]. That leaves as the last step en route to carbon and energy autonomy, the origin of the first pumping reactions. Though some will argue that pumping mechanisms that are dependent upon cytochromes, quinones, and high potential oxidants generated by lightning came first [71], it seems more likely to us that cytochrome-independent pumping mechanisms operating under highly reducing conditions came first. Congruent with our other inferences, so far, there are acetogens and methanogens that have cytochrome-independent pumping mechanisms.

In acetogens that lack cytochromes, *Acetobacterium woodii* [31] serving as an example, ion pumping is achieved with the help of a single protein complex called Rnf [72] that pumps sodium ions from the inside of the cell to the outside in the process of transferring electrons in an exergonic reaction from a low-potential reduced ferredoxin to NAD^+ [73]. The evolution of such a complex would have paved the way to the free-living lifestyle (▶ Fig. 9.1 E). Neither cytochromes nor quinones are involved in this pumping. The reduced ferredoxin required at the Rnf reaction is generated with electrons from H_2 with the help of flavin-based electron bifurcation at the iron only hydrogenase used by *A. woodii*: the electron pair of H_2 is split (bifurcated), one electron going energetically uphill to generate the low-potential reduced ferredoxin, the other electron going energetically downhill to NAD^+ [74] so that the overall reaction

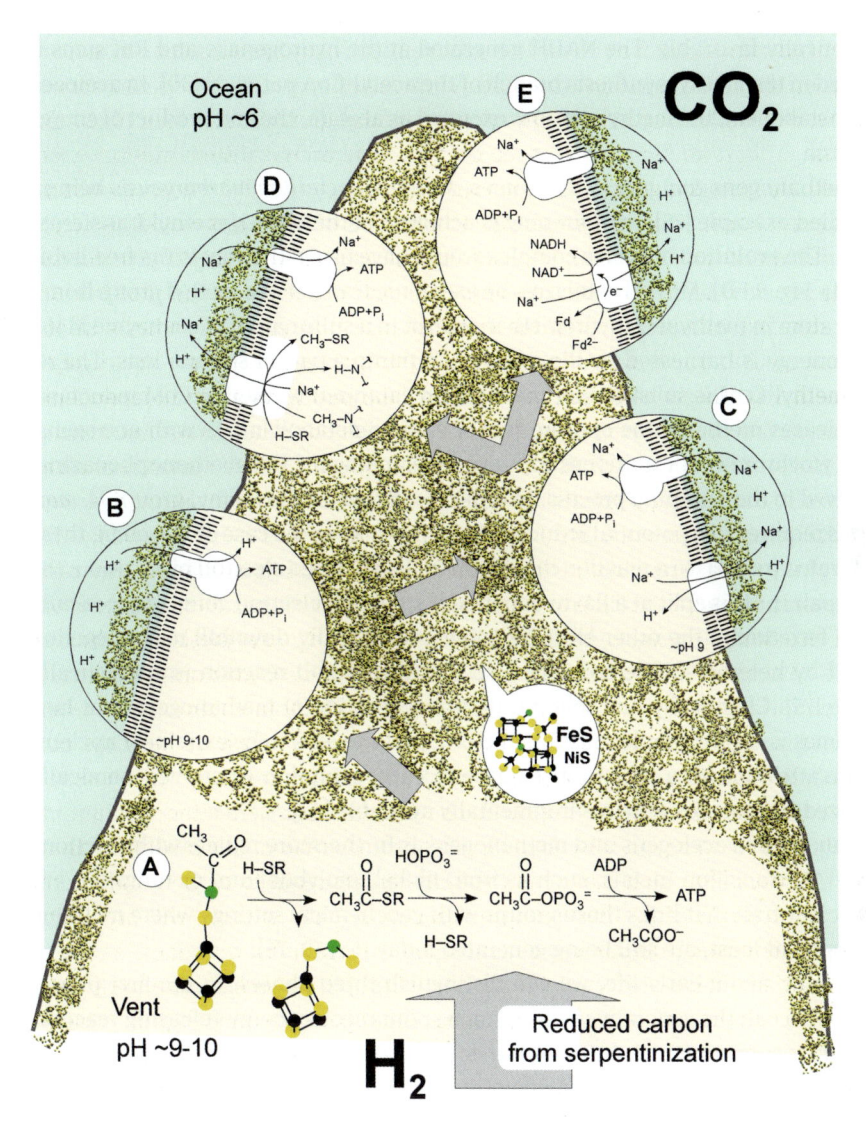

Fig. 9.1. Possible early bioenergetic evolution in an alkaline hydrothermal vent. **A** Substrate-level phosphorylation is thermodynamically feasible under mild hydrothermal conditions [2]. **B** The pH difference between ocean and vent fluids results in a stable, geochemically sustained pH gradient of the polarity and magnitude utilized by modern cells. With the advent of translation and proteins, this could have been harnessed by the universal rotor-stator type ATPase. **C** Membranes are far more permeable to protons than sodium ions. An antiporter, as found in methanogens and acetogens, could convert the geochemical proton gradient into a sodium gradient [40]. The promiscuity of some ATPases for H^+ and Na^+ [70] permits continuity of function. **D, E** The origin of biological ion pumping. In primitive methanogens **D**, Na^+ pumping is powered by the exergonic transfer of a methyl group from a nitrogen atom in methanopterin to a sulfur atom in coenzyme M [6]. In primitive acetogens **E**, Na^+ pumping is powered by the exergonic transfer of electrons from iron-sulfur clusters in ferredoxin to NAD^+ [73]. Pumping reactions of this type involve simple chemicals and might be ancient.

is energetically favorable. The NADH generated at the hydrogenase and Rnf steps is reoxidized in the methyl synthesis branch of the acetyl-CoA pathway [30]. In acetogen energy metabolism, the methyl group is excreted as acetate, the end product of energy metabolism.

In methanogens that lack cytochromes, *Methanobacterium marburgensis* being a well-studied example [75], ion pumping is achieved at the MtrA-H methyl transferase complex. The evolution of such a complex would have paved the way to the free-living lifestyle (▶ Fig. 9.1 **D**). MtrA-H catalyzes the exergonic transfer of a methyl group from a nitrogen atom in methyltetrahydromethanopterin to a sulfur atom in coenzyme M [6], the free energy is harnessed by the complex to pump a pair of sodium ions. The resulting methyl-CoM is substrate for the reaction catalyzed by methyl-CoM reductase, which releases methane, the end product of energy metabolism. As with acetogens, neither cytochromes nor quinones (nor quinone analogues like methanophenazine) are involved in the pumping process. In order to synthesize the methyl group, *M. marburgensis* requires low potential reduced ferredoxins. As in the case of *A. woodii*, these are generated from H_2 in a reaction that involves flavin-based electron bifurcation: the electron pair in H_2 is split at a flavin dependent step, one electron going energetically uphill to ferredoxin, the other electron going energetically downhill to the reaction catalysed by heterodisulfide reductase, so that the overall reaction is energetically favorable [75]. Chemically, the sole ion pumping reaction of methanogens that lack cytochromes is a very simple reaction and one that appears to be extremely ancient; it has the attributes of a relic from a phase in early evolution where geochemically synthesized C1 moieties were environmentally available [47].

Metabolism in acetogens and methanogens is furthermore replete with reactions catalyzed by transition metals, such as iron, nickel, molybdenum, or tungsten, another ancient trait that links these groups with geochemical settings where methane and formate, at least, are still being generated today [58, 59, 76].

In talking about early life, we can distinguish three phases. In the first phase, which we can call the generative phase, some spontaneous energy-releasing reaction was going on somewhere in the environment, and this reaction led to the accumulation of reduced carbon compounds, the building blocks of life. This reaction fostered and financed, in the energetic sense, the synthesis of molecular constituents and their organization into the first free-living cells. In the second phase, which we can call emergence, there were free-living cells that were able to foster, by virtue of genetic instructions, their own organization into likenesses of themselves, and energetically finance that organization by tapping environmentally available energy sources with the help of chemical tools that they synthesized by themselves from CO_2, electrons, and nitrogen. That is a way of saying that in early bioenergetic evolution, carbon and energy conversion using only inorganic or spontaneously-formed catalysts was followed by energy conversion using proteins [47]. In the third phase of early life, free-living cells began to diversify and harness environmentally available redox couples.

If acetogens and methanogens are the founders of the bacterial and archaeal lineages, what would be the next evolutionary steps after the origin of free-living cells? The evolution of cytochromes, quinones, and sulfur reduction as manifest in both high and low cytochrome c-containing sulfate reducers [77, 78] would be one possible next step [77], not only because they are strict anaerobes [13], but also because many of them can grow autotrophically, using the acetyl-CoA pathway for carbon metabolism and sulfur reduction for energy metabolism. Moreover, the similarity of subunits of the dissimilatory sulfite reductase module (DsrMK) and the quinone membrane-bound oxidoreductase complex (QmoABC) with key enzymes involved in the last step of methanogenesis tightens their close relationship. DsrMK and QmoABC are highly conserved complexes among sulfate reducers and both interact with quinones that have heme b in their membranar subunit [77]. Nevertheless, the cytoplasmatic DsrK subunit is homologous to the catalytic HdrD subunit of membranar heterodisulfide reductase HdrED present in methanogens [79] while the soluble QmoA and B subunits have homology with the soluble heterodisulfide reductase subunit HdrA [77]. However it is not yet clear whether archaea and bacteria independently evolved sulfate reduction pathways, or whether one lineage evolved the pathway with subsequent distribution to other lineages. Once standard bioenergetic electron transport chains using cytochromes and quinones were in place, then the evolution of electron accepting complexes, like complex I [52], and various terminal oxidases would follow [80, 81].

It has been suggested that electrochemical gradients at the vent-ocean interface might have been required for early CO_2 reduction [40], but Schuchmann and Müller [82] recently characterized a soluble molybdoenzyme with several FeS clusters from *A. woodii* that reduces CO_2 with H_2 to formate: a hydrogen-dependent carbon dioxide reductase [83]. This is a clear biological demonstration that the concerns expressed by some that there might not be enough energy in H_2 to reduce CO_2 [84, 85] are not founded. The enzymatic reaction shows that CO_2 reduction with H_2 is a matter of limited catalysis, not limited free energy release.

What do phylogenetics say about these issues? Newer trees of early evolution show the archaeal component of the eukaryotes (i.e. information processing systems) branching within the archaea, not as their sisters [17, 86–89]. Furthermore, although those studies – with the exception of one [86] – were aimed specifically at determining the position of the archaeal component of eukaryotes among the archaea, they had to employ bacterial outgroups in order to root the trees. If we look at the position of the bacterial root within the archaeal tree in those analyses, what we see is the nature of the most ancient archaeal lineages emerging from those analyses: methanogens are very close to the root [88], methanogens are on the first branch emerging from the root [17], or the archaeal tree roots within the methanogens, specifically within the hydrogenotrophic forms [86, 87] that lack cytochromes and quinones, as some views of early microbial evolution would have it [13, 39, 40, 46]. Despite many uncertainties [90], the foregoing is one possible route that early life might have taken.

References

[1] R. K. Thauer, K. Jungermann, K. Decker. Energy-conservation in chemotropic anaerobic bacteria. Bacteriol Rev 1977, 41, 100–180.

[2] J. P. Amend, E. L. Shock. Energetics of overall metabolic reactions of thermophilic and hyperthermophilic Archaea and Bacteria. FEMS Microbiol Rev 2001, 25, 175–243.

[3] L. D. Hansen, R. S. Criddle, E. H. Battley. Biological calorimetry and the thermodynamics of the origination and evolution of life. Pure Appl Chem 2009, 81, 1843–1855.

[4] A. H. Stouthamer. Energy-yielding pathways. In: The Bacteria Vol VI: Bacterial diversity, vol 6 Bacterial Diversity (ed. I. C. Gunsalus) Academic Press, New York. 1978

[5] G. Herrmann, J. Elamparithi, M. Galina, W. Buckel. Energy conservation via electron-transferring flavoprotein in anaerobic bacteria. J Bacteriol 2008, 190, 784–791.

[6] R. K. Thauer, A. K. Kaster, H. Seedorf, W. Buckel, R. Hedderich, Methanogenic archaea: ecologically relevant differences in energy conservation. Nature Rev Microbiol 2008, 6, 579–591.

[7] N. T. Arndt, E. G. Nisbet. Processes on the young Earth and the habitats of early life. Annu Rev Earth Planet Sci 2012, 40, 521–549.

[8] Y. Ueno, K. Yamada, N. Yoshida, S. Maruyama, Y. Isozaki. Evidence from fluid inclusions for microbial methanogenesis in the early archaean era. Nature 2006, 440, 516–519.

[9] N. V. Grassineau, P. Abell, P. W. U. Appel, D. Lowry, E. G. Nisbet. Early life signatures in sulfur and carbon isotopes from Isua, Barberton, Wabigoon (Steep Rock), and Belingwe greenstone belts (3.8 to 2.7 ga). Geol Soc Am Mem 2006, 198, 33–52.

[10] J. E. Johnson, S. M. Webb, K. Thomas, S. Ono, J. L. Kirschvink, W. W. Fischer. Manganese-oxidizing photosynthesis before the rise of cyanobacteria, Proc Natl Acad Sci USA, 2013, 108, 11238–11243.

[11] E. G. Nisbet, N. H. Sleep. The habitat and nature of early life. Nature 2001, 409, 1083–1091.

[12] W. Martin. On the ancestral state of microbial physiology. In: R. Amann, W. Goebel, B. Schink, F. Widdel (Ed) Life Strategies of Microorganisms in the Environment and in Host Organisms. Nova Acta Leopoldina 2008, 96, 53–60.

[13] K. Decker, K. Jungerman, R. K. Thauer. Energy production in anaerobic organisms. Angew Chem Int Ed 1970, 9, 138–158.

[14] R. V. Eck, M. O. Dayhoff. Evolution of the structure of ferredoxin based on living relics of primitive amino acid sequences. Science 1966, 152, 363–366.

[15] T. M. Embley, W. Martin: Eukaryotic evolution, changes and challenges. Nature 2006, 440, 623–630.

[16] N. Lane, W. Martin. The energetics of genome complexity. Nature 2010, 467, 929–934.

[17] T. Williams, P. G. Foster, C. J. Cox, T. M. Embley. An archaeal origin of eukaryotes supports only two primary domains of life. Nature 2013, 504, 231–236.

[18] J. O. McInerney, M. O'Connell, D. Pisani. The hybrid nature of the eukaryota and a consilient view of life on Earth. Nature Rev Microbiol 2014, 12, 449–455.

[19] G. Wächtershäuser. Groundworks for an evolutionary biochemistry – The iron-sulfur world. Prog Biophys Mol Biol 1992, 58, 85–201.

[20] B. E. H. Maden. No soup for starters? Autotrophy and the origins of metabolism. Trends Biochem Sci 1995, 20, 337–341.

[21] D. O. Hall, R. Cammack, K. K. Rao. Role of ferredoxins in the origin of life and biological evlution. Nature 1971, 233, 136–138.

[22] T. Dagan, M. Roettger, D. Bryant, W. Martin. Genome networks root the tree of life between prokaryotic domains. Genome Biol Evol 2010, 2, 379–392.

[23] E. V. Koonin, W. Martin. On the origin of genomes and cells within inorganic compartments. Trends Genet 2005, 21, 647–654.

[24] V. Sojo, A. Pomiankowski, N. Lane. A bioenergetic basis for membrane divergence in Archaea and Bacteria. PloS Biology 2014, 12, e1001926.
[25] G. Fuchs. Alternative pathways of carbon dioxide fixation: insights into the early evolution of life? Annu Rev Microbiol 2011, 65, 631–658.
[26] I. A. Berg, D. Kockelkorn, W. H. Ramos-Vera, S. F. Say, J. Zarzycki, M. Hugler, B. E. Alber, G. Fuchs. Autotrophic carbon fixation in archaea. Nature Rev Microbiol 2010, 8, 447–460.
[27] G. Fuchs, E. Stupperich. Evolution of autotrophic CO_2 fixation. In: Evolution of prokaryotes (eds KH Schleifer & E Stackebrandt) FEMS Symposium. London, UK: Academic Press, 1985, 29, 235–251.
[28] G. Fuchs. Alternative pathways of autotrophic CO_2 fixation. In: Autotrophic Bacteria, (eds Schlegel, H.G. & Bowien, B.) Madison, WI, Science Tech, 1989, 365–382.
[29] S. W. Ragsdale, E. Pierce. Acetogenesis and the Wood-Ljungdahl pathway of CO_2 fixation. Biochim Biophys Acta 2008, 1784, 1873–1898.
[30] W. Buckel, R. K. Thauer. Energy conservation via electron bifurcating ferredoxin reduction and proton/Na^+ translocating ferredoxin oxidation. Biochim Biophys Acta 2013, 1827, 94–113.
[31] A. Poehlein, S. Schmidt, A. K. Kaster, M. Goenrich, J. Vollmers, A. Thürmer, J. Bertsch, K. Schuchmann, B. Voigt, M. Hecker, D. Daniel, R. K. Thauer, G. Gottschalk, V. Müller. An ancient pathway combining carbon dioxide fixation with the generation and utilization of a sodium ion gradient for ATP synthesis. PLoS One 2012, 7, e33439.
[32] M. A. Lever et al. Acetogenesis in deep subseafloor sediments of the Juan de Fuca Ridge flank: a synthesis of geochemical, thermodynamic, and gene-based evidence. Geomicrobiol J 2010, 27, 183–211.
[33] M. A. Lever. Frontiers Acetogenesis in the energy-starved deep biosphere. Frontiers Microbiol 2012, 2, 284.
[34] Takami et al. A deeply branching thermophilic bacterium with an ancient acetyl-CoA pathway dominates a subsurface ecosystem. PLoS one 2008, 7, e30559.
[35] F. H. Chapelle, K. O'Neill, P. M. Bradley, B. A. Methé, S. A. Ciufo, L. L. Knobel, D. R. Lovley. A hydrogen-based subsurface microbial community dominated by methanogens. Nature 2002, 415, 312–315.
[36] D. Chivian, E. L. Brodie, E. J. Alm et al. Environmental genomics reveals a single-species ecosystem deep within earth. Science 2008, 322, 275–278.
[37] F. Mayer, V. Müller. Adaptations of anaerobic archaea to life under extreme energy limitation. FEMS Microbiol Rev 2013, DOI: 10.1111/1574-6976.12043.
[38] T. M. Hoehler, B. B. Jörgensen. Microbial life under extreme energy limitation. Nat Rev Microbiol 2013,11, 83–94.
[39] W. F. Martin. Hydrogen, metals, bifurcating electrons, and proton gradients: The early evolution of biological energy conservation. FEBS Lett 2012, 586, 485–493.
[40] N. Lane, W. F. Martin. The origin of membrane bioenergetics. Cell 2012, 151, 1406–1416.
[41] J. G. Ferry, C. H. House. The stepwise evolution of early life driven by energy conservation. Mol Biol Evol 2006, 23, 1286–1292.
[42] P. Mitchell. Coupling of phosphorylation to electron and hydrogen transfer by a chemi-osmotic type of mechanism. Nature 1961, 191, 144–148
[43] A. I. Oparin, J. B. S. Haldane. The origin of life. Rationalist Annual 1929, 3, 3–10.
[44] R. F. Say, G. Fuchs. Fructose 1,6-bisphosphate aldolase/phosphatase may be an ancestral gluconeogenic enzyme. Nature 2010, 464, 1077–1081.
[45] N. Lane, J. F. Allen, W. Martin. How did LUCA make a living? Chemiosmosis in the origin of life. BioEssays 2010, 32, 271–280.
[46] W. Martin, M. J. Russell. On the origin of biochemistry at an alkaline hydrothermal vent. Phil Trans Roy Soc Lond B 2007, 367, 1887–1925.

[47] F. Sousa, W. F. Martin. Biochemical fossils of the ancient transition from geoenergetics to bioenergetics in prokaryotic one carbon compound metabolism. Biochim Biophys Acta 2014, 1837, 964–981.

[48] M. J. Russell, R. M. Daniel, A. J. Hall, J. Sherringham. A hydrothermally precipitated catalytic iron sulphide membrane as a first step toward life. J Mol Evol 1994, 39, 231–243.

[49] M. J. Russell, A. J. Hall. The emergence of life from iron monosulphide bubbles at a submarine hydrothermal redox and pH front. J Geol Soc Lond 1997, 154, 377–402.

[50] W. Martin, M. Russell. On the origins of cells: a hypothesis for the evolutionary transitions from abiotic geochemistry to chemoautotrophic prokaryotes, and from prokaryotes to nucleated cells. Phil Trans Roy Soc Lond B 2003, 358, 59–85.

[51] F. Baymann, E. Lebrun, M. Brugna, B. Schoepp-Cothenet, M. T. Giudici-Orticoni, W. Nitschke. The redox protein construction kit: pre-last universal common ancestor evolution of energy-conserving enzymes. Phil Trans Roy Soc Lond 2003, 358B, 267–274.

[52] B. C. Marreiros, A. P. Batista, A. M. S. Duarte, M. M. Pereira. A missing link between complex I and group 4 membrane-bound [NiFe]-hydrogenases, BBA – Bioenergetics 2013, 1827, 198–209.

[53] J. A. Baross, S. E. Hoffman. Submarine hydrothermal vents and associated gradient environments as sites for the origin and evolution of life. Origins Life Evol Biosph 1985, 15, 327–345.

[54] D. S. Kelley, J. A. Baross, J. R. Delaney. Volcanoes, fluids, and life at mid-ocean ridge spreading centers. Annu Rev Earth Planet Sci 2002, 30, 385–491.

[55] W. Martin, J. Baross, D. Kelley, M. J. Russell. Hydrothermal vents and the origin of life. Nature Rev Microbiol 2008, 6, 805–814.

[56] W. Bach, H. Paulick, C. J. Garrido, B. Ildefonse, W. P. Meurer, S. E. Humphris. Unraveling the sequence of serpentinization reactions: petrography, mineral chemistry, and petrophysics of serpentinites from MAR 15°N (ODP Leg 209, Site 1274), Geophys Res Lett 2006, 33, L13306.

[57] M. J. Russell, A. J. Hall, W. Martin. Serpentinization as a source of energy at the origin of life. Geobiol 2010, 8, 355–371.

[58] T. M. McCollom, J. S. Seewald, Serpentinites, hydrogen, and life. Elements 2013, 9, 129–134.

[59] M. O. Schrenk, W. J. Brazelton, S. Q. Lang. Serpentinization, carbon and deep life. Rev Mineral Geochem 2013, 75, 575–606.

[60] D. S. Kelley, J. A. Karson, D. K. Blackman, G. L. Fruh-Green, D. A. Butterfield, M. D. Lilley, E. J. Olson, M. O. Schrenk, K. K. Roe, G. T. Lebon, P. Rivizzigno. An off-axis hydrothermal vent field near the Mid-Atlantic Ridge at 30°N. Nature 2001, 412, 145–149.

[61] D. S. Kelley, J. A. Karson, G. L. Früh-Green et al. A serpentinite-hosted ecosystem: The Lost City hydrothermal field. Science 2005, 307, 1428–1434.

[62] G. L. Früh-Green, D. S. Kelley, S. M. Bernasconi, J. A. Karson, K. A. Ludwig, D. A. Butterfield, C. Boschi. 30,000 years of hydrothermal activity at the Lost City vent field. Science 2003, 301, 495–498.

[63] G. Proskurowski, M. D. Lilley, J. S. Seewald, G. L. Früh-Green, E. J. Olson, J. E. Lupton, S. P. Sylva, D. S. Kelley. Abiogenic hydrocarbon production at Lost City Hydrothermal Field. Science 2008, 319, 604–607.

[64] G. Wächtershäuser. Pyrite fomation, the first energy source for life: A hypothesis. Syst Appl Microbiol 1988, 10, 207–210.

[65] J. L. Bada. How life began on Earth: a status report. Earth Planet Sci Lett 2004, 226, 1–15.

[66] A. Y. Mulkidjanian, A. Y. Bychkov, D. V. Dibrova, M. Y. Galperin, E. V. Koonin. Origin of first cells at terrestrial, anoxic geothermal fields. Proc Natl Acad Sci USA 2012, 10.1073/pnas.1117774109.

[67] G. Etiope, M. Schoell, H. Hosgörmez. Abiotic methane flux from the Chimaera seep and Tekirova ophiolites (Turkey): Understanding gas exhalation from low temperature serpentinization and implications for Mars. Earth Planet Sci Lett 2011, 310, 96–104.

[68] F. L. Sousa, T. Thiergart, G. Landan, S. Nelson-Sathi, I. A. C. Pereira, J. F. Allen, N. Lane, W. F. Martin. Early bioenergetic evolution. Phil Trans R Soc Lond B 2013, 368, 20130088.

[69] A. Y. Mulkidjanian, M. Y. Galperin, K. S. Makarova, Y. I. Wolf, E. V. Koonin. Evolutionary primacy of sodium bioenergetics. Biol Direct 2008, 3, 13.

[70] K. Schlegel, V. Leone, J. D. Faraldo-Gómez, V. Müller. 2012 Promiscuous archaeal ATP synthase concurrently coupled to Na^+ and H^+ translocation. Proc Natl Acad Sci USA 2012, 109, 947–952.

[71] W. Nitschke, M. J. Russell. Beating the acetyl coenzyme A-pathway to the origin of life. Phil Trans Roy B. Soc-Biol Sci 2013, 368, 1622, 20120258.

[72] E. Biegel, S. Schmidt, J. M. Gonzalez, V. Müller. 2011. Biochemistry, evolution and physiological function of the Rnf complex, a novel ion-motive electron transport complex in prokaryotes. Cell Mol Life Sci 2011, 68, 613–634.

[73] E. Biegel, V. Müller. Bacterial Na^+-translocating ferredoxin: NAD^+ oxidoreductase. Proc Natl Acad Sci USA 2010, 107, 18138–18142.

[74] K. Schuchmann, V. Müller. A bacterial electron-bifurcating hydrogenase. J Biol Chem 2012, 287, 31165–31171.

[75] A. K. Kaster, J. Moll, K. Parey, R. K. Thauer. Coupling of ferredoxin and heterodisulfide reduction via electron bifurcation in hydrogenotrophic methanogenic Archaea. Proc Natl Acad Sci USA 2011, 108, 2981–6298.

[76] S. Q. Lang, D. A. Butterfield, M. Schulte, D. S. Kelley, M. D. Lilley. Elevated concentrations of formate, acetate and dissolved organic carbon found at the Lost City hydrothermal field. Geochim Cosmochim Ac 2010, 74, 941–952.

[77] I. A. C. Pereira, A. R. Ramos, F. Grein, M. C. Marques, S. M. da Silva, S. S. Venceslau. A comparative genomic analysis of energy metabolism in sulfate reducing bacteria and archaea. Front Microbiol 2011, 2, 69.

[78] Romão CV, M. Archer, S. A. Lobo, R. O. Louro, I. A. C. Pereira, L. M. Saraiva, M. Teixeira, P. M. Matias. Diversity of heme proteins in sulfate reducing bacteria. In: Handbook of Porphyrin Science – With Applications to Chemistry, Physics, Materials Science, Engineering, Biology and Medicine, Singapore, World Scientific Publishing Co., Pte. Ltd., Singapore, 2012,139–230.

[79] R. H. Pires, S. S. Venceslau, F. Morais, M. Teixeira, A. V. Xavier, I. A. Pereira. Characterization of the *Desulfovibrio desulfuricans* ATCC 27774 DsrMKJOP complex – A membrane-bound redox complex involved in the sulfate respiratory pathway. Biochemistry 2006, 45, 249–262.

[80] G. Schäfer, M. Engelhard, V. Müller. Bioenergetics of the Archaea. Microbiol Mol Biol Rev 1999, 63, 570–620.

[81] M. M. Pereira, M. Santana, M. Teixeira. A novel scenario for the evolution of haem-copper oxygen reductases. Biochim Biophys Acta 2001, 1505, 185–208.

[82] K. Schuchmann, V. Müller. Direct and reversible hydrogenation of CO_2 to formate by a bacterial carbon dioxide reductase. Science 2013, 342, 1382–1385.

[83] I. A. C. Pereira. An enzymatic route to H_2 storage. Science 2013, 342, 1329–1330.

[84] G. Wächtershäuser. Before enzymes and templates: Theory of surface metabolism. Microbiol Rev 1988b, 52, 452–484.

[85] E. Branscomb, M. J. Russell. Turnstiles and bifurcators: The disequilibrium converting engines that put metabolism on the road. BBA-Bioenergetics 2013, 1827, 62–78.

[86] S. Kelly, B. Wickstead, K. Gull. Archaeal phylogenomics provides evidence in support of a methanogenic origin of the archaea and a thaumarchaeal origin for the eukaryotes. Proc Roy Soc Lond B 2011, 278, 1009–1018.

[87] T. A. Williams, T. M. Embley. Archaeal "Dark Matter" and the origin of eukaryotes Genome Biol Evol 2014, 6, 474–481.
[88] N. Yutin, Puigbò P, E. V. Koonin, Y. I. Wolf. Phylogenomics of prokaryotic ribosomal proteins. PLoS One 2012, 7(5), e36972.
[89] L. Guy, J. H. Saw, T. J. Ettema. The archaeal legacy of Eukaryotes: A phylogenomic perspective. Cold Spring Harb Perspect Biol. 2014, doi: 10.1101/cshperspect.a016022
[90] W. F. Martin, F. L. Sousa, N. Lane. Energy at life's origin. Science 2014, 344, 1092–1093.

Cene Gostinčar, Nina Gunde-Cimerman, and Martin Grube
10 Polyextremotolerance as the fungal answer to changing environments

10.1 Introduction

It has been just a few decades ago, when extreme environments were thought of as virtually sterile. Physicochemical conditions in these environments are so different from those that are tolerable for a majority of organisms that extreme environments were considered unable to support metabolism, growth, or development of most organisms. This view has changed substantially after it became clear that species tolerating such conditions comprise a variety of archaea and bacteria, but also of eukaryotes. Some fungal species have specialized for survival in a set of similar habitats, either with constant extreme conditions or with drastically changing conditions. Others have evolved into highly adaptable and versatile species, which are able to tolerate a wide variety or combinations of different stressors. Such polyextremotolerant fungi may colonize a broad range of ecologically similar habitats, often by using the same cellular mechanisms to survive conditions such as desiccation, high radiation levels, acidity or salinity, or low nutrient availability. They also seize the opportunities, which emerged with the ubiquity of anthropogenic habitats, leading to their frequent occurrence in and around our homes, with potentially negative consequences to human well-being.

We review some of the possible mechanisms contributing to the pronounced persistence of polyextremotolerant fungi, with special reference to the black yeast *Aureobasidium pullulans*. This fungus is a potential model organism for studies of polyextremotolerance and is a good representative of the other species discussed here. We discuss the genomic footprint of extremotolerance that enables polyextremotolerant species to efficiently colonize the varied habitats created by humans and, in certain unfortunate circumstance, the human body.

10.2 Extremes in nature

Snow, glacial ice, and sea ice support active growth and reproduction of microbes [1–4], including fungi which have been isolated in considerable numbers from subglacial ice of polythermal glaciers [5, 6]. Similar fungi were found in habitats of hypersaline water, such as salterns [7–9]. The isolated species are characterised by extensive and complex molecular adaptations to low water activities and high concentrations of toxic ions (for reviews see e.g. [10–12]).

In many cases, shared patterns of morphology, phylogeny, and population characteristics exist between fungi from extreme environments (e.g. extremely cold or salty [13]). These patterns can be explained by several factors. On the one hand, there is a large overlap of the mechanisms of tolerance to different kinds of stress in polyextremotolerant fungi, resulting in the occurrence of individuals representing the same species in many different extreme environments. On the other hand, the fragmentation and isolation of extreme habitats at the geographical or ecological peripheries creates small populations, with specific selection pressures, resulting in populations with various degrees of adaptation and specialization. These peripheral populations can shelter unique adaptations [14, 15].

Possibly as a result of the conditions described above, a large diversity of closely related fungal species has been reported from stressful and often quite unusual environments [13, 16]. Some species are specialized, while others retained (different degrees of) flexibility in also inhabiting other habitats. A good example of the latter is *A. pullulans*, an osmotolerant black yeast [17] with substantial biotechnological importance due to its production of pullulan [18, 19]). *A. pullulans* is widespread and frequently found in the phyllosphere, in hypersaline and fresh water, in glacial habitats, and many other habitats, including indoors [7, 20, 21]. *A. pullulans* has a relatively broad amplitude of ecological tolerance, and can survive hypersaline [7], acidic, and basic [22, 23] conditions, as well as low temperatures and low nutrient concentrations [24]. Although in recent years, it has become clear that the degree of specialization within the species is larger than previously thought, good stress tolerance is (with some differences between the strains) a characteristic of all taxonomic groups to which the species was divided. A multilocus phylogenetic comparison of *A. pullulans*-like isolates has revealed the existence of at least four clusters in the species. These were initially described as *A. pullulans* varieties [21], but have been raised to the species level on the basis of genome sequencing [25]. Despite their close phylogenetic proximity, three of these taxa exhibit different ecological preferences (because only one isolate of *Aureobasidium namibiae* is known to date, no trends can be established for this taxon). *A. pullulans* s.str. mostly occurs in mildly osmotic and plant associated habitats and *Aureobasidium melanogenum* in freshwater habitats, although in both cases this preference is far from exclusive. *Aureobasidium subglaciale* on the other hand was only isolated from subglacial ice and the immediate surroundings of glaciers. While the ecological preferences of *A. subglaciale* appear to be very narrowly specialized, *A. pullulans* and *A. melanogenum* retain a large share of the generalist character which was attributed to *A. pullulans* before it was split into four separate taxa. Due to a recent description of the taxa, most published works do not yet distinguish between them, therefore in the rest of the chapter they will be discussed under the name of *A. pullulans* s.l. unless explicitly stated otherwise.

Not all fungi from extreme environments are ubiquitous. Some of them occur only in a very narrow set of habitats. A good example is the black yeast-like fungus *Hortaea werneckii*. It is one of the most halotolerant fungi, with a broad growth optimum from

1.0 M to 3.0 M NaCl [7], and the ability to grow in nearly saturated salt solutions as well as in standard mycological media without additional salt. Hypersaline waters (such as those created by the evaporation of sea water) appear to be its preferred habitat in nature [7, 26]. Even more extreme is *Wallemia ichthyophaga*, the most halophilic fungus known to date [13]. It cannot grow in media with NaCl concentrations lower than 1.5 M. Such a narrow ecological amplitude was previously thought to be a characteristic of archaeal halophiles only, and in the fungal kingdom it still remains an exception. Even the most salt-tolerant fungal species normally grow successfully at salt concentrations that are available in standard mycological media, and any further addition of salt frequently decreases their growth rates. With only two other species in addition to *W. ichthyophaga* in the same genus, order, and class, this group of species is an extremely distinct sister group of *Agaricomycotina* [27, 28]. Despite extensive sampling efforts, so far, only a few isolates of the species are available, thus the occurrence of the species appears to be extremely rare or scattered.

Which factors contribute to the existence of different life strategies that are recognizable in species from extreme habitats? One possibility is that the generalist species represent a pool of genetic resources for the evolution of highly adapted specialists [13]. *In-vitro* studies have shown that generalists are much more likely to diversify than specialists [29]. They commonly persist across varied environments, often on metabolic "low flame" or even in cryptobiosis, without necessarily adapting to local conditions. Initial proteomic analyses of fungal growth under Mars-like conditions indicate that the tested species of fungi do not even show typical stress responses on the level of proteins [30]. Similarly, during desiccation some fungi respond with the production of small proteins, while the extremotolerant *Cryomyces antarcticus* appears to accumulate sugar osmolytes and simply down-regulates its metabolism [31]. Further studies should be performed to improve our understanding of survival in such extreme conditions. For example it would be interesting to know to what extent the cellular proteins remain functional after prolonged desiccation. Are all of them preserved to the same extent or are the ones that are crucial for resuming the metabolism upon the improvement of conditions more stable or even specifically protected?

The distributional limits of generalistic fungal species will frequently grade into more or less extreme environments. Fragmentation of permissive habitats at the ecological edge can create small bottleneck populations [32] that are exposed to severe selection pressures due to extreme physicochemical parameters. At this point, dispersal efficiency will be an important determinant of gene flow between the main and edge populations and will influence the amount of differences that accumulate between them over time. In small populations that have managed to adapt to extreme habitats the importance of genetic drift will also be pronounced. Adaptation at the level of phenotypic plasticity [33] and "ecological fitting" [34] may be aided by the expression of cryptic genetic variation, which is more likely to occur when an organism encounters a rare or novel environment [35] and may prove to have an adaptive value. This will be followed by selection for genetic adaptations [33], and possibly evolutionary radiation [34]. Gene copy number variations in populations appear to contribute to the ecological fitting of fungi [36]. In this light it is interesting to note the great genetic redundancy that was uncovered by the genome sequencing of *A. pullulans* [25]. *H. werneckii* even experienced recent whole-genome duplication [37].

Genetic variation may be provided by sexual reproduction, but avoiding the additional costs of forming sexual structures could be beneficial in nutrient-poor and stressful conditions. No sexual reproduction has been observed for many fungi thriving in hostile habitats. Some of them, like *H. werneckii* [37] and *A. pullulans* [25] – but not all, e.g. *W. ichthyophaga* [28] – maintain recognizable genes presumably encoding proteins of mating systems, however their function has not been proven so far. Since mating loci have been found through genomic sequencing in many species, which were previously thought to be asexual, further experiments should confirm (or deny) the ability of these species to undergo sexual recombination. Laboratory mating of strains and the sequencing of presumable mating loci in a wider selection of wild-type isolates are just two possible approaches.

Even in the absence of sexual recombination genetic variation can be generated through other ways of diversification, either by increasing mutation rates or genome dynamics. Increased mutation rate induced by stress is a general trait of organisms that has been observed in fungi as well [38–40]. The inherited spontaneous mutation rates in *Penicillium lanosum* and *Aspergillus niger* even correlate to the levels of stress in the environment they were isolated from [41]. Several studies have revealed strong positive correlations of diversity and stress at conditions near the edge of life, and it has been suggested that increases in mutation rate, recombination, and gene conversion ensure higher levels of genetic diversity, providing greater potential for genetic adaptation [42, 43]. Another source of genetic variation may be large scale genome restructuring. In the asexual plant pathogen *Verticillium dahliae* extensive chromosomal rearrangements were observed and were frequently associated with retrotransposons [44]. Similar genome restructuring was found in *Fusarium oxysporum* [45]. This is an important finding relevant to adaptation to extreme environments, since many studies report a connection between stress and the activation of transposable elements – not exclusively in fungi, but also in animals and plants [46]. Furthermore, stress-responsive transposable elements may act in *cis* and *trans* to regulate the host stress response. This mechanism, including associated epigenetic processes, renders stress-induced changes in gene expression heritable [46]. In the light of these findings, some authors believe that transposable elements have a beneficial effect on the evolution of their host genome and the environmental adaptation of the host organism [47]. Genome expansions have been observed in *H. werneckii* and *A. pullulans* and genome contractions have been observed in *W. ichthyophaga* [25, 28, 37]. All of these phenomena could be important for rapid adaptation in cases where limited or absent sexual reproduction is advantageous [48].

Such processes may form a basis for the typical behaviour of polyextremotolerant species, which can be described as 'antifragility'. The term originates from the economic realm and marks the ability to gain from uncertain, changing, and stressful events. Biological systems are antifragile when they accumulate functional redundancy [49]. This can happen by maintaining cryptic genetic variation, by canalisation, genetic redundancy, or alternative splicing [35]. Additionally, various stress-related mechanisms for generating genetic diversity can perhaps explain the great diversity of some groups of closely related stress-tolerant species. Two endemic species of *Thelebolus* for example appear to have evolved in the Antarctic, where they lost the ability to undergo sexual reproduction [50].

Strains identified as *Cladosporium sphaerospermum* have been consistently isolated from hypersaline environments around the globe [51]. *C. sphaerospermum* is one of the most common air-borne, cosmopolitan *Cladosporium* species, and it has been frequently isolated from indoor and outdoor air [52] and dwellings [53], and occasionally from humans [54] and plants [55]. Although this species has also been reported from osmotically non-stressed niches, strains that were morphologically identified as *C. sphaerospermum* can grow at a very low water activity (contrary to other species in *Cladosporium* [56]). Molecular analysis has shown that *C. sphaerospermum*-like isolates were actually a complex of species. *C. sphaerospermum* was thus redefined, and seven new species from hypersaline environments have been described. Some of these have to date been found exclusively in hypersaline waters of solar salterns [51]. Similarly, several new species taken exclusively from hypersaline environments have been described for the genera *Eurotium* [57], *Trimmatostroma* [58], *Emericella* [59], and *Phaeotheca* [60].

Such divergence of species has been reported as characteristic of the initial phase of adaptive radiation [61]. The message resulting from studying extreme environments is that these do not harbor only specialized fungal extremophiles, but also polyextremotolerant generalists with an exceptional versatility and adaptability. These species are not hemerophobic and limited to exotic habitats such as saline lakes or Arctic glaciers. They readily occupy domestic habitats, since the essential microecological parameters there are quite similar to those in extreme natural environments, as will be shown below.

How do extremotolerant fungi thrive in the conditions that are detrimental to the majority of other species? Apart from the already mentioned high degree of phenotypic plasticity extremotolerant fungi are able to employ efficient stress-response mechanisms such as the abundance of alkali metal cation transporters and enzymes for degrading a variety of substrates [25, 37]. They also produce an array of protective small molecules [62]. Melanin is the polymer product of one of these small molecules (e.g. dihydroxynaphthalene) and is known to have a role in the ability of melanized fungi to survive excessive heat or cold, extreme pH or osmotic conditions, polychromatic UV-radiation, simulated space and Martian conditions [63–66] and even appears to have a role in survival of ionizing radiation [67]. Other fungal protective molecules include mycosporines and mycosporine-like amino acids (which protect against UV radiation and oxidative stress [68] and possibly act as compatible solutes [69, 70]), trehalose [71,

72], polyalcohols, betaine and proline [71, 73, 74], and carotenoid pigments (e.g. torularhodin or astaxanthin) [75]. In addition to these features, polyextremotolerant fungi are often oligotrophic and slow-growing. Several unusual sources of energy and carbon that might eventually supplement the often scarce resources of extreme environments have been proposed so far, from aerial CO_2 uptake to harvesting visible light with rhodopsins and ionizing radiation with melanin (reviewed in [62]), but the existence and significance of these mechanisms needs to be investigated in more detail.

10.3 Anthropogenic extremes: indoor habitats

Bacterial abundance in the indoor environments (especially contamination hot spots such as the kitchens and the bathrooms) has been known for a long time [76, 77]. These airborne communities are influenced by ventilation, occupancy, and outdoor air [78]. They may include pathogens or opportunists [79–81], which may reside in certain niches under appropriate conditions. Compared to bacteria, the public awareness of the high abundance of fungi in anthropogenic habitats is lower. However, recent studies have uncovered a surprising diversity of species including pathogenic species in hospitals [82]. Opportunistic pathogens are present also in residential buildings, where habitats with the regular presence of liquid water (e.g. plumbing, sinks or drains) support active growth and accumulation of high fungal biomass, while dry surfaces appear to contain mostly fungi deposited from air [83, 84]. Furthermore, polyextremotolerant oligotrophic fungi are particularly common indoors [84–86]. With the growing human population and advances in technology we are exposing microbes to new conditions similar to those encountered in nature, but nevertheless different in important details. So far, we have very little knowledge about how this exposure might impact the survival, biodiversity, and even evolution of microbial species. These aspects should be investigated carefully, because consequences are difficult to predict and are not necessarily harmless. For example, the opportunistic pathogen *Candida albicans* grown in spaceflight conditions expresses several traits with medical relevance, from increased cell aggregation and biofilm formation to altered expression of genes involved in resistance to pathogenesis-related stresses and antifungal drugs [87]. Despite this, increased virulence of the changed strains was not observed in a mouse model [87].

The occurrence of polyextremotolerant fungi in indoor habitats is the result of specific selection pressures. Even if our homes are designed in ways to limit the survival of microbes, this leaves ample room for those species that are able to adapt and thrive in such suboptimal conditions. A metagenomic study of indoor air organisms reported enrichment of the genes involved in overcoming iron limitation, oxidative damage, and desiccation [88]. Low water activity can be permanent, such as in the case of dry library books and documents, which are nevertheless conducive for the growth of the xerophilic *Eurotium halophilicum* [89]. However, occasional desiccation is a problem even in environments where large quantities of liquid water and high humidity are usually present (such as bathrooms and kitchens). Humid habitats in bathrooms and kitchens usually dry out for at least short periods of time, creating a strongly fluctuating environment. Besides water availability, microbial colonization is also determined by the nutritional substances that can be derived from the building materials or environment [90] and the presence of specific chemicals, such as surfactants from soap and shampoo [91]. Co-occurring species also play an important role and can actively change the conditions for microbial growth. For example when a substrate is attacked by a fungus, this leads to changes in its water activity and enables the growth of other fungi and bacteria [92]. As a case study of ecological facilitation, *E. halophilicum* and *Aspergillus penicillioides* were proposed to stimulate the growth of house dust mites [93].

Each home contains a large number of different habitats with profoundly different selection pressures, which lead to specific and in most cases limited diversity. For example, the showerhead environment is enriched in microbes that form biofilms in water systems, possibly due to their mechanic resilience and chlorine resistance [79]. Microbial growth in dishwashers, on the other hand, is determined by fluctuations in pH, temperature, salinity, and humidity [84, 94]. It should also be considered that with the tendency for energy efficiency (with decreased operating temperatures) and the substitution of aggressive chemical detergents with enzymatic mixtures, dishwashers are becoming more prone to microbial growth than ever. In addition, the widespread use of synthetic chemicals, some of them used specifically due to their antimicrobial activity, represents a relatively novel parameter of selection. Substances, such as triclosan, are released in the environment in ever larger amounts [95, 96]. Household cleaners often contain oxidative chemicals, such as bleach or hydrogen peroxide, thus adding to the secondary oxidative stress microbes encounter due to other stress factors. The presence of aromatic pollutants, detergents, biocides, and other household chemicals might expose the cells to water stress, which can be mediated by their chaotropicity [97] or hydrophobicity [98]. Both types of compounds trigger similar cellular responses, which are also used to counteract the consequences of other types of water stress.

Another novel niche is provided with the increasing use of synthetic materials and chemicals, such as plastics and aromatic pollutants. Although these substances are not present in nature, they are prone to microbial growth, often in the form of biofilms, on the surface of the material. It has been shown that *Exophiala dermatitidis* can colonize the rubber seals in dishwashers [84]. The novel substrate can even serve as a source of nutrients, when it is degraded by biologically generated free radicals or enzymes [99]. Aromatic pollutants can be metabolized by several extremotolerant fungi from the group of black yeasts; at least some of them can also use complex phenolic hydrocarbons as the sole source of carbon and energy [100]. However, black yeasts can also grow on glass, metals or silicon, and on a range of organic surfaces such as plastic materials and other polymers ([101], reviewed in [102]), some of which they might help to degrade.

There are many other habitats that were created or were profoundly changed due to human activity and the fungal diversity in many of them is poorly researched. Gas recovery by hydraulic fracturing subjects microorganisms to high pressures, elevated temperatures, chemical additives, high salinity, and of course various hydrocarbons, but published studies on microbial diversity in these sites are limited to prokaryotes [103]. Creosote-treated oak railway ties and concrete sleepers stained with petroleum oil were recently found to be contaminated with *E. dermatitidis* [104]. Soils polluted with extra-heavy crude oil contain several fungal species, including the already mentioned *C. sphaerospermum* [105]. The investigation of municipal landfill leachates revealed a relatively low fungal load, but included several emerging or established (opportunistic) pathogens, including *Candida parapsilosis* and *Aspergillus fumigatus*. Many of these species are able to grow on hydrocarbons and aromatic compounds, as well as tolerate oxygen limitation and hypersaline conditions [106]. Emerging pathogens of the complex *Pseudallescheria boydii* were found at the highest densities in human-impacted areas, such as in agricultural areas, fluids from wastewater treatment plants, and industrial areas [107].

Radiation encountered in and around the Chernobyl Nuclear Power Plant led to an increased abundance of pigmented fungi [108], among them *C. sphaerospermum* and *A. pullulans* [109], some of them showing unusual adaptations such as radiotropism [110]. Surface biofilms in areas exposed to ionizing radiation have a similar composition to biofilms in UV-exposed areas in other parts of the world – presumably as a result of their adaptation to UV and desiccation, here serving as a pre-adaptation [111].

Although man-made and natural outdoor extreme environments may appear very different at first glance, they may seem quite similar from the perspective of the microbes. In both coastal hypersaline environments and Arctic glacial ice, as well as on bathroom surfaces organisms for example experience periods of extremely low water activity. This explains the overlap of microbial diversity between these habitats. Sequences suggestive of opportunistic pathogens as well as virulence-associated genes are common in the indoor air [88]. Some of the species that persist in (either natural or indoor) stressful habitats, can also cause opportunistic human infections. Thus, while the design of indoor spaces may prevent good growth of the majority of microbes, this does not necessarily eliminate the most troublesome ones that may even be selected for and enriched for due to their adaptability and polyextremotolerance.

10.4 Coincidental opportunities: opportunistic infections

With an increasing frequency of opportunistic mycoses, many of the fungi that were previously thought to be non-pathogenic are now recognized causes of invasive fungal infections in compromised patients [112]. This phenomenon has been attributed to several factors, from changed hygienic measures and chemical pollution, to global en-

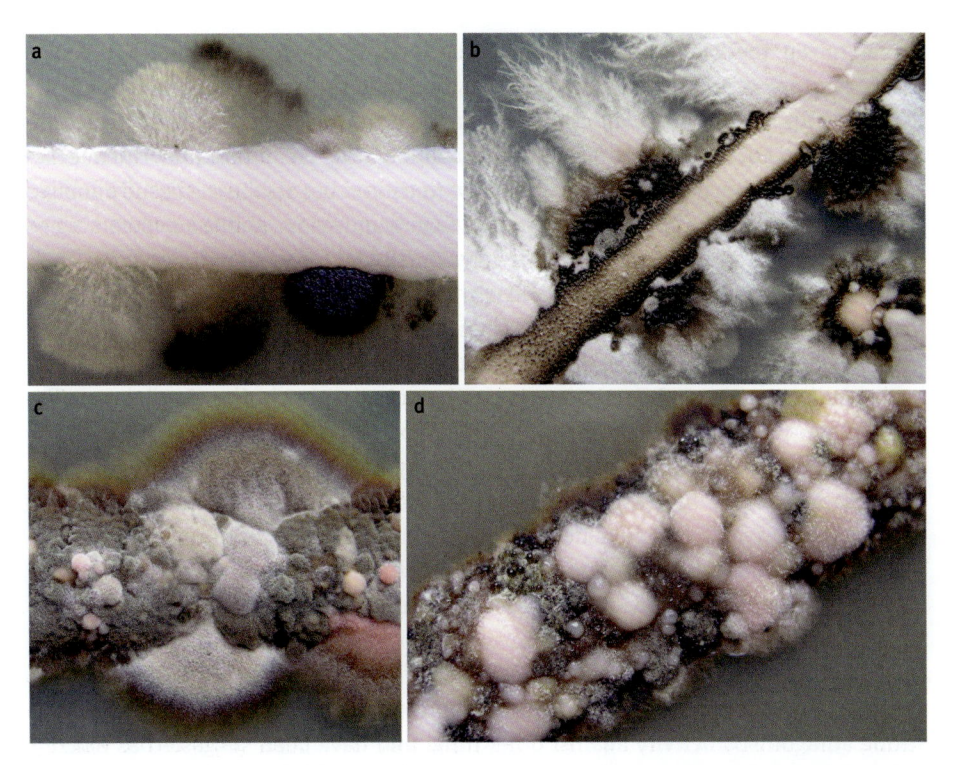

Fig. 10.1. Plastic morphology of *Aureobasidium* spp. on different media. (a) *Aureobasidium pullulans* after one month on a minimal medium. (b) *Aureobasidium subglaciale* after one month on a malt extract agar. (c) *Aureobasidium melanogenum* after four months on a high-glucose malt extract agar. (d) *Aureobasidium namibiae* after four months on a high-glucose malt extract agar. All cultures were inoculated from two-day-old liquid cultures (each species grown from a single phenotypically uniform colony) and all were grown at 24 °C.

vironmental changes, increased numbers of susceptible hosts, global travel, urbanisation, and novel extreme environments in our immediate surroundings [102, 113, 114]. Certain types of environments may act as a training ground that prepares the species for invasion of a human body [115], which may result in "accidental" infections [116], especially if the fungal abundance is high. Also, for effective survival in their host those fungi also require general stress tolerance [117]. We argue that polyextremotolerant fungi, which are often also characterised by exceptional adaptability, are excellent candidates for such "troublemakers".

The black yeast *A. pullulans* s.l., as already mentioned, is a group of taxa found in a variety of natural habitats. In addition, it was isolated from tap water, bathroom surfaces, and dishwashers (mainly *A. melanogenum* [21, 84]); refrigerated, frozen, salt-preserved, and dry food [118, 119]; and numerous other man-made habitats [25]. Besides the already described mechanisms, its ability to survive might be supported by the rapid dimorphic switching from small colorless yeast cells to thick-walled, heavily melanized, meristematic forms [120] and a very plastic morphology (▸ Fig. 10.1) – it has even been proposed as a model for investigating fungal phenotypic plasticity [121]. Additionally, *A. pullulans* produces a large number of extracellular hydrolytic enzymes [25, 122]. Proteases, lipases, and phospholipases are important virulence factors. These enzymes not only have roles in the nutrition, but also in the tissue damage, dissemination within the human organism, iron acquisition, and the overcoming of the host immune system, which all strongly contribute to fungal pathogenicity [123], as reported for *A. fumigatus* and *C. albicans* [124]. Importantly, *A. pullulans* s.l. (specifically, *A. melanogenum*; [25]) has also been reported to cause a variety of localized infections, as well as rare systemic infections (reviewed in [125]).

A. pullulans is not the only species with such a wide choice of environments. Some other, phylogenetically distant fungi can be just as adaptable. *C. parapsilosis* and *Meyerozyma guilliermondii* from *Saccharomycetales* (previously *Pichia guilliermondii*, anamorph *Candida guilliermondii*) and basidiomycetous yeast *Rhodotorula mucilaginosa* have all been isolated from glacial ice [126, 127], hypersaline water [8], salted meat products [128–130], and dishwashers [84], to name just a few habitats. All species have been described as the causes of human infections, and are considered by some authors as emerging opportunistic pathogens [131–133]. *C. parapsilosis* is an increasingly important cause of *Candida* blood-stream infections [134]. Interestingly, similar to *A. pullulans*, both *R. mucilaginosa* and *M. guilliermondii* species also exhibit strong antagonistic activity against other fungi and have been suggested as biocontrol agents of postharvest diseases of plant-derived products [131, 135]. There are other similar examples; a large share of the oligotrophic black fungi encountered in humid indoor environments has significant potential to cause human infection [86]. We believe that this is tightly linked to their polyextremotolerant nature, which enables them to survive the stressful conditions.

Many of the virulence factors identified for human pathogenic fungi appear to allow both the establishment of the microbe in a mammalian host, and its survival in the environment, e.g. by preventing predation by amoeba, slime molds, and nematodes (reviewed in [136]). Thus traits that enable pathogenesis, including thermotolerance, need not have evolved primarily as traits selected during the development of pathogenicity, but can be considered to be pre-existing adaptations to relatively extreme environmental conditions outside the natural host [137]. *Cryptococcus neoformans*, for example, copes efficiently with several stresses during infection: pH fluctuations, anoxia and nutrient deprivation, and reactive oxidative, nitrosative, and chlorinating species. However, it is considered unlikely that *C. neoformans* stress response pathways were developed specifically for survival in a mammalian host; they are more likely to be the result of the stress that the fungus encounters in its primary ecological niche, bird manure [138]. Since opportunistic infections do not spread from host to host, they are most likely an evolutionary dead end from a fungal perspective. To preserve any adaptations gained during the infection, the pathogen would have to escape into the environment and the genetic basis of the adaptation would have to avoid being drowned in the much larger gene pool of environmental strains. Traits that enable an opportunistic fungus to survive in its host are therefore in most if not all cases pre-adaptations or exaptations.

In such cases the number of human infections is relatively small compared to the out-of-host population of the fungus and probably has no major impact on the biological success of the species. Thereby preventive measures are complicated and the species growth rate is decoupled from host densities (since the fungus can survive without the host), i.e. the spread of the infection causing fungus is not self-limiting by host availability. This phenomenon has already led to catastrophic declines in several animal species in recent years [139], and an increasing incidence of infections with black yeasts was already observed more than a decade ago [140]. A further increase is predicted [141].

As described above, certain generalist species, such as the black yeasts, are particularly good at adapting to a variety of different stressful environments. For example according to phylogenetic studies, the ancestors of *Chaetothyriales* black fungi were rock-inhabiting species, and the traits that they evolved for life in this extreme environment are the basis of the numerous independent shifts to pathogenicity that have occurred in this group of fungi [142]. It has also been proposed previously that the human pathogenicity of fungi is associated with moderate osmotolerance at the order level [143]. Black fungi are wide-spread in nature, but can also successfully colonize various indoor habitats, where they come in contact with their potential hosts on a daily basis.

Habitats with fluctuating and occasionally high temperatures for example appear to be particularly good at harboring problematic fungal species. Examination of bathwater and sludge in drainpipes that is warmed daily to over 42 °C correlates with the occurrence of the medically important genus *Exophiala* [144]. Several species from this genus are known to cause various diseases, which include fatal systemic and brain infections, especially when caused by *E. dermatitidis* [145, 146]. This species has also been isolated in large numbers from steam baths [147], sink drains [85], and water taps [148]. Interestingly, despite its good stress-tolerance, *E. dermatitidis* is rarely found in nature, which distinguishes it from the ubiquitous polyextremotolerant species described above. Unless sampling has so far missed *E. dermatitidis*'s preferred natural habitat, this could mean that anthropogenic habitats are relatively important for its success. A new and unexpected indoor habitat of *E. dermatitidis* and *Exophiala phaeomuriformis* was reported by Zalar et al. [84]. Both species have consistently been found to form a stable community in dishwashers, together with some other human opportunists belonging to the genera *Aspergillus, Candida, Dipodascus, Fusarium, Penicillium, Pichia,* and *Rhodotorula*. In addition to their thermotolerance, the majority of *E. dermatitidis* isolates in dishwashers belong to the most virulent genotype A [84].

This preference for high temperatures could be linked to the fact that temperatures in a warm-blooded animal body are above those that are favoured by most fungi. For every 1 °C gain in body temperature in the range of 30 °C to 42 °C, approximately 6% of the fungal species are excluded as potential pathogens [113]. The ability to grow at 37 °C thus appears to be the decisive factor for pathogenesis. It has to be noted, however, that this does not mean that opportunistic pathogens cannot tolerate low temperatures. *A. pullulans*, *R. mucilaginosa* and *M. guilliermondii* are all commonly found in glacial environments. Even *E. dermatitidis*, despite its pronounced thermotolerance, has been isolated from glacial ice [149]. Additionally, several other opportunistic pathogens, such as *Trichophyton* and *Malassezia* have been found in extremely cold polar and alpine environments [149–151].

The ability to assimilate aromatic hydrocarbons was suggested to be related to certain patterns of mammalian infection [100]. Neurochemistry features a distinctive array of phenolic and aliphatic compounds that are related to molecules involved in the metabolism of aromatic hydrocarbons. Many of the volatile-hydrocarbon-degrading strains are closely related to, or in some cases apparently conspecific with, human-pathogenic fungal species, which can cause severe mycoses in immunocompetent people, especially neurological infections.

Tolerance to oxidative stress and oligotrophism is another possible virulence factor of polyextremotolerant fungi. Oxidative stress is triggered by various abiotic stressors (e.g. light, high salinity, extreme temperatures, starvation, mechanical damage) as a primary or secondary stress [152–154]. Oxidative burst is also an ancient part of the animal and plant immune response [155] and is crucial for defense against fungi [156]. It has been shown that antioxidant pathways of fungi are important for surviving the attack by neutrophils [157]. As described above, polyextremotolerant fungi produce a plethora of compounds, which protect the cells from oxidative damage: from melanin, to mycosporines, mycosporine-like amino acids, carotenoid pigments and others. During infection, phagocytes can restrict fungal growth by releasing mediators that sequester iron [156]. Since many generalistic polyextremotolerant species grow in oligotrophic environments, they are able to employ various mechanisms that overcome the starvation imposed by the immune system. In the case of iron this can be achieved for example by producing siderophores, high-affinity iron chelating compounds [159–162]. The ability of oligotrophic growth (which includes growth at limited iron availability), possessed by many black fungi [25, 86, 142], may be beneficial in such conditions.

The stressful conditions of domestic environments not only lead to enrichment of the adaptable and stress-tolerant species, but may also select for traits that serve as pre-adaptations for an opportunistic potential and tolerance to antimicrobials. Subinhibitory concentrations of commonly used biocides, such as benzalkonium chloride, and even stress, can lead to a significant decrease in sensitivity to biocides as well as antibiotics. This is achieved through several mechanisms, for example, by changes in the expression or structure of cellular efflux pumps or by biofilm formation [163–167]. Especially, biofilms can confer additional tolerance to other stresses [168]. Embedded in extracellular polysaccharides, such as those produced by black fungi [169], the cells are protected from a variety of hostile conditions. Apart from the high evolvability of certain groups of organisms [170], adaptation to (extreme) indoor environments and human hosts can be facilitated by several other factors, such as horizontal gene transfer (which occurs more frequently under stressful conditions, reviewed in [171], and is enhanced in biofilms [172]), high expression noise [173], phenotypic plasticity [33], ecological fitting [34], and even stress-triggered increase of mutation rates [40]. Increases in mutation rates have previously been proposed to fuel the evolution of microbial pathogenesis and antibiotic resistance, both of which occur under stress [174]. Stressful conditions also mobilize transposons and lead to genetic rearrangements, a phenomenon, which is involved in adaptation of plant-pathogenic and other fungi [175, 176]. Finally, fluctuating conditions can contribute to the evolution of stress-induced mutagenesis mechanisms, which appear to be under short-term environment-specific selection. They are likely selected for in rapidly changing environments and toned down or lost in static ones [174]. Periodic high temperatures, desiccation, and the presence of cleaning chemicals in dishwashers are good examples of a changing environment. Dormancy is another beneficial trait in a fluctuating environment [177], and it has also been suggested to be an important factor in *C. neoformans* infections [178]. Persistence in a desiccated state can substantially improve the ability of an organism to survive (temporary) hostile conditions, as is demonstrated by lichens [179].

The above described *A. pullulans* is just one example of polyextremotolerant and potentially problematic species, with which we share our homes. Some of these species may have closer relationships with human opportunists or even pathogens, while others are involved in allergic or other immunomodulatory reactions [180]. Even though it may not benefit their long-term survival, these polyextremotolerant fungi will seize the opportunity presented by the human hosts, especially when the obstacles presented by otherwise efficient defensive mechanisms of the host are lowered due to a compromised immune system, accessible entry points, or abuse of drugs (including alcohol and nicotine).

10.5 Conclusions: polyextremotolerance

Polyextremotolerance appears to be an efficient strategy for organisms to allow their occurrence in more than one type of stressful environment. This strategy also lowers the effects of competition posed by other species and thus increases the potential to radiate by evolutionary progression. We recognized several factors that are involved in the polyextremotolerant strategy.

We argue that environmental heterogeneity and stress could contribute to the maintenance of genetic polymorphisms, especially in dynamically cycling environments. Several ways to quickly generate the genetic and/or phenotypic variation even in the absence of (or in addition to) sexual reproduction were discovered, from stress-induced increases in mutation rates, genome restructuring associated with transposable elements, and epigenetic mechanisms. All of them could contribute to the good adaptability of polyextremotolerant species.

The wealth of stress tolerance mechanisms is enabled not only by highly specialized mechanisms evolved for combating a specific type of stress (e.g. hypersaline), but also, and perhaps even more importantly, by general mechanisms, which confer tolerance to more than one type of stress. Examples of such mechanisms are the production of protective small molecules, melanization of the cell wall, production of extracellular polysaccharides, pleomorphic growth, and growth in biofilms.

Finally, by maintaining the redundancy of the stress-tolerance mechanisms instead of fine-tuning them for a specific habitat or type of stress, polyextremotolerant organisms avoid specialization and maintain their potential antifragility. Specialization may be prevented by migration and exchange of genetic material, thus continually drowning local adaptations in the larger gene pool of individuals from other environment types. Moreover, adaptation may not necessarily lead to speciation if it is primarily based on non-genetic changes (e.g. phenotypic plasticity, epigenetic changes). In this case gene-flow-reducing selection pressure (leading to local accumulation of adaptive mutations) will be smaller. Another possibility is that the specialization of local populations is occurring all the time, but the resulting specialists are being repeatedly wiped out by each major change of the environmental conditions. The individual species of *A. pullulans* s.l., which show varying degrees of specialization, could be a result of such an ongoing process.

Polyextremotolerance has an important advantage: it keeps the organism prepared for occupying novel habitats, especially if these are stressful enough to limit competition by other species. This may also be true for habitats such as those found indoors and in animal bodies [102]. The overlap in types of stress the species are exposed to in all these situations facilitates the habitat shifts: tolerance to stress is crucial for persistence in extreme natural environments, but also in many domestic habitats and during an infection. The pervasive colonization of human products (ranging from dishwashers to grout between bathroom tiles) and opportunistic pathogenicity is thus a side effect of the extreme adaptability of polyextremotolerant fungi.

Acknowledgments

The authors acknowledge the financial support from the state budget by the Slovenian Research Agency (Infrastructural Centre Mycosmo, MRIC UL, Applied Research Project L4-5533, and Postdoctoral Project Z4-5531 to C. Gostinčar). The scientific work was also partly financed via operation "Centre of excellence for integrated approaches in chemistry and biology of proteins" number OP13.1.1.2.02.0005, financed by European Regional Development Fund (85% share of financing) and by the Slovenian Ministry of Higher Education, Science and Technology (15% share of financing).

References

[1] B. C. Christner, E. Mosley-Thompson, L. G. Thompson, V. Zagorodnov, K. Sandman, J. N. Reeve. Recovery and identification of viable bacteria immured in glacial ice. Icarus 2000,479–485.

[2] P. B. Price, T. Sowers. Temperature dependence of metabolic rates for microbial growth, maintenance, and survival. Proc Natl Acad Sci USA 2004,101,4631–4636.

[3] R. A. Rohde, P. B. Price. Diffusion-controlled metabolism for long-term survival of single isolated microorganisms trapped within ice crystals. Proc Natl Acad Sci USA 2007,104,16592–16597.

[4] P. B. Price. Microbial genesis, life and death in glacial ice. J. Can Microbiol 2009,55,1–11.

[5] N. Gunde-Cimerman, S. Sonjak, P. Zalar, J. C. Frisvad, B. Diderichsen, A. Plemenitaš. Extremophilic fungi in Arctic ice: a relationship between adaptation to low temperature and water activity Phys Chem Earth 2003,28,1273–1278.

[6] L. Butinar, T. Strmole, N. Gunde-Cimerman. Relative incidence of ascomycetous yeasts in Arctic coastal environments. Antonie van Leeuwenhoek 2011,61,832–843.

[7] N. Gunde-Cimerman, P. Zalar, S. de Hoog, A. Plemenitaš. Hypersaline waters in salterns – natural ecological niches for halophilic black yeasts. FEMS Microbiol Ecol 2000,32,235–240.

[8] L. Butinar, S. Santos, I. Spencer-Martins, A. Oren, N. Gunde-Cimerman. Yeast diversity in hypersaline habitats. FEMS Microbiol Lett 2005,244,229–234.

[9] P. Zalar, M. A. Kocuvan, A. Plemenitas, N. Gunde-Cimerman. Halophilic black yeasts colonize wood immersed in hypersaline water. Bot Mar 2005,48,323–326.

[10] N. Gunde-Cimerman, J. Ramos, A. Plemenitas. Halotolerant and halophilic fungi. Mycol Res 2009,113,1231–1241.

[11] A. Plemenitaš, Vaupotič T, M. Lenassi, T. Kogej, N. Gunde-Cimerman. Adaptation of extremely halotolerant black yeast *Hortaea werneckii* to increased osmolarity: a molecular perspective at a glance. Stud Mycol 2008,67–75.

[12] Gostinčar C, M. Lenassi, N. Gunde-Cimerman, A. Plemenitaš. Fungal adaptation to extremely high salt concentrations. Adv Appl Microbiol 2011,77,71–96.

[13] Gostinčar C, M. Grube, G. S. de Hoog, P. Zalar, N. Gunde-Cimerman. Extremotolerance in fungi: evolution on the edge. FEMS Microbiol Ecol 2010,71,2–11.

[14] M. Alleaume-Benharira, I. R. Pen, O. Ronce. Geographical patterns of adaptation within a species' range: interactions between drift and gene flow. J Evol Biol 2006,19,203–215.

[15] K. Johannesson, C. Andre. Life on the margin: genetic isolation and diversity loss in a peripheral marine ecosystem, the Baltic Sea. Mol Ecol 2006,15,2013–2029.

[16] S. A. Cantrell, J. C. Dianese, J. Fell, N. Gunde-Cimerman, P. Zalar. Unusual fungal niches. Mycologia 2011,103,1161–1174.

[17] T. Kogej, J. Ramos, A. Plemenitas, N. Gunde-Cimerman. The halophilic fungus *Hortaea werneckii* and the halotolerant fungus *Aureobasidium pullulans* maintain low intracellular cation concentrations in hypersaline environments. Appl Environ Microbiol 2005,71,6600–6605.

[18] T. D. Leathers. Biotechnological production and applications of pullulan. Appl Microbiol Biotechnol 2003,62,468–473.

[19] R. S. Singh, G. K. Saini, J. F. Kennedy. Pullulan: microbial sources, production and applications. Carbohydr Polym 2008,6986–6997.

[20] J. H. Andrews, R. N. Spear, E. V. Nordheim. Population biology of *Aureobasidium pullulans* on apple leaf surfaces. Can J Microbiol 2002,48,500–513.

[21] P. Zalar, Gostinčar C, G. S. de Hoog, Uršič V, M. Sudhadham, N. Gunde-Cimerman. Redefinition of *Aureobasidium pullulans* and its varieties. Stud Mycol 2008,61,21–38.

[22] H. M. Ranta. Effect of simulated acid rain on quantity of epiphytic microfungi on Scots pine (*Pinus sylvestris* L.) needles. Environ Pollut 1990,67,349–359.
[23] N. Shiomi, T. Yasuda, Y. Inoue, et al. Characteristics of neutralization of acids by newly isolated fungal cells. J Biosci Bioeng 2004,97,54–58.
[24] S. Onofri. Antarctic Microfungi. In: J. Seckbach, ed. Enigmatic microorganisms and life in extreme environments. Dordrecht; London: Kluwer Academic; 1999:323–336.
[25] Gostinčar C, R. A. Ohm, T. Kogej, et al. Genome sequencing of four *Aureobasidium pullulans* varieties: biotechnological potential, stress tolerance, and description of new species. BMC Genomics 2014, 5:549.
[26] L. Butinar, S. Sonjak, P. Zalar, A. Plemenitaš, N. Gunde-Cimerman. Melanized halophilic fungi are eukaryotic members of microbial communities in hypersaline waters of solar salterns. Bot Mar 2005,48,73–79.
[27] M. Padamsee, T. K. A. Kumar, R. Riley, et al. The genome of the xerotolerant mold *Wallemia sebi* reveals adaptations to osmotic stress and suggests cryptic sexual reproduction. Fungal Genet Biol 2012,49,217–226.
[28] J. Zajc, Y. Liu, W. Dai, et al. Genome and transcriptome sequencing of the halophilic fungus *Wallemia ichthyophaga*: haloadaptations present and absent. BMC Genomics 2013,14,617.
[29] A. Buckling, M. A. Wills, N. Colegrave. Adaptation limits diversification of experimental bacterial populations. Science 2003,302,2107–2109.
[30] K. Zakharova, G. Marzban, J. P. de Vera, A. Lorek, K. Sterflinger. Protein patterns of black fungi under simulated Mars-like conditions. Scientific Reports 2014,4.
[31] K. Zakharova, D. Tesei, G. Marzban, J. Dijksterhuis, T. Wyatt, K. Sterflinger. Microcolonial fungi on rocks: a life in constant drought? Mycopathologia 2013,175,537–547.
[32] J. R. Bridle, T. H. Vines. Limits to evolution at range margins: when and why does adaptation fail? Trends Ecol Evol 2007,22,140–147.
[33] M. J. West-Eberhard. Developmental plasticity and the origin of species differences. Proc Natl Acad Sci USA 2005,102,6543–6549.
[34] S. J. Agosta, J. A. Klemens. Ecological fitting by phenotypically flexible genotypes: implications for species associations, community assembly and evolution. Ecology Letters 2008,11,1123–1134.
[35] C. D. Schlichting. Hidden reaction norms, cryptic genetic variation, and evolvability. Ann N Y Acad Sci 2008,1133,187–203.
[36] N. Corradi, I. R. Sanders. Evolution of the P-type II ATPase gene family in the fungi and presence of structural genomic changes among isolates of *Glomus intraradices*. BMC Evol Biol 2006,6,21.
[37] M. Lenassi, Gostinčar C, S. Jackman, et al. Whole genome duplication and enrichment of metal cation transporters revealed by de novo genome sequencing of extremely halotolerant black yeast *Hortaea werneckii*. PLoS ONE 2013,8,e71328.
[38] S. M. Rosenberg, C. Thulin, R. S. Harris. Transient and heritable mutators in adaptive evolution in the lab and in nature. Genetics 1998,148,1559–1566.
[39] E. Heidenreich, R. Novotny, B. Kneidinger, V. Holzmann, U. Wintersberger. Non-homologous end joining as an important mutagenic process in cell cycle-arrested cells. EMBO J 2003,22,2274–2283.
[40] S. M. Rosenberg, P. J. Hastings. Genomes: worming into genetic instability. Nature 2004,430,625–626.
[41] B. C. Lamb, S. Mandaokar, B. Bahsoun, I. Grishkan, E. Nevo. Differences in spontaneous mutation frequencies as a function of environmental stress in soil fungi at "Evolution Canyon," Israel. Proc Natl Acad Sci USA 2008,105,5792–5796.

[42] T. Kis-Papo, V. Kirzhner, S. P. Wasser, E. Nevo. Evolution of genomic diversity and sex at extreme environments: fungal life under hypersaline Dead Sea stress. Proc Natl Acad Sci USA 2003,100,14970–14975.
[43] E. Nevo. Evolution of genome-phenome diversity under environmental stress. Proc Natl Acad Sci USA 2001,98,6233–6240.
[44] R. de Jonge, M. D. Bolton, A. Kombrink, G. C. M. van den Berg, K. A. Yadeta, B. P. H. J. Thomma. Extensive chromosomal reshuffling drives evolution of virulence in an asexual pathogen. Genome Res 2013,23,1271–1282.
[45] S. M. Schmidt, P. M. Houterman, I. Schreiver, et al. MITEs in the promoters of effector genes allow prediction of novel virulence genes in *Fusarium oxysporum*. BMC Genomics 2013,14.
[46] B. Wheeler. Small RNAs, big impact: small RNA pathways in transposon control and their effect on the host stress response. Chromosome Res 2013,21,587–600.
[47] E. Casacuberta, J. Gonzalez. The impact of transposable elements in environmental adaptation. Mol Ecol 2013,22,1503–1517.
[48] S. Sun, J. Heitman. Is sex necessary? BMC Biol 2011,9.
[49] Antifragility – or – the property of disorder-loving systems. 2011. (Accessed 13. 12. 2013, at www.edge.org/q2011/q11_3.html#taleb.)
[50] G. S. de Hoog, E. Göttlich, G. Platas, O. Genilloud, G. Leotta, J. van Brummelen. Evolution, taxonomy and ecology of the genus *Thelebolus* in Antarctica. Stud Mycol 2005,33–76.
[51] P. Zalar, G. S. de Hoog, H. J. Schroers, P. W. Crous, J. Z. Groenewald, N. Gunde-Cimerman. Phylogeny and ecology of the ubiquitous saprobe *Cladosporium sphaerospermum*, with descriptions of seven new species from hypersaline environments. Stud Mycol 2007,58,157–183.
[52] H. G. Park, J. R. Managbanag, E. K. Stamenova, S. C. Jong. Comparative analysis of common indoor *Cladosporium* species based on molecular data and conidial characters. 2004,441–451.
[53] M. Aihara, T. Tanaka, K. Takatori. *Cladosporium* as the main fungal contaminant of locations in dwelling environments. Biocontrol Science 2001,6,49–52.
[54] G. Badillet, Bièvre Cd, S. Spizajzen. Isolement de dématiées à partir d'ongles et de squames. Bulletin de la Société Française de Mycologie 1982,69–72.
[55] P. T. Pereira, M. M. de Carvalho, F. M. Girio, J. C. Roseiro, M. T. Amaral-Collaco. Diversity of microfungi in the phylloplane of plants growing in a Mediterranean ecosystem. J Basic Microbiol 2002,42,396–407.
[56] A. D. Hocking, B. Miscamble, J. Pitt. Water relations of *Alternaria alternata*, *Cladosporium cladosporioides*, *Cladosporium sphaerospermum*, *Curvularia lunata* and *Curvularia pallescens*. Mycol Res 1994,91–94.
[57] L. Butinar, P. Zalar, J. C. Frisvad, N. Gunde-Cimerman. The genus *Eurotium* – members of indigenous fungal community in hypersaline waters of salterns. FEMS Microbiol Ecol 2005,51,155–166.
[58] P. Zalar, G. S. de Hoog, N. Gunde-Cimerman. *Trimmatostroma salinum*, a new species from hypersaline water. Stud Mycol 1999,43,57–62.
[59] P. Zalar, J. C. Frisvad, N. Gunde-Cimerman, J. Varga, R. A. Samson. Four new species of *Emericella* from the Mediterranean region of Europe. Mycologia 2008,100,779–795.
[60] P. Zalar, G. S. de Hoog, N. Gunde-Cimerman. Taxonomy of the endoconidial black yeast genera *Phaeotheca* and *Hyphospora*. Stud Mycol 1999,43,49–56.
[61] S. Gavrilets, J. B. Losos. Adaptive radiation: contrasting theory with data. Science 2009,323,732–737.
[62] Gostinčar C, L. Muggia, M. Grube. Polyextremotolerant black fungi: oligotrophism, adaptive potential, and a link to lichen symbioses. Front Microbiol 2012,3,390.

[63] T. Kogej, M. H. Wheeler, Lanisnik T. Rizner, N. Gunde-Cimerman. Evidence for 1,8-dihydroxynaphthalene melanin in three halophilic black yeasts grown under saline and non-saline conditions. FEMS Microbiol Lett 2004,232,203–209.

[64] S. Onofri, D. Barreca, L. Selbmann, et al. Resistance of Antarctic black fungi and cryptoendolithic communities to simulated space and Martian conditions. Stud Mycol 2008,61,99–109.

[65] K. Sterflinger, D. Tesei, K. Zakharova. Fungi in hot and cold deserts with particular reference to microcolonial fungi. Fungal Ecol 2012,5,453–462.

[66] S. Onofri, R. de la Torre, de J. P. Vera, et al. Survival of rock-colonizing organisms after 1.5 years in outer space. Astrobiology 2012,12,508–516.

[67] E. Dadachova, R. A. Bryan, X. Huang, et al. Ionizing radiation changes the electronic properties of melanin and enhances the growth of melanized fungi. PLoS ONE 2007,2,e457.

[68] A. Oren, N. Gunde-Cimerman. Mycosporines and mycosporine-like amino acids: UV protectants or multipurpose secondary metabolites? FEMS Microbiol Lett 2007,269,1–10.

[69] A. A. Gorbushina, K. Whitehead, T. Dornieden, A. Niesse, A. Schulte, J. Hedges. Black fungal colonies as units of survival: hyphal mycosporines synthesized by rock dwelling microcolonial fungi. Can J Bot 2003,2,131–138.

[70] T. Kogej, Gostinčar C, M. Volkmann, A. A. Gorbushina, N. Gunde-Cimerman. Mycosporines in extremophilic fungi – novel complementary osmolytes? Environ Chem 2006,3,105–110.

[71] J. Shima, H. Takagi. Stress-tolerance of baker's-yeast (*Saccharomyces cerevisiae*) cells: stress-protective molecules and genes involved in stress tolerance. Biotechnol Appl Biochem 2009,53,155–164.

[72] A. Ocon, R. Hampp, N. Requena. Trehalose turnover during abiotic stress in arbuscular mycorrhizal fungi. New Phytol 2007,174,879–891.

[73] A. Blomberg, L. Adler. Physiology of osmotolerance in fungi. Adv Microb Physiol 1992,33,145–212.

[74] H. Takagi. Proline as a stress protectant in yeast: physiological functions, metabolic regulations, and biotechnological applications. Appl Microbiol Biotechnol 2008,81,211–223.

[75] A. Madhour, H. Anke, A. Mucci, P. Davoli, R. W. S. Weber. Biosynthesis of the xanthophyll plectaniaxanthin as a stress response in the red yeast *Dioszegia* (Tremellales, Heterobasidiomycetes, Fungi). Phytochemistry 2005,66,2617–2626.

[76] M. Ojima, Y. Toshima, E. Koya, et al. Hygiene measures considering actual distributions of microorganisms in Japanese households. J Appl Microbiol 2002,93,800–809.

[77] R. R. Beumer, H. Kusumaningrum. Kitchen hygiene in daily life. Int Biodeterior Biodegradation 2003,51,299–302.

[78] J. F. Meadow, A. E. Altrichter, S. W. Kembel, et al. Indoor airborne bacterial communities are influenced by ventilation, occupancy, and outdoor air source. Indoor Air 2014,24,41–48.

[79] L. M. Feazel, L. K. Baumgartner, K. L. Peterson, D. N. Frank, J. K. Harris, N. R. Pace. Opportunistic pathogens enriched in showerhead biofilms. Proc Natl Acad Sci USA 2009,106,16393–16398.

[80] S. D. Perkins, J. Mayfield, V. Fraser, L. T. Angenent. Potentially pathogenic bacteria in shower water and air of a stem cell transplant unit. Appl Environ Microbiol 2009,75,5363–5372.

[81] E. Scott, S. Duty, M. Callahan. A pilot study to isolate *Staphylococcus aureus* and methicillin-resistant *S. aureus* from environmental surfaces in the home. Am J Infect Control 2008,36,458–460.

[82] C. P. Garcia-Cruz, M. J. N. Aguilar, O. E. Arroyo-Helguera. Fungal and bacterial contamination on indoor surfaces of a hospital in Mexico. Jundishapur Journal of Microbiology 2012,5,460–464.

[83] R. I. Adams, M. Miletto, J. W. Taylor, T. D. Bruns. The diversity and distribution of Fungi on residential surfaces. PLoS One 2013,8.

[84] P. Zalar, M. Novak, G. S. De Hoog, N. Gunde-Cimerman. Dishwashers – A man-made ecological niche accommodating human opportunistic fungal pathogens. Fungal Biol 2011,115,997–1007.

[85] N. Hamada, N. Abe. Comparison of fungi found in bathrooms and sinks. Biocontrol Science 2010,15,51–56.

[86] X. Lian, G. S. de Hoog. Indoor wet cells harbour melanized agents of cutaneous infection. Med Mycol 2010,48,622–628.

[87] A. Crabbe, S. M. Nielsen-Preiss, C. M. Woolley, et al. Spaceflight enhances cell aggregation and random budding in *Candida albicans*. PLoS One 2013,8,e80677.

[88] S. G. Tringe, T. Zhang, X. G. Liu, et al. The airborne metagenome in an indoor urban environment. PLoS One 2008,3,e1862.

[89] M. Montanari, V. Melloni, F. Pinzari, G. Innocenti. Fungal biodeterioration of historical library materials stored in Compactus movable shelves. Int Biodeterior Biodegradation 2012,75,83–88.

[90] R. L. Gorny. Filamentous microorganisms and their fragments in indoor air – A review. Ann Agric Environ Med 2004,11,185–197.

[91] N. Hamada, N. Abe. Growth characteristics of four fungal species in bathrooms. Biocontrol Science 2010,15,111–115.

[92] R. A. Samson. Health implications of fungi in indoor environments. Amsterdam; Oxford: Elsevier; 1994.

[93] R. A. Samson, B. V. D. Lustgraaf. *Aspergillus penicilloides* and *Eurotium halophilicum* in association with house-dust mites. Mycopathologia 1978,64,13–16.

[94] Döğen A, E. Kaplan, Öksüz, M. S. Serin, M. Ilkit, G. S. de Hoog. Dishwashers are a major source of human opportunistic yeast-like fungi in indoor environments in Mersin, Turkey. Med Mycol 2013,51,493–498.

[95] S. B. Levy. Antibacterial household products: cause for concern. Emerg Infect Dis 2001,7,512–515.

[96] J. L. Fang, R. L. Stingley, F. A. Beland, W. Harrouk, D. L. Lumpkins, P. Howard. Occurrence, efficacy, metabolism, and toxicity of triclosan. Journal of Environmental Science and Health – Part C: Environmental Carcinogenesis & Ecotoxicology Reviews 2010,28,147–171.

[97] J. E. Hallsworth, S. Heim, K. N. Timmis. Chaotropic solutes cause water stress in *Pseudomonas putida*. Environ Microbiol 2003,5,1270–1280.

[98] P. Bhaganna, R. J. M. Volkers, A. N. W. Bell, et al. Hydrophobic substances induce water stress in microbial cells. Microb Biotechnol 2010,3,1751–7915.

[99] S. Wallstrom, S. Karlsson. Biofilms on silicone rubber insulators; microbial composition and diagnostics of removal by use of ESEM/EDS – Composition of biofilms infecting silicone rubber insulators. Polymer Degradation and Stability 2004,85,841–846.

[100] F. X. Prenafeta-Boldu, R. Summerbell, G. S. de Hoog. Fungi growing on aromatic hydrocarbons: biotechnology's unexpected encounter with biohazard? FEMS Microbiol Rev 2006,30,109–130.

[101] D. Isola, L. Selbmann, G. S. de Hoog, et al. Isolation and screening of black fungi as degraders of volatile aromatic hydrocarbons. Mycopathologia 2013,175,369–379.

[102] Gostinčar C, M. Grube, N. Gunde-Cimerman. Evolution of fungal pathogens in domestic environments? Fungal Biol 2011,115,1008–1018.

[103] M. A. Cluff, A. Hartsock, J. D. MacRae, K. Carter, P. J. Mouser. Temporal changes in microbial ecology and geochemistry in produced water from hydraulically fractured marcellus shale gas wells. Environ Sci Technol 2014,48,6508–6517.

[104] Dögen A, E. Kaplan, M. Ilkit, G. S. de Hoog. Massive contamination of *Exophiala dermatitidis* and *E. phaeomuriformis* in railway stations in subtropical Turkey. Mycopathologia 2013,175,381–386.
[105] L. Naranjo, H. Urbina, A. De Sisto, V. Leon. Isolation of autochthonous non-white rot fungi with potential for enzymatic upgrading of Venezuelan extra-heavy crude oil. Biocatalysis and Biotransformation 2007,25,341–349.
[106] V. Tigini, V. Prigione, G. C. Varese. Mycological and ecotoxicological characterisation of landfill leachate before and after traditional treatments. Sci Total Environ 2014,487,335–341.
[107] A. Rougeron, G. Schuliar, J. Leto, et al. Human-impacted areas of France are environmental reservoirs of the *Pseudallescheria boydii/Scedosporium apiospermum* species complex. Environ Microbiol 2014,http://onlinelibrary.wiley.com/doi/10.1111/1462-2920.12472/abstract.
[108] N. N. Zhdanova, A. I. Vasilevskaya, L. V. Artyshkova, et al. Changes in micromycete communities in soil in response to pollution by long-lived radionuclides emitted in the chernobyl accident. Mycol Res 1994,98,789–795.
[109] N. N. Zhdanova, V. A. Zakharchenko, V. V. Vember, L. T. Nakonechnaya. Fungi from Chernobyl: mycobiota of the inner regions of the containment structures of the damaged nuclear reactor. Mycol Res 2000,104,1421–1426.
[110] N. N. Zhdanova, T. Tugay, J. Dighton, V. Zheltonozhsky, P. McDermott. Ionizing radiation attracts soil fungi. Mycol Res 2004,108,1089–1096.
[111] M. Ragon, G. Restoux, D. Moreira, A. P. Moller, P. Lopez-Garcia. Sunlight-exposed biofilm microbial communities are naturally resistant to Chernobyl ionizing-radiation levels. PLoS One 2011,6.
[112] M. A. Pfaller, D. J. Diekema. Epidemiology of invasive mycoses in North America. Crit Rev Microbiol 2010,36,1–53.
[113] V. A. Robert, A. Casadevall. Vertebrate endothermy restricts most fungi as potential pathogens. J Infect Dis 2009,200,1623–1626.
[114] A. Casadevall, F. C. Fang, L. A. Pirofski. Microbial virulence as an emergent property: consequences and opportunities. PLoS Pathog 2011,7.
[115] J. Heitman. Microbial pathogens in the fungal kingdom. Fungal Biol Rev 2011,25,48–60.
[116] A. Casadevall, L. A. Pirofski. Accidental virulence, cryptic pathogenesis, Martians, lost hosts, and the pathogenicity of environmental microbes. Eukaryot Cell 2007,6,2169–2174.
[117] C. A. Kumamoto. Niche-specific gene expression during *C. albicans* infection. Curr Opin Microbiol 2008,11,325–330.
[118] J. I. Pitt, A. D. Hocking. Fungi and food spoilage. 2nd ed. ed. Gaithersburg, Maryland: Aspen Publishers, Inc.; 1999.
[119] A. A. Nisiotou, N. Chorianopoulos, G. J. E. Nychas, E. Z. Panagou. Yeast heterogeneity during spontaneous fermentation of black *Conservolea* olives in different brine solutions. J Appl Microbiol 2010,108,396–405.
[120] J. M. Bermejo, J. B. Dominguez, F. M. Goni, F. Uruburu. Influence of pH on the transition from yeast-like cells to chlamydospores in *Aureobasidium pullulans*. Antonie van Leeuwenhoek 1981,47,385–392.
[121] R. A. Slepecky, W. T. Starmer. Phenotypic plasticity in fungi: a review with observations on *Aureobasidium pullulans*. Mycologia 2009,101,823–832.
[122] Z. Chi, F. Wang, L. Yue, G. Liu, T. Zhang. Bioproducts from *Aureobasidium pullulans*, a biotechnologically important yeast. Appl Microbiol Biotechnol 2009,82,793–804.
[123] J. Karkowska-Kuleta, M. Rapala-Kozik, A. Kozik. Fungi pathogenic to humans: molecular bases of virulence of *Candida albicans*, *Cryptococcus neoformans* and *Aspergillus fumigatus*. Acta Biochim Pol 2009,56,211–224.

[124] P. van Baarlen, van A. Belkum, R. C. Summerbell, P. W. Crous, B. P. H. J. Thomma. Molecular mechanisms of pathogenicity: how do pathogenic microorganisms develop cross-kingdom host jumps? FEMS Microbiol Rev 2007,31,239–277.
[125] M. Hawkes, R. Rennie, C. Sand, W. Vaudry. *Aureobasidium pullulans* infection: fungemia in an infant and a review of human cases. Diagn Microbiol Infect Dis 2005,51,209–213.
[126] L. Butinar, I. Spencer-Martins, N. Gunde-Cimerman. Yeasts in high Arctic glaciers: the discovery of a new habitat for eukaryotic microorganisms. Antonie van Leeuwenhoek 2007,91,277–289.
[127] P. Singh, M. Tsuji, S. M. Singh, U. Roy, T. Hoshino. Taxonomic characterization, adaptation strategies and biotechnological potential of cryophilic yeasts from ice cores of Midre Lovenbreen glacier, Svalbard, Arctic. Cryobiology 2013,66,167–175.
[128] F. Gardini, G. Suzzi, A. Lombardi, et al. A survey of yeasts in traditional sausages of southern Italy. FEMS Yeast Res 2001,1,161–167.
[129] A. SaldanhaDaGama, M. MalfeitoFerreira, V. Loureiro. Characterization of yeasts associated with Portuguese pork-based products. Int J Food Microbiol 1997,37,201–207.
[130] V. M. Dillon, R. G. Board. Yeasts associated with red meats. J Appl Bacteriol 1991,71,93–108.
[131] N. Papon, V. Savini, A. Lanoue, et al. *Candida guilliermondii*: biotechnological applications, perspectives for biological control, emerging clinical importance and recent advances in genetics. Curr Genet 2013,59,73–90.
[132] M. Desnos-Ollivier, M. Ragon, V. Robert, D. Raoux, J. C. Gantier, F. Dromer. *Debaryomyces hansenii (Candida famata)*, a rare human fungal pathogen often misidentified as *Pichia guilliermondii (Candida guilliermondii)*. J Clin Microbiol 2008,46,3237–3242.
[133] F. F. Tuon, S. F. Costa. *Rhodotorula* infection. A systematic review of 128 cases from literature. Rev Iberoam Micol 2008,25,135–140.
[134] S. Silva, M. Negri, M. Henriques, R. Oliveira, D. W. Williams, J. Azeredo. *Candida glabrata*, *Candida parapsilosis* and *Candida tropicalis*: biology, epidemiology, pathogenicity and antifungal resistance. FEMS Microbiol Rev 2012,36,288–305.
[135] A. Robiglio, M. C. Sosa, M. C. Lutz, C. A. Lopes, M. P. Sangorrin. Yeast biocontrol of fungal spoilage of pears stored at low temperature. Int J Food Microbiol 2011,147,211–216.
[136] A. Casadevall. Determinants of virulence in the pathogenic fungi. Fungal Biol Rev 2007,21,130–132.
[137] J. A. H. van Burik, P. T. Magee. Aspects of fungal pathogenesis in humans. Annu Rev Microbiol 2001,55,743–772.
[138] S. M. Brown, L. T. Campbell, J. K. Lodge. *Cryptococcus neoformans*, a fungus under stress. Curr Opin Microbiol 2007,10,320–325.
[139] M. C. Fisher, D. A. Henk, C. J. Briggs, et al. Emerging fungal threats to animal, plant and ecosystem health. Nature 2012,484,186–194.
[140] F. Silveira, M. Nucci. Emergence of black moulds in fungal disease: epidemiology and therapy. Current Opinion in Infectious Diseases 2001,14,679–684.
[141] M. A. Garcia-Solache, A. Casadevall. Global warming will bring new fungal diseases for mammals. MBio 2010,1.
[142] C. Gueidan, C. R. Villasenor, G. S. de Hoog, A. A. Gorbushina, W. A. Untereiner, F. Lutzoni. A rock-inhabiting ancestor for mutualistic and pathogen-rich fungal lineages. Stud Mycol 2008,61,111–119.
[143] G. S. de Hoog, P. Zalar, A. H. G. Gerrits van den Ende, N. Gunde-Cimerman. Relation of halotolerance to human-pathogenicity in the fungal tree of life: an overview of ecology and evolution under stress. In: N. Gunde-Cimerman, A. Oren, A. Plemenitaš, eds. Adaptation to life at high salt concentrations in Archaea, Bacteria, and Eukarya. Dordrecht, The Netherlands: Springer; 2005:373–395.

[144] K. Nishimura, M. Miyaji, H. Taguchi, R. Tanaka. Fungi in bathwater and sludge of bathroom drainpipes. Frequent isolation of *Exophiala* species. Mycopathologia 1987,97,17–23.
[145] Li DM, Li RY, G. S. de Hoog, M. Sudhadham, D. L. Wang. Fatal *Exophiala* infections in China, with a report of seven cases. Mycoses 2011,4,136–142.
[146] J. S. Zeng, D. A. Sutton, A. W. Fothergill, M. G. Rinaldi, M. J. Harrak, G. S. de Hoog. Spectrum of clinically relevant *Exophiala* species in the United States. J Clin Microbiol 2007,45,3713–3720.
[147] T. Matos, G. S. de Hoog, A. G. de Boer, I. de Crom, G. Haase. High prevalence of the neurotrope *Exophiala dermatitidis* and related oligotrophic black yeasts in sauna facilities. Mycoses 2002,45,373–377.
[148] G. Heinrichs, I. Hubner, C. K. Schmidt, G. S. de Hoog, G. Haase. Analysis of black fungal biofilms occurring at domestic water taps (I): Compositional analysis using tag-encoded FLX amplicon pyrosequencing. Mycopathologia 2013,175,387–397.
[149] E. Branda, B. Turchetti, G. Diolaiuti, M. Pecci, C. Smiraglia, P. Buzzini. Yeast and yeast-like diversity in the southernmost glacier of Europe (Calderone Glacier, Apennines, Italy). FEMS Microbiol Ecol 2010,72,354–369.
[150] P. D. Bridge, K. K. Newsham. Soil fungal community composition at Mars Oasis, a southern maritime Antarctic site, assessed by PCR amplification and cloning. Fungal Ecol 2009,2,66–74.
[151] J. W. Fell, G. Scorzetti, L. Connell, S. Craig. Biodiversity of micro-eukaryotes in Antarctic Dry Valley soils with < 5% soil moisture. Soil Biology & Biochemistry 2006,38,3107–3119.
[152] Petrovič U. Role of oxidative stress in the extremely salt-tolerant yeast *Hortaea werneckii*. FEMS Yeast Res 2006,6,816–822.
[153] E. Garre, F. Raginel, A. Palacios, A. Julien, E. Matallana. Oxidative stress responses and lipid peroxidation damage are induced during dehydration in the production of dry active wine yeasts. Int J Food Microbiol 2010,136,295–303.
[154] N. N. Gessler, Aver'yanov AA, T. A. Belozerskaya. Reactive oxygen species in regulation of fungal development. Biochemistry-Moscow 2007,72,1091–1109.
[155] A. J. Nappi, E. Ottaviani. Cytotoxicity and cytotoxic molecules in invertebrates. Bioessays 2000,22,469–480.
[156] M. Hamad. Antifungal immunotherapy and immunomodulation: A double-hitter approach to deal with invasive fungal infections. J. Scand Immunol 2008,67,533–543.
[157] S. M. Leal, C. Vareechon, S. Cowden, et al. Fungal antioxidant pathways promote survival against neutrophils during infection. J Clin Invest 2012,122,2482–2498.
[158] A. Oren. Mycosporine-like amino acids as osmotic solutes in a community of halophilic cyanobacteria. J. Geomicrobiol 1997,231–240.
[159] W. L. Wang, Z. M. Chi, Z. Chi, Li J, X. H. Wang. Siderophore production by the marine-derived *Aureobasidium pullulans* and its antimicrobial activity. Bioresour Technol 2009,100,2639–2641.
[160] M. Holzberg, W. M. Artis. Hydroxamate siderophore production by opportunistic and systemic fungal pathogens. Infect Immun 1983,40,1134–1139.
[161] C. L. Atkin, J. B. Neilands, H. J. Phaff. Rhodotorulic acid from species of *Leucosporidium, Rhodosporidium, Rhodotorula, Sporidiobolus,* and *Sporobolomyces,* and a new alanine-containing ferrichrome from *Cryptococcus melibiosum*. J Bacteriol 1970,103,722–733.
[162] T. Cairns, F. Minuzzi, E. Bignell. The host-infecting fungal transcriptome. FEMS Microbiol Lett 2010,307,1–11.
[163] Mc P. H. Cay, A. A. Ocampo-Sosa, G. T. Fleming. Effect of subinhibitory concentrations of benzalkonium chloride on the competitiveness of *Pseudomonas aeruginosa* grown in continuous culture. Microbiol 2010,156,30–38.

[164] S. Langsrud, G. Sundheim, A. L. Holck. Cross-resistance to antibiotics of *Escherichia coli* adapted to benzalkonium chloride or exposed to stress-inducers. J Appl Microbiol 2004,96,201–208.
[165] M. Braoudaki, A. C. Hilton. Adaptive resistance to biocides in *Salmonella enterica* and *Escherichia coli* O157 and cross-resistance to antimicrobial agents. J Clin Microbiol 2004,42,73–78.
[166] A. D. Russell. Introduction of biocides into clinical practice and the impact on antibiotic-resistant bacteria. J Appl Microbiol 2002,92,121s–35s.
[167] B. Szomolay, I. Klapper, J. Dockery, P. S. Stewart. Adaptive responses to antimicrobial agents in biofilms. Environ Microbiol 2005,7,1186–1191.
[168] A. K. Mangalappalli-Illathu, D. R. Korber. Adaptive resistance and differential protein expression of *Salmonella enterica* serovar Enteritidis biofilms exposed to benzalkonium chloride. Antimicrob Agents Chemother 2006,50,3588–3596.
[169] L. Selbmann, G. S. de Hoog, A. Mazzaglia, E. I. Friedmann, S. Onofri. Fungi at the edge of life: cryptoendolithic black fungi from Antarctic desert. Stud Mycol 2005,51,1–32.
[170] J. A. Draghi, T. L. Parsons, G. P. Wagner, J. B. Plotkin. Mutational robustness can facilitate adaptation. Nature 2010,463,353–355.
[171] F. Baquero. Environmental stress and evolvability in microbial systems. Clin Microbiol Infect 2009,15,5–10.
[172] S. J. Sorensen, M. Bailey, L. H. Hansen, N. Kroer, S. Wuertz. Studying plasmid horizontal transfer in situ: A critical review. Nat Rev Microbiol 2005,3,700–710.
[173] Z. H. Zhang, W. F. Qian, J. Z. Zhang. Positive selection for elevated gene expression noise in yeast. Mol Syst Biol 2009,5,299.
[174] R. S. Galhardo, P. J. Hastings, S. M. Rosenberg. Mutation as a stress response and the regulation of evolvability. Crit Rev Biochem Mol Biol 2007,42,399–435.
[175] K. Ikeda, H. Nakayashiki, M. Takagi, Y. Tosa, S. Mayama. Heat shock, copper sulfate and oxidative stress activate the retrotransposon MAGGY resident in the plant pathogenic fungus *Magnaporthe grisea*. Mol Genet Genomics 2001,266,318–325.
[176] H. Ogasawara, H. Obata, Y. Hata, S. Takahashi, K. Gomi. Crawler, a novel Tc1/mariner-type transposable element in *Aspergillus oryzae* transposes under stress conditions. Fungal Genet Biol 2009,46,441–449.
[177] J. T. Lennon, S. E. Jones. Microbial seed banks: the ecological and evolutionary implications of dormancy. Nat Rev Microbiol 2011,9,119–130.
[178] Z. Moranova, S. Kawamoto, V. Raclavsky. Hypoxia sensing in *Cryptococcus neoformans*: biofilm-like adaptation for dormancy? Biomedical Papers-Olomouc 2009,153,189–193.
[179] I. Kranner, W. J. Cram, M. Zorn, et al. Antioxidants and photoprotection in a lichen as compared with its isolated symbiotic partners. Proc Natl Acad Sci USA 2005,102,3141–3146.
[180] M. Breitenbach, B. Simon-Nobbe. The allergens of *Cladosporium herbarum* and *Alternaria alternata*. Fungal Allergy and Pathogenicity 2002,81,48–72.

Alexander I. Culley, Migun Shakya, and Andrew S. Lang
11 Viral evolution at the limits

11.1 Introduction

Viral evolution is a complex process driven by a variety of forces including the extracellular environment, the intracellular milieu, the interaction of viruses with one another, and is inextricably linked with the evolution of the host organism. There have been a number of recent interesting discussions of the potential roles of viruses in the evolution of cellular organisms on Earth (e.g. [1–4]), which include the possibility of a pre-cellular viral world. Regardless of the exact timing of their evolution, the effects of viruses on the evolution of cellular organisms are indisputable. These effects range from direct exertion of selection pressure [5, 6] and mediating genetic exchange [7] to more indirect effects such as shaping of microbial community structure through their killing of community members (reviewed in [8]). It is perfectly reasonable to assume that every cellular organism in every environment on the planet is a host for viruses, and even viruses are "infected" by other viruses [9, 10]. Here we cover issues of viruses and evolution in extreme environments as examples. What is an "extreme" environment? This is a subjective term, but we have taken a fairly traditional view and examined virus evolutionary processes in polar, deep-sea, acidic hot spring, and hypersaline environments as examples. We have examined some of what is known about viruses from these environments to look for information about virus evolution in these extreme conditions.

11.2 Acidic hot springs and hypersaline environments

Although bacteria are present, archaea dominate hypersaline environments and acidic hot springs. In consequence, the majority of viruses isolated and studied from these environments infect archaea. In this light, another relevant factor for virus evolution in these environments is the presence of clustered regularly inter-spaced short palindromic repeat (CRISPR) systems in almost all archaea [11], which is certain to exert specific selective pressures for archaeal viruses to counteract these defense systems. This dynamic arms race between viruses and CRISPR defenses has resulted in evolution of viral mechanisms to evade the system. Some of the ways in which viruses escape this defense system include the use of minimal mutation or recombinational changes in proto-spacers [12], total inactivation of the CRISPR system through production of a transcriptional repressor [13], and hijacking the CRISPR system and turning the defense system into an assault weapon against the host [14]. Anti-CRISPR defenses have been described in bacteriophages [15] and will presumably be discovered in archaeal viruses in due time. A recent comparative genomics study across prokaryotes suggests that there is an escalated arms race between viruses and hosts in extreme environments, with a pronounced enrichment of defense systems in the genomes of archaea compared with bacteria, and in hyperthermophiles compared to the mesophiles and psychrophiles [16]. Further genomic analyses showed that archaea and hyperthermophiles contained a greater distribution of toxin-antitoxin and CRISPR defense systems with a higher fraction of their genomes dedicated to these anti-viral systems compared to the average [17]. Toxin-antitoxin systems are part of abortive infection (Abi) systems that can cause cell death when activated. These systems, which are activated during stress or phage infection, consist of toxin and corresponding antitoxin proteins that neutralize each other during normal cell growth. However, when triggered, the more labile antitoxin is degraded resulting in dormancy or cell death, thus preventing the dissemination of phages in a microbial community [18]. In return, phages have developed mechanisms to promote dissemination by preventing cell death through production of proteins that function as antitoxins [19, 20], harboring mutations at specific regions to escape detection by Abi systems [21], harboring or hijacking the toxin-antitoxin system through recombination with plasmids that carry analogous systems [19]. Although much remains to be understood about the mechanism of toxin-antitoxin systems from extreme environments, it is likely that the viruses from these environments have developed similar, if not more complex, mechanisms to escape host Abi systems.

While the characterized viruses infecting bacteria are dominated by the *Caudovirales* (tailed phages), viruses infecting archaea show a much greater diversity of morphologies, as comprehensively covered in several recent reviews [22–24]. Another relevant observation is that there is a higher prevalence of non-lytic viruses in archaea, and one possible explanation for this is that these extreme environments make survival of extracellular virions more difficult [25]. It has been proposed that living in conditions of chronic energy stress particularly shapes the evolution of archaea [26]; because viruses are dependent on the physiology of their hosts for replication, this selection pressure will act similarly on archaeal viruses. In accordance with the chronic energy stress lifestyle, the majority of known archaeal viruses establish a chronic "carrier state" infection. In this state, the host remains alive with ongoing virion production [27].

One of the many interesting examples of viral novelties found in extreme environments is *Acidianus* two-tailed virus (ATV), a virus isolated from an acidic hot spring that infects the archaeon *Acidianus convivator* [28]. ATV is released into the environment from infected cells as a lemon-shaped particle, where it then undergoes morphological development by "growing" tails on each end of its virion. Tail development is temperature-dependent, only occurring at elevated temperatures (> 75 °C) and quickest at the natural hot spring temperature (85–90 °C). Unlike many of the other viruses from this type of environment that show more of a temperate lifestyle, ATV is a lytic virus, and it was proposed this structural feature might make the virions more stable outside of host cells in the hot acidic environment [28].

Viruses isolated from high temperature environments are able to withstand exposure to high temperatures. One thermostable virus, *Sulfolobus* turreted icosahedral virus (STIV) infecting the archaeon *Sulfolobus*, has been particularly well characterized structurally [29–33]. As summarized by Ortmann and colleagues [32], the STIV virion features several structural features believed to contribute to thermostability, including a capsid with tightly packed subunits and reduced volume, high proline content, and increased polar surface areas. Glycosylation of the major capsid protein has also been suggested as a stabilizing feature [29]. Interestingly, conservation of certain structural features between STIV and viruses infecting bacteria and eukaryotes support a 3-billion year evolutionary link between viruses found in the three domains [31, 33, 34].

Structural adaptations to the extreme environment are not limited to the archaea-specific virus structures. An archaeal virus with tailed-phage morphology, ϕH, infecting the halophile *Halobacterium halobium*, was found to be dependent on high salt for infectivity [35]. Incubation of the virus in < 2 M NaCl resulted in loss of infectious particles, which could be partially abrogated by addition of 0.1 M Mg^{2+}. Similar findings were observed with ϕCh1 infecting the haloalkaliphile *Natronobacterium magadii*, where incubation of the virus in ≤ 2 M NaCl resulted in a loss of > 4 orders of magnitude in plaque-forming units, which was not relieved with the addition of Mg^{2+} [36], and *Salisaeta* icosahedral phage 1 (SSIP-1), a virus that infects the halophile *Salisaeta* sp., where incubation in reduced salt concentrations lowered infectivity of the particles and plaque formation was lost when salt concentrations were relatively low [37]. Other viruses that infect halophiles have been shown to tolerate limited exposure to lowered salinity if subsequently restored into higher salinity conditions [38].

The recent use of metagenomic techniques to characterize viruses from extreme environments has greatly expanded our understanding of viral diversity in these locations. For example, the analysis of the viral community from an acidic geothermal lake uncovered sequences from a virus that appears to be an RNA-DNA virus hybrid [39]. This finding is of importance because it suggests that gene exchange between highly unrelated groups of viruses can occur. The authors propose that a deeper understanding of this virus may lead to greater insight into the ancient transition from an RNA-based world to a DNA-based world and lends support to the hypothesis that some types of DNA viruses emerged when DNA plasmids acquired structural genes from RNA viruses [40]. Also quite striking is the finding of two haloarchaeal viruses that have different genome structures and infect different species but which share remarkable protein sequence and genome organization similarities [41]. HHPV-1 infects *Haloarcula hispanica* and has a circular dsDNA genome [41] while HRPV-1 infects *Halorubrum* sp. and has a circular ssDNA genome [42]. Despite this difference in genomic structure, six of the eight HHPV-1 protein sequences show recognizable homology to HRPV-1 proteins, and one non-homologous protein in each viral genome is predicted to be the replication protein, Rep. Therefore, these viruses represent close relatives that have diverged via exchange of replication genes to generate ssDNA and dsDNA entities. Comparative genomics of *Sulfolobus* turreted icosahedral viruses (STIVs), *Thermococcus kodakarensis* virus 4 (TKV4), and *Methanococcus voltae* virus (MVV) revealed that these viruses have exchanged genes and also contain host genes [24]. The presence of host genes in viruses can be important for viral propagation as demonstrated in some phages infecting cyanobacterial genera *Synechococcus* and *Prochlorococcus*, which carry photosynthesis genes [43, 44]. During infection, these genes are expressed and provide increased energy to support virus replication [45, 46]. It is likely that host-derived metabolism genes are similarly being carried by viruses in extreme environments and used to gain a competitive edge in their host organisms.

11.3 The deep sea

The fluids within the ocean crust constitute a water mass that is estimated to be the largest aquifer system on Earth [47]. Driven by hydrothermal circulation, water enters the aquifer from the deep ocean through exposed crustal outcroppings on the seafloor and circulates through the porous, permeable volcanic crustal layer. It is estimated that the entire volume of the ocean passes through the subseafloor every one hundred thousand to one million years and the chemical exchange that occurs between crust and water during this period of time is an essential component of marine biogeochemical cycles [48]. Despite the extreme chemical and temperature gradients and energy limitations that are characteristic of this environment, a growing body of research has now established that the habitats of this system, including the sediments, hydrothermal vents, cold seeps, mud volcanos, seamounts, and even subseafloor, support a great diversity of microbial life, including archaea, bacteria, and some protists [49].

Our understanding of the viruses in this environment is limited. A majority of the viral work thus far represents reports of the total abundance of viruses in and around hydrothermal vents [50–54]. It is clear from this research that viruses are present in these habitats and that total viral abundance can vary over an order of magnitude, nevertheless, the relationships between viral abundance, host diversity, and geochemical conditions remain unclear. The first direct investigation of virus-host interactions at hydrothermal vents demonstrated that the proportion of lysogenic hosts in the vents was significantly greater than in the surrounding seawater [55]. A subsequent study based on sequences generated from excised pulsed-field gel electrophoresis bands from vent viral communities also suggested that a high proportion of vent viruses are temperate [56]. These data indicate that vent conditions may favor the prophage condition and that these intimate interactions between the viral and host genomes could result in a high rate of virus-mediated horizontal gene transfer (HGT). An innovative approach that capitalized on the CRISPR antiviral defense systems of archaea and bacteria has successfully linked viral sequences with hosts from a hydrothermal vent system [57], indicating that diverse taxa of vent-associated prokaryotes are susceptible to viral infection. A recent study of viruses associated with a mud volcano concluded that viral activity accounted for approximately a third of the total prokaryotic mortality, which is markedly higher than the 10–20% estimated for bacterial communities in surface waters [53], suggesting that viruses play an important role in the top-down control of prokaryotes in this ecosystem [58]. Subsequently, viral lysis of a greater percentage of the cellular population in this environment could result in a more diverse viral gene pool and an accelerated arms race between host and virus, ultimately producing a more diverse and rapidly evolving viral community. Several viruses have been isolated from prokaryotes cultured from hydrothermal vents and characterized [59–62]. Although several of these viruses had some sequence similarity to phages isolated from Yellowstone hot springs, they were otherwise distantly related to other characterized viruses. Hydrothermal vents are an exit point of subseafloor fluids so it is likely that there is some relationship between vent viral communities and the viruses in the crust.

The role of viruses in the evolution of the microbes that inhabit the subseafloor remains largely undetermined. However, given the paucity of grazers in this environment, it is likely that viruses are one of the primary agents of mortality. If this is the case, the host specificity of viruses has important implications for the overall community structure. It has been proposed that the nature of the interaction between virus and host is likely to be dependent on the degree of fluid flux and sedimentation present in a particular habitat [63]. In hydrologically active regions where the volume of fluid flux is high and host density and metabolic activity are greater, viruses are more likely to drive host evolution via mechanisms of HGT associated with lytic infection (i.e. generalized transduction and transformation after release of DNA from lysed host cells), while in low flux, nutrient-depleted regions with low cell abundance, lysogenic conversion is expected to be the predominant mechanism of HGT [63].

11.4 Polar environments

Polar environments are characterized by extreme cold, dramatic changes in light exposure, and ice, yet microbial life persists in a variety of habitats throughout the Arctic and Antarctic. Some of the organisms that inhabit these environments are psychrophiles, which have evolved a variety of strategies to overcome the effects of low temperatures on cellular processes [64]. Viruses infecting these organisms presumably face strong selection pressure for replication ability at lowered temperatures and possibly enhanced resistance to freeze-thaw cycles. Similar to the deep sea, polar viruses remain largely unstudied. Viruses are abundant in a diversity of polar aquatic habitats, including the marine water column and sediments [65, 66], sea ice [67], glacial cryoconite holes [68], and microbial mats in melt water streams and lakes [69, 70]. In the Arctic, the abundance of viruses in aquatic ecosystems is an order of magnitude lower than in temperate areas, with some startling exceptions such as in sea ice [67, 71]. A comprehensive analysis of the microbial communities of a meromictic lake in Antarctica using metagenomic methods indicated that the lake harbored a dynamic assemblage of viruses, whose community composition co-varied with changes in the cellular host community [72]. A global metagenomic survey of viral communities (viromes) pooled from four distinct oceanic regions, including the Arctic, resulted in the first comprehensive assessment of Arctic dsDNA viruses [73]. Comparison of these four viromes suggested that Arctic phage communities have a higher incidence of prophage-like sequences, similar to what was discussed above for hydrothermal vents and the subseafloor. These data are commensurate with results that indicate the percentage of lysogenic bacteria in the Beaufort Sea can at times reach 38% [74], and that viral sequences are three times more prevalent in the genomes of Arctic marine bacteria than in temperate bacterial genomes [75]. Although there are some reports to the contrary [76], viral mediated mortality is generally a substantial percentage of the total mortality of aquatic Arctic prokaryotes [65, 68, 74, 77–79], indicating that viruses play an important role in the cycling of nutrients in this environment.

There has been some limited work on viruses of eukaryotes in polar systems. One study [80] found that a group of viruses that infect important protistan primary producers in an Antarctic lake, are themselves preyed upon by virophages, satellite viruses that adversely affect the reproduction of the helper virus [10]. It was predicted that virophage activity reduces the impact of viral lysis on the host, leading to a higher frequency of blooms and greater overall primary production in this environment [80]. The apparent widespread prevalence of, in essence, a viral predator introduces an additional evolutionary pressure that must ultimately mold the viruses infected by virophages. It remains unclear what mechanisms viruses have developed to resist virophage infection and whether the maintenance of these defenses results in a reduction in viral productivity. An additional examination of an Antarctic lake virus community uncovered diverse populations of eukaryotic viruses, including viruses with ssDNA genomes, distantly related to any known viral taxa [81].

11.5 Viruses and their effects on host organisms and communities

Viruses undoubtedly play an important role in the evolution of their hosts, mediating horizontal gene transfer, as genetic reservoirs, determining community composition, and affecting fitness [8]. It remains to be determined if the conditions in extreme environments enhance or reduce the role of viruses in host evolution. In some extreme environments, the communities are comprised primarily of prokaryotes, and in these it could be predicted that the role of viruses as agents of mortality is enhanced because larger predators are generally less abundant [82]. Moreover, the high frequency of infection, relatively low host growth rates, and the increased incidence of "latency" that appear to characterize some extreme environments (low temperature environments in particular) may prolong the residency time of viruses in this environment, increasing the frequency of gene exchange between virus and host and enhancing the importance of viruses as a source and mediator of genetic diversity [82].

HGT is an important mechanism for natural variation and it plays an important role for prokaryotes in adapting to their environments. Analysis of genome sequences has made it clear that viruses have played an important role in the acquisition of novel genes in many prokaryotes [83]. Although there remains much to be discovered about viruses and HGT in extreme environments, genomic studies have suggested that HGT is rampant in many extreme environments. HGT can occur across species boundaries and even between different domains of life, which has contributed greatly to the evolution of prokaryotes in, and for adaptation to, extreme environments [84]. For example, a high level of gene transfer between archaea and hyperthermophilic bacteria, *Aquifex aeolicus* and *Thermatoga maritima*, has been documented in a hyperthermophilic environment [85, 86]. Similarly, genomes of deep-sea archaea have been shown to contain many horizontally acquired genes [87], psychrophilic diatoms have acquired prokaryotic ice-binding proteins (IBPs) from polar bacteria [88], and *Methanococcoides burtonii*, a psychrophilic archaeon isolated from the Antarctic, obtained CRISPRs from pathogenic bacteria [89]. It can be hypothesized that in extreme environments where physical constraints, such as pH and temperature, compromise the survival of naked DNA, virus-mediated gene transfer will be one of the dominant means of HGT between extremophiles.

One of the best-studied virus host systems in extreme environments involves viruses infecting hyperthermophilic archaea in the genus *Sulfolobus*. Starting with *Sulfolobus* spindle shaped virus 1 (SSV1) [90], the genome sequences of > 10 SSVs, in the family *Fuselloviridae*, have been published. SSVs are generally not host species-specific and can infect different species of *Sulfolobus*, which provides increased chances for exchange of genetic material between species [91]. SSV1 can also integrate into the *Sulfolobus* chromosome to form a provirus [92], which could also increase the likelihood that it packages host DNA. Additionally, SSVs might be involved in HGT by encapsidating non-conjugative plasmids into virus-like particles [93]. Similar modes of plasmid transfer have been found for *Acidianus* filamentous virus 1 (AFV1) [94] and *Pyrococcus abyssi* virus 1 (PAV1) [95].

The effects of HGT on viruses in extreme environments are not yet clear. Genome level biases in extremophiles for defense system genes could result in high selection pressures for viruses in these environments. However, the combined evidence of high structural diversity of archaeal viruses and novel defense-evading mechanisms and the importance of HGT in maintenance of microbial communities in extreme environments suggests there are strong effects on evolution of viruses in extreme environments.

11.6 Future perspectives

Viruses from extreme environments, in particular those infecting archaea, have been the subjects of greatly increased attention over the past decade. This has revealed a tremendous diversity of virion structures and viral genomes, greatly expanding our view of the virosphere. The processes of virus evolution in extreme environments are not different from those occurring under non-extreme conditions, but the selective forces are different. One trend that recurs in these systems is the observation of increased tendency for temperate lifestyles (▶ Fig. 11.1) in polar and deep-sea environments and increased release through non-lytic mechanisms in hot springs. However, in particular for environments such as polar and the deep sea, there is clearly a need to more comprehensively sample these environments. This is now being facilitated by technological advances including better and cleaner sampling platforms, cheaper and more robust sequencing and high-throughput approaches, cultivation-indepen-

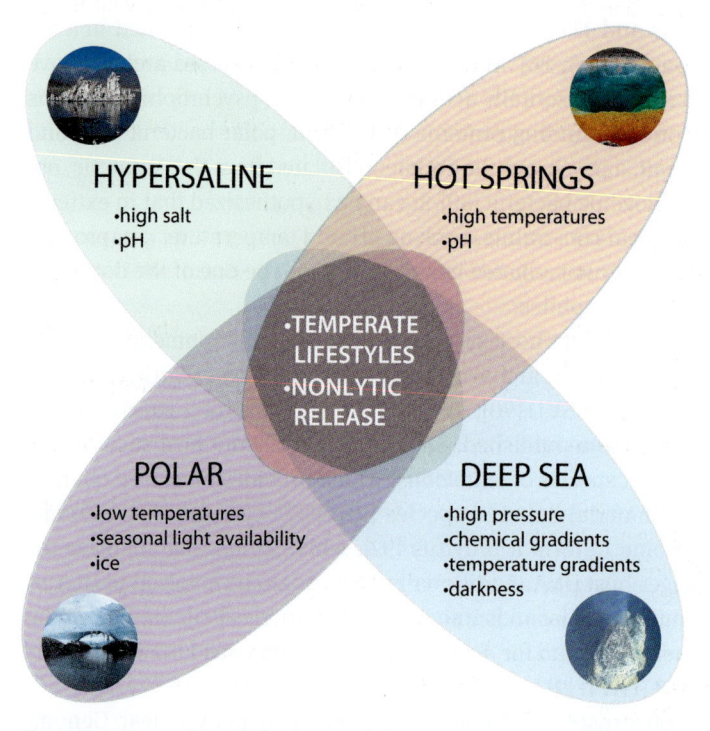

Fig. 11.1. Forces and trends in virus evolution in extreme environments. Physical and chemical forces acting on evolutionary processes in the different environments are indicated. Common themes discovered in these environments are tendencies towards temperate lifestyles and/or predominance of virus release through non-lytic mechanisms (center). Image credits: Deep Sea, Submarine Ring of Fire 2006 Exploration, U.S. National Oceanic and Atmospheric Administration Vents Program, Wikimedia Commons; Hot Springs, Jon Sullivan, Wikimedia Commons; Hypersaline, Carol M. Highsmith Archive, Wikimedia Commons; Polar, Alexander Culley.

dent single cell and virus sequencing, greater computer power, and more sophisticated bioinformatics and statistical approaches.

Acknowledgments

We thank Olga Zhaxybayeva for helpful discussions and Amanda Toperoff for graphical assistance with the figure. A.S.L. and A.I.C. thank the Natural Sciences and Engineering Research Council (NSERC) of Canada for funding to their research groups.

References

[1] P. Forterre, D. Prangishvili. The major role of viruses in cellular evolution: facts and hypotheses. Curr Opin Virol 2013, 3, 558–565.
[2] E. V. Koonin, V. V. Dolja. A virocentric perspective on the evolution of life. Curr Opin Virol 2013, 3, 546–557.
[3] E. V. Koonin, T. G. Senkevich, V. V. Dolja. The ancient Virus World and evolution of cells. Biol Direct 2006, 1, 29.
[4] A. Nasir, K. M. Kim, G. Caetano-Anolles. Viral evolution: Primordial cellular origins and late adaptation to parasitism. Mob Genet Elem 2012, 2, 247–252.
[5] C. Pal, M. D. Macia, A. Oliver, I. Schachar, A. Buckling. Coevolution with viruses drives the evolution of bacterial mutation rates. Nature 2007, 450, 1079–1081.
[6] F. Rodriguez-Valera, A.-B. Martin-Cuadrado, B. Rodriguez-Brito, L. Pasic, T. F. Thingstad, F. Rohwer, A. Mira. Explaining microbial population genomics through phage predation. Nat Rev Microbiol 2009, 7, 828–836.
[7] C. Canchaya, G. Fournous, S. Chibani-Chennoufi, M.-L. Dillmann, H. Brussow. Phage as agents of lateral gene transfer. Curr Opin Microbiol 2003, 6, 417–424.
[8] C. A. Suttle. Marine viruses – major players in the global ecosystem. Nat Rev Microbiol 2007, 5, 801–812.
[9] M. G. Fischer, C. A. Suttle. A virophage at the origin of large DNA transposons. Science 2011, 332, 231–234.
[10] B. La Scola, C. Desnues, I. Pagnier, C. Robert, L. Barrassi, G. Fournous, M. Merchat, M. Suzan-Monti, P. Forterre, E. Koonin, D. Raoult. The virophage as a unique parasite of the giant mimivirus. Nature 2008, 455, 100–104.
[11] R. A. Garrett, G. Vestergaard, S. A. Shah. Archaeal CRISPR-based immune systems: exchangeable functional modules. Trends Microbiol 2011, 19, 549–556.
[12] H. Deveau, R. Barrangou, J. E. Garneau, J. Labonte, C. Fremaux, P. Boyaval, D. A. Romero, P. Horvath, S. Moineau. Phage Response to CRISPR-Encoded Resistance in Streptococcus thermophilus. J Bacteriol 2008, 190, 1390–1400.
[13] C. I. Skennerton, F. E. Angly, M. Breitbart, L. Bragg, He S, K. D. McMahon, P. Hugenholtz, G. W. Tyson. Phage encoded H-NS: a potential Achilles heel in the bacterial defence system. PLoS ONE 2011, 6, e20095.
[14] K. D. Seed, D. W. Lazinski, S. B. Calderwood, A. Camilli. A bacteriophage encodes its own CRISPR/Cas adaptive response to evade host innate immunity. Nature 2013, 494, 489–491.
[15] J. Bondy-Denomy, A. Pawluk, K. L. Maxwell, A. R. Davidson. Bacteriophage genes that inactivate the CRISPR/Cas bacterial immune system. Nature 2013, 493, 429–432.

[16] K. S. Makarova, Y. I. Wolf, S. Snir, E. V. Koonin. Defense islands in bacterial and archaeal genomes and prediction of novel defense systems. J Bacteriol 2011, 193, 6039–6056.
[17] K. S. Makarova, Y. I. Wolf, E. V. Koonin. Comparative genomics of defense systems in archaea and bacteria. Nucleic Acids Res 2013, 41, 4360–4377.
[18] J. E. Samson, A. H. Magadan, M. Sabri, S. Moineau. Revenge of the phages: defeating bacterial defences. Nat Rev Microbiol 2013, 11, 675–687.
[19] T. R. Blower, T. J. Evans, R. Przybilski, P. C. Fineran, G. P. C. Salmond. Viral evasion of a bacterial suicide system by RNA-based molecular mimicry enables infectious altruism. PLoS Genet 2012, 8, e1003023.
[20] Y. Otsuka, T. Yonesaki. Dmd of bacteriophage T4 functions as an antitoxin against *Escherichia coli* LsoA and RnlA toxins. Mol Microbiol 2012, 83, 669–681.
[21] E. Bidnenko, A. Chopin, S. D. Ehrlich, M. C. Chopin. Activation of mRNA translation by phage protein and low temperature: the case of *Lactococcus lactis* abortive infection system AbiD1. BMC Mol Biol 2009, 10, 4.
[22] M. K. Pietilä, T. A. Demina, N. S. Atanasova, H. M. Oksanen, D. H. Bamford. Archaeal viruses and bacteriophages: comparisons and contrasts. Trends Microbiol 2014, 22, 334–344.
[23] M. Pina, A. Bize, P. Forterre, D. Prangishvili. The archeoviruses. FEMS Microbiol Rev 2011, 35, 1035–1054.
[24] D. Prangishvili, P. Forterre, R. A. Garrett. Viruses of the Archaea: a unifying view. Nat Rev Microbiol 2006, 4, 837–848.
[25] J. C. Snyder, K. Stedman, G. Rice, B. Wiedenheft, J. Spuhler, M. J. Young. Viruses of hyperthermophilic Archaea. Res Microbiol 2003, 154, 474–482.
[26] D. L. Valentine. Adaptations to energy stress dictate the ecology and evolution of the Archaea. Nat Rev Microbiol 2007, 5, 316–323.
[27] D. Prangishvili, R. A. Garrett. Viruses of hyperthermophilic Crenarchaea. Trends Microbiol 2005, 13, 535–542.
[28] M. Haring, G. Vestergaard, R. Rachel, L. Chen, R. A. Garrett, D. Prangishvili. Independent virus development outside a host. Nature 2005, 436, 1101–1102.
[29] R. Khayat, Fu C-y, A. C. Ortmann, M. J. Young, J. E. Johnson. The architecture and chemical stability of the archaeal *Sulfolobus* turreted icosahedral virus. J Virol 2010, 84, 9575–9583.
[30] R. Khayat, L. Tang, E. T. Larson, C. M. Lawrence, M. Young, J. E. Johnson. Structure of an archaeal virus capsid protein reveals a common ancestry to eukaryotic and bacterial viruses. Proc Natl Acad Sci U S A 2005, 102, 18944–18949.
[31] W. S. A. Maaty, A. C. Ortmann, M. Dlakic, K. Schulstad, J. K. Hilmer, L. Liepold, B. Weidenheft, R. Khayat, T. Douglas, M. J. Young, B. Bothner. Characterization of the archaeal thermophile *Sulfolobus* turreted icosahedral virus validates an evolutionary link among double-stranded DNA viruses from all domains of life. J Virol 2006, 80, 7625–7635.
[32] A. C. Ortmann, B. Wiedenheft, T. Douglas, M. Young. Hot crenarchaeal viruses reveal deep evolutionary connections. Nat Rev Microbiol 2006, 4, 520–528.
[33] D. Veesler, Ng T-S, A. K. Sendamarai, B. J. Eilers, C. M. Lawrence, S.-M. Lok, M. J. Young, J. E. Johnson, Fu C-y. Atomic structure of the 75 MDa extremophile *Sulfolobus* turreted icosahedral virus determined by CryoEM and X-ray crystallography. Proc Natl Acad Sci U S A 2013, 110, 5504–5509.
[34] G. Rice, L. Tang, K. Stedman, F. Roberto, J. Spuhler, E. Gillitzer, J. E. Johnson, T. Douglas, M. Young. The structure of a thermophilic archaeal virus shows a double-stranded DNA viral capsid type that spans all domains of life. Proc Natl Acad Sci U S A 2004, 101, 7716–7720.
[35] H. Schnabel, W. Zillig, M. Pfaffle, R. Schnabel, H. Michel, H. Delius. *Halobacterium halobium* phage øH. EMBO J 1982, 1, 87–92.

[36] A. Witte, U. Baranyi, R. Klein, M. Sulzner, C. Luo, G. Wanner, D. H. Krüger, W. Lubitz. Characterization of *Natronobacterium magadii* phage ΦCh1, a unique archaeal phage containing DNA and RNA. Mol Microbiol 1997, 23, 603–616.

[37] A. P. Aalto, D. Bitto, J. J. Ravantti, D. H. Bamford, J. T. Huiskonen, H. M. Oksanen. Snapshot of virus evolution in hypersaline environments from the characterization of a membrane-containing *Salisaeta* icosahedral phage 1. Proc Natl Acad Sci U S A 2012, 109, 7079–7084.

[38] M. K. Pietilä, P. Laurinmaki, D. A. Russell, Ko CC, D. Jacobs-Sera, S. J. Butcher, D. H. Bamford, R. W. Hendrix. Insights into head-tailed viruses infecting extremely halophilic Archaea. J Virol 2013, 87, 3248–3260.

[39] G. S. Diemer, K. M. Stedman. A novel virus genome discovered in an extreme environment suggests recombination between unrelated groups of RNA and DNA viruses. Biol Direct 2012, 7, 13.

[40] M. Krupovič, J. Ravantti, D. Bamford. Geminiviruses: a tale of a plasmid becoming a virus. BMC Evol Biol 2009, 9, 1–11.

[41] E. Roine, P. Kukkaro, L. Paulin, S. Laurinavicius, A. Domanska, P. Somerharju, D. H. Bamford. New, closely related haloarchaeal viral elements with different nucleic acid types. J Virol 2010, 84, 3682–3689.

[42] M. K. Pietilä, E. Roine, L. Paulin, N. Kalkkinen, D. H. Bamford. An ssDNA virus infecting archaea: a new lineage of viruses with a membrane envelope. Mol Microbiol 2009, 72, 307–319.

[43] N. H. Mann, A. Cook, A. Millard, S. Bailey, M. Clokie. Bacterial photosynthesis genes in a virus. Nature 2003, 424, 741.

[44] M. B. Sullivan, D. Lindell, J. A. Lee, L. R. Thompson, J. P. Bielawski, S. W. Chisholm. Prevalence and evolution of core photosystem II genes in marine cyanobacterial viruses and their hosts. PLoS Biol 2006, 4, e234.

[45] D. Lindell, J. D. Jaffe, Z. I. Johnson, G. M. Church, S. W. Chisholm. Photosynthesis genes in marine viruses yield proteins during host infection. Nature 2005, 438, 86–89.

[46] N. H. Mann, M. R. J. Clokie, A. Millard, A. Cook, W. H. Wilson, P. J. Wheatley, A. Letarov, H. M. Krisch. The genome of S-PM2, a "photosynthetic" T4-type bacteriophage that infects marine *Synechococcus* strains. J Bacteriol 2005, 187, 3188–3200.

[47] H. P. Johnson, M. J. Pruis. Fluxes of fluid and heat from the oceanic crustal reservoir. Earth Plan Sci Lett 2003, 216, 565–574.

[48] C. G. Wheat. Oceanic phosphorus imbalance: Magnitude of the mid-ocean ridge flank hydrothermal sink. Geophys Res Lett 2003, 30, 1895.

[49] B. N. Orcutt, J. B. Sylvan, N. J. Knab, K. J. Edwards. Microbial ecology of the dark ocean above, at, and below the Seafloor. Microbiol Mol Biol Rev 2011, 75, 361–422.

[50] C. Geslin, M. Le Romancer, G. Erauso, M. Gaillard, G. Perrot, D. Prieur. PAV1, the first virus-like particle isolated from a hyperthermophilic euryarchaeote,"*Pyrococcus abyssi*". J Bacteriol 2003, 185, 3888–3894.

[51] S. K. Juniper, M.-A. Cambon, F. Lesongeur, G. Barbier. Extraction and purification of DNA from organic rich subsurface sediments (ODP Leg 169S). Mar Geol 2001, 174, 241–247.

[52] A. C. Ortmann, C. A. Suttle. High abundances of viruses in a deep-sea hydrothermal vent system indicates viral mediated microbial mortality. Deep Sea Res I 2005, 52, 1515–1527.

[53] K. E. Wommack, R. R. Colwell. Virioplankton: Viruses in aquatic ecosystems. Microbiol Mol Biol Rev 2000, 64, 69–114.

[54] Y. Yoshida-Takashima, T. Nunoura, H. Kazama, T. Noguchi, K. Inoue, H. Akashi, T. Yamanaka, T. Toki, M. Yamamoto, Y. Furushima, Y. Ueno, H. Yamamoto, K. Takai. Spatial Distribution of Viruses Associated with Planktonic and Attached Microbial Communities in Hydrothermal Environments. Appl Environ Microbiol 2012, 78, 1311–1320.

[55] S. J. Williamson, D. B. Rusch, S. Yooseph, A. L. Halpern, C. Andrews-Pfannkoch, G. Sutton, M. Frazier, J. C. Venter. The Sorcerer II Global Ocean Sampling Expedition: metagenomic characterization of viruses within aquatic microbial samples. PLoS ONE 2008, 3, e1456.

[56] J. Ray, M. Dondrup, S. Modha, I. H. Steen, R.-A. Sandaa, M. Clokie. Finding a needle in the virus metagenome haystack – micro-metagenome analysis captures a snapshot of the diversity of a bacteriophage armoire. PLoS ONE 2012, 7, e34238.

[57] R. E. Anderson, W. J. Brazelton, J. A. Baross. Using CRISPRs as a metagenomic tool to identify microbial hosts of a diffuse flow hydrothermal vent viral assemblage. FEMS Microbiol Ecol 2011, 77, 120–133.

[58] C. Corinaldesi, Anno ADa, R. Danovaro. Viral infections stimulate the metabolism and shape prokaryotic assemblages in submarine mud volcanoes. ISME J 2011, 6, 1250–1259.

[59] C. Geslin, M. Gaillard, D. Flament, K. Rouault, M. Le Romancer, D. Prieur, G. Erauso. Analysis of the first genome of a hyperthermophilic marine virus-like particle, PAV1, isolated from *Pyrococcus abyssi*. J Bacteriol 2007, 189, 4510–4519.

[60] B. Liu, Wu S, Q. Song, X. Zhang, L. Xie. Two novel bacteriophages of thermophilic bacteria Isolated from deep-sea hydrothermal fields. Curr Microbiol 2006, 53, 163–166.

[61] Y. Wang, X. Zhang. Genome analysis of deep-sea thermophilic phage D6E. Appl Environ Microbiol 2010, 76, 7861–7866.

[62] Y. Yoshida-Takashima, Y. Takaki, S. Shimamura, T. Nunoura, K. Takai. Genome sequence of a novel deep-sea vent epsilonproteobacterial phage provides new insight into the co-evolution of *Epsilonproteobacteria* and their phages. Extremophiles 2013, 17, 405–419.

[63] R. E. Anderson, W. J. Brazelton, J. A. Baross. The deep viriosphere: assessing the viral impact on microbial community dynamics in the deep subsurface. Rev Mineral Geochem 2013, 75, 649–675.

[64] K. S. Siddiqui, T. J. Williams, D. Wilkins, S. Yau, M. A. Allen, M. V. Brown, F. M. Lauro, R. Cavicchioli. Psychrophiles. Ann Rev Earth Plan Sci 2013, 41, 87–115.

[65] G. F. Steward, D. C. Smith, F. Azam. Abundance and production of bacteria and viruses in the Bering and Chukchi Seas. Mar Ecol Prog Ser 1996, 131, 287–300.

[66] C. Winter, B. Matthews, C. A. Suttle. Effects of environmental variation and spatial distance on bacteria, archaea and viruses in sub-polar and arctic waters. ISME J 2013, 7, 1507–1518.

[67] R. Maranger, D. F. Bird, S. K. Juniper. Viral and bacterial dynamics in arctic sea ice during the spring algal bloom near Resolute, NWT, Canada. Mar Ecol Prog Ser 1994, 111, 121–127.

[68] C. Säwström, J. Laybourn-Parry, A. M. Anesio. High viral infection rates in Antarctic and Arctic bacterioplankton. Environ Microbiol 2007, 9, 250–255.

[69] C. Säwström, J. Laybourn-Parry. Influence of environmental conditions, bacterial activity and viability on the viral component in 10 Antarctic lakes. FEMS Microbiol Ecol 2008, 63, 12–22.

[70] W. F. Vincent, J. A. Gibson, R. Pienitz, V. Villeneuve, P. A. Broady, P. B. Hamilton, C. Howard-Williams. Ice shelf microbial ecosystems in the high arctic and implications for life on snowball earth. Die Naturwissenschaften 2000, 87, 137–141.

[71] L. E. Wells, J. W. Deming. Modelled and measured dynamics of viruses in Arctic winter sea-ice brines. Environ Microbiol 2006, 8, 1115–1121.

[72] F. M. Lauro, M. Z. Demaere, S. Yau, M. V. Brown, Ng C, D. Wilkins, M. J. Raftery, J. A. Gibson, C. Andrews-Pfannkoch, M. Lewis, J. M. Hoffman, T. Thomas, R. Cavicchioli. An integrative study of a meromictic lake ecosystem in Antarctica. ISME J 2010, 5, 879–895.

[73] F. E. Angly, B. Felts, M. Breitbart, P. Salamon, R. A. Edwards, C. Carlson, A. M. Chan, M. Haynes, S. Kelley, H. Liu, J. M. Mahaffy, J. E. Mueller, J. Nulton, R. Olson, R. Parsons, S. Rayhawk, C. A. Suttle, F. Rohwer. The marine viromes of four oceanic regions. PLoS Biol 2006, 4, e368.

[74] J. P. Payet, C. A. Suttle. To kill or not to kill: The balance between lytic and lysogenic viral infection is driven by trophic status. Limnol Oceanogr 2013, 58, 465–474.

[75] M. T. Cottrell, D. L. Kirchman. Virus genes in Arctic marine bacteria identified by metagenomic analysis. Aquat Microb Ecol 2012, 66, 107–116.
[76] G. F. Steward, L. B. Fandino, J. T. Hollibaugh, T. E. Whitledge, F. Azam. Microbial biomass and viral infections of prokaryotes in the mid-waters of the central Arctic Ocean. Deep-Sea Res I 2007, 54, 1744–1757.
[77] A. I. Kopylov, D. B. Kosolapov, E. A. Zabotkina, P. V. Boyarskii, V. N. Shumilkin, N. A. Kuznetsov. Planktonic viruses, heterotrophic bacteria, and nanoflagellates in fresh and coastal marine waters of the Kara Sea Basin (the Arctic). Inland Water Biol 2012, 5, 241–249.
[78] M. Middelboe. Bacterial growth rate and marine virus-host dynamics. Microb Ecol 2000, 40, 114–124.
[79] L. E. Wells, J. W. Deming. Significance of bacterivory and viral lysis in bottom waters of Franklin Bay, Canadian Arctic, during winter. Aquat Microb Ecol 2006, 43, 209–221.
[80] S. Yau, F. M. Lauro, M. Z. Demaere, M. V. Brown, T. Thomas, M. J. Raftery, C. Andrews-Pfannkoch, M. Lewis, J. M. Hoffman, J. A. Gibson, R. Cavicchioli. Virophage control of antarctic algal host-virus dynamics. Proc Natl Acad Sci U S A 2011, 108, 6163–6168.
[81] A. López-Bueno, J. Tamames, D. Velázquez, A. Moya, A. Quesada, A. Alcamí. High diversity of the viral community from an Antarctic lake. Science 2009, 326, 858–861.
[82] A. M. Anesio, C. M. Bellas. Are low temperature habitats hot spots of microbial evolution driven by viruses? Trends Microbiol 2011, 19, 52–57.
[83] H. Ochman, J. G. Lawrence, E. A. Groisman. Lateral gene transfer and the nature of bacterial innovation. Nature 2000, 405, 299–304.
[84] M. Wolferen, M. Ajon, A. M. Driessen, S.-V. Albers. How hyperthermophiles adapt to change their lives: DNA exchange in extreme conditions. Extremophiles 2013, 17, 545–563.
[85] L. Aravind, R. L. Tatusov, Y. I. Wolf, D. R. Walker, E. V. Koonin. Evidence for massive gene exchange between archaeal and bacterial hyperthermophiles. Trends Genet 1998, 14, 442–444.
[86] C. L. Nesbo, S. L'Haridon, K. O. Stetter, W. F. Doolittle. Phylogenetic analyses of two "archaeal" genes in *Thermotoga maritima* reveal multiple transfers between Archaea and Bacteria. Mol Biol Evol 2001, 18, 362–375.
[87] C. Brochier-Armanet, P. Deschamps, P. Lopez-Garcia, Y. Zivanovic, F. Rodriguez-Valera, D. Moreira. Complete-fosmid and fosmid-end sequences reveal frequent horizontal gene transfers in marine uncultured planktonic archaea. ISME J 2011, 5, 1291–1302.
[88] J. A. Raymond, H. J. Kim. Possible role of horizontal gene transfer in the colonization of sea ice by algae. PLoS ONE 2012, 7, e35968.
[89] N. F. Saunders, A. Goodchild, M. Raftery, M. Guilhaus, P. M. Curmi, R. Cavicchioli. Predicted roles for hypothetical proteins in the low-temperature expressed proteome of the antarctic archaeon *Methanococcoides burtonii*. J Proteome Res 2005, 4, 464–472.
[90] P. Palm, C. Schleper, B. Grampp, S. Yeats, P. McWilliam, W.-D. Reiter, W. Zillig. Complete nucleotide sequence of the virus SSV1 of the archaebacterium *Sulfolobus shibatae*. Virology 1991, 185, 242–250.
[91] R. M. Ceballos, C. D. Marceau, J. O. Marceau, S. Morris, A. J. Clore, K. M. Stedman. Differential virus host-ranges of the *Fuselloviridae* of hyperthermophilic Archaea: implications for evolution in extreme environments. Frontiers Microbiol 2012, 3, 295.
[92] N. L. Held, R. J. Whitaker. Viral biogeography revealed by signatures in *Sulfolobus islandicus* genomes. Environ Microbiol 2009, 11, 457–466.
[93] Y. Wang, Z. Duan, H. Zhu, X. Guo, Z. Wang, J. Zhou, Q. She, L. Huang. A novel *Sulfolobus* non-conjugative extrachromosomal genetic element capable of integration into the host genome and spreading in the presence of a fusellovirus. Virology 2007, 363, 124–133.
[94] T. Basta, J. Smyth, P. Forterre, D. Prangishvili, X. Peng. Novel archaeal plasmid pAH1 and its interactions with the lipothrixvirus AFV1. Mol Microbiol 2009, 71, 23–34.

[95] M. Krupovič, P. Forterre, D. H. Bamford. Comparative analysis of the mosaic genomes of tailed archaeal viruses and proviruses suggests common themes for virion architecture and assembly with tailed viruses of Bacteria. J Mol Biol 2010, 397, 144–160.

Eva C. M. Nowack and Arthur R. Grossman

12 Evolutionary pressures and the establishment of endosymbiotic associations

12.1 Introduction

In natural habitats, organisms of different species interact in many different capacities, for example, as competitors, prey, predators, or partners in metabolic exchange. In a broad sense, interactions between unlike organisms are referred to as 'symbioses'. These interactions shape the genetic and physiological capacities of organisms and have a strong impact on characteristics of ecosystems. However, since interspecific interactions shape both the physiologies of individuals and the characteristics of the environments in which they live, their effects on genome evolution are usually not easy to discern.

A situation in which partner organisms mutually benefit from interacting with one another is termed mutualism, or symbiosis in a narrower sense, which is the phenomenon that we discuss in this review. Often, symbiotic partners form close knit associations in which one partner (symbiont) populates the inter- or intracellular space of the other (host). In looser assemblies, referred to as 'ectosymbioses', the symbiont may live attached to the host or populate its body cavities. There are many fascinating examples of this latter type of association. (i) A widespread strategy among insects to protect themselves from potential pathogens exploits antimicrobial compounds synthesized by symbiotic bacteria [1]. For example, female solitary beewolves (*Philanthus triangulum*) cultivate the actinobacterium '*Ca.* Streptomyces philanthi' in specialized antennal gland reservoirs and then distribute these bacteria, which produce a cocktail of different antibiotic substances, into their subterranean larval brood cells (▶ Fig. 12.1 i). The incorporation of the bacteria into the cocoon silk of the developing larvae [2] provides the larvae with chemical protection against microbial infestation during maturation and hibernation [3]. (ii) The bioluminescent gammaproteobacterium *Vibrio fischeri* populates the light organ of squids. The bacteria produce light that can be modulated by the squid to match the down-welling illumination from the moon and stars, masking the silhouette of the squid, which minimizes predation (▶ Fig. 12.1 j and [4]). (iii) Trillions of symbionts of many different types (microbiota) are housed in the intestines of coelomate animals (▶ Fig. 12.1 k). These multi-species consortia help degrade complex nutrients and influence the health and well-being of the host organism, including humans, in numerous ways [5].

Symbiont-host integrations go a step further in 'endosymbiosis'; the endosymbiont resides within the tissue or intracellularly ('endocytobiosis') within a – typically eukaryotic – host. In multicellular hosts, endosymbionts are often confined to specialized cells or tissue types (e.g. bacteriocytes of insects (▶ Fig. 12.1 A), root nodules of leguminous plants (▶ Fig. 12.1 B), trophosomes of tubeworms (▶ Fig. 12.1 C), or endodermal cells of corals (▶ Fig. 12.1 D)). Endosymbionts can be derived from any domain of life (Eukaryota, Bacteria, Archaea) and confer a multitude of new biochemical abilities to the host that enable exploitation of otherwise inaccessible niches (discussed in 12.2.1). The different endosymbiotic associations use various strategies to establish

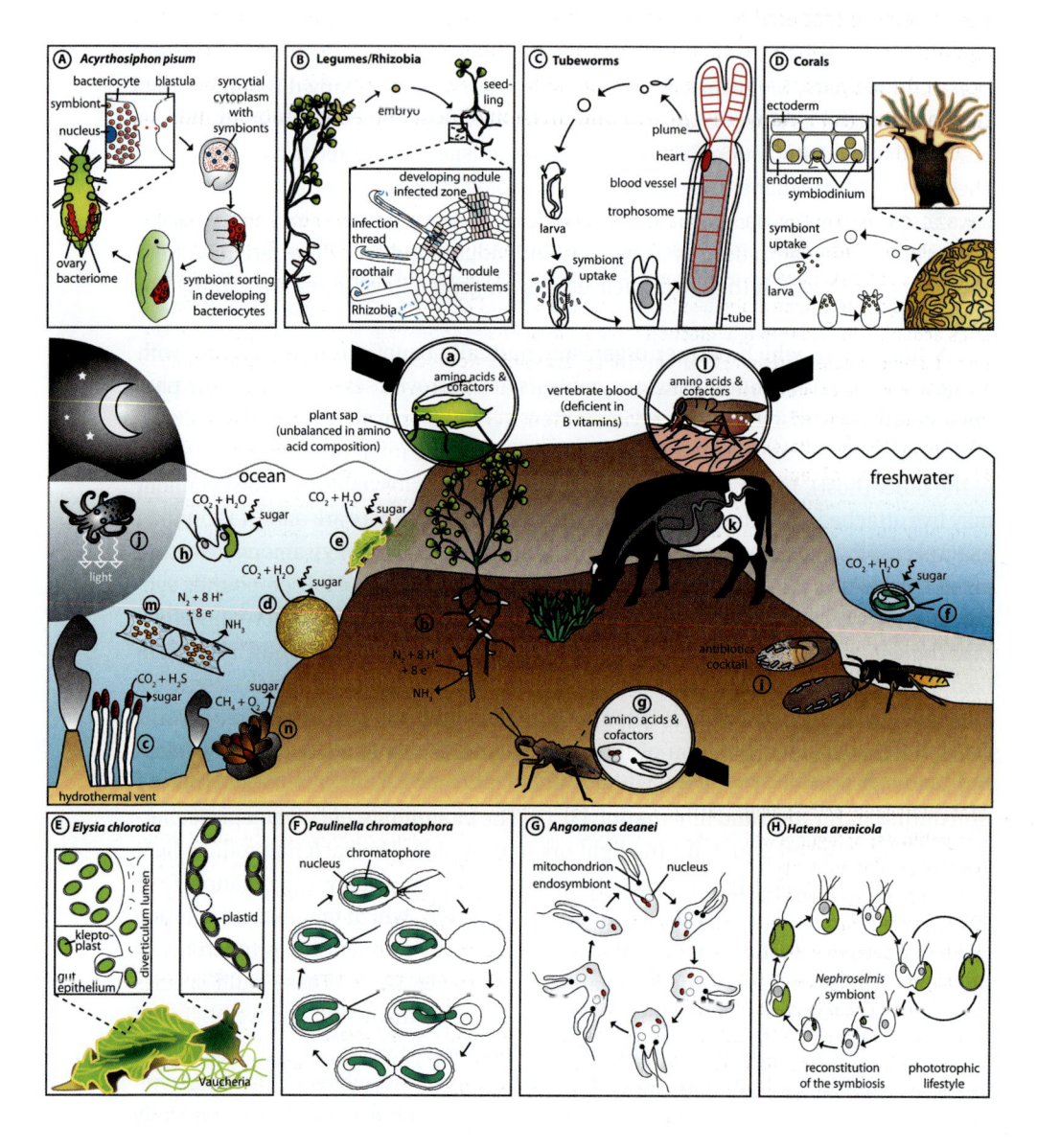

and maintain endosymbiosis (discussed in 12.2.2), often resulting in a robust association that can be stable for hundreds of millions of years (discussed in 12.2.3).

There are four key challenges related to establishing and maintaining an endosymbiotic association. (i) Endosymbionts have to be captured, overcome digestion, and interact with the host in a way that prevents immunological attack. (ii) Host and endosymbiont physiologies have to be integrated in a way that minimizes redundancy and energetic costs of bidirectional maintenance and control. (iii) A set of transporters have to evolve that enables exchange of metabolites between partner organisms, and signaling mechanisms have to evolve that coordinate transport processes. The evolution of these systems would facilitate the beneficial distribution of nutrients between the partners. (iv) Finally, the growth and division of host and endosymbiont must be coordinated to create a sustainable interaction.

Fig. 12.1. Diversity of Symbiotic Associations. *Small upper and lower panels*: Endosymbiont acquisition and/or transmission strategies: A) in the parthenogenic lifecyle of the aphid *Acyrthosiphon pisum Buchnera* symbionts are exocytosed from maternal bacteriocytes, and endocytosed by adjacent syncytial blastulae at the ovariole tip (adapted from [33]); B) in legumes aposymbiotic seedlings acquire Rhizobia through infection threads in root hairs and bacterial signals elicit the formation of a root nodule meristem; C) after settling, aposymbiotic tubeworm larvae take up symbiotic bacteria from the environment and sequester them in bacteriocytes that form the trophosome in the gut-less adult (adapted from [34]); D) endodermis cells of aposymbiotic coral larvae capture *Symbiodinium* (dinoflagellate) symbionts endocytotically from the environment, the adult polyps are also still capable of capturing symbionts; E) the sea slug *Elysia chlorotica* feeds on the xanthophyte alga *Vaucheria litorea* and the algal plastids are deposited as kleptoplasts in gland cells of finely branching gut diverticula and can be maintained for several months; F) in *Paulinella chromatophora*, chromatophore and host cell divisions are synchronized and upon cell division, one of the chromatophores is squeezed through the mouth opening and transferred to the new daughter cell; G) in trypanosomatids host and endosymbiont cell division are also synchronized and upon cell division, one symbiont is partitioned into each daughter cell (adapted from [35]); H) in dividing *Hatena arenicola* cells only one of the daughter cells inherits the symbiont while the other daughter cell can reestablish the symbiotic lifestyle by phagocytosis of a *Nephroselmis* cell (adapted from [36]). *Large central panel*: diverse symbiotic associations in their natural environment with information on the physiological functions of the symbionts. The lower case letters correspond to the upper case letters used in small panels when the association is depicted in both (more detail is generally shown in the individual small panels). a) Aphids with amino acid- and cofactor-producing *Buchnera* endosymbionts; b) legume with N_2-fixing rhizobial endosymbionts; c) tubeworms with thioautotrophic bacterial endosymbionts; d) corals with photosynthetic *Symbiodinium* endosymbionts; e) *E. chlorotica* with photosynthetic kleptoplasts; f) *P. chromatophora* with photosynthetic chromatophores; g) the trypanosomatid *Angomonas deanei* (gut-dwelling in insects) with amino acid- and cofactor-producing betaproteobacterial endosymbiont; h) *H. arenicola* with photosynthetic *Nephroselmis* endosymbiont; i) beewolf with antibiotics-producing '*Ca.* Streptomyces philanthi' ectosymbionts; j) squid with chemiluminescent *Vibrio fischeri* ectosymbionts; k) coelomate animals with diverse gut microbiota; l) tsetse fly with amino acid- and cofactor-producing *Wigglesworthia glossinidia* endosymbionts; m) the diatom *Hemiaulus* with N_2-fixing *Richelia intracellularis* endosymbionts; n) mytilid deep-sea mussels with methanotrophic bacterial endosymbionts.

As the processes mentioned above are shaped by evolutionary pressures, endosymbiosis can result in the complete merger of the two partners to form a novel organism with new biochemical and biosynthetic properties. The endosymbiont has become integral to the host's physiology, while at the same time losing its ability to function independently because of a massive loss of its gene content; the import of proteins to sustain endosymbiont functions marks the transition to an organelle (discussed in 12.3.2). However, every step in the evolution of the symbiotic partnership impacts both endosymbiont and host and can leave traces in their genomes (discussed in 12.3.1 and 12.4). Since the endosymbiont's environment is largely defined by interactions with a single host genotype rather than the countless organismic interactions associated with the free-living life style, genetic adaptations associated with the host-endosymbiont relationship are relatively easily distinguished in the endosymbiont genome sequence. Therefore, endosymbiosis is a universally important process that is providing us with insights into the merging of genetic systems in response to interspecies interactions and dependencies.

Glossary

aposymbiotic: host organisms in which their symbionts have been eliminated either by experimental treatments or natural conditions

bacteriocytes: specialized cells found in some insects including aphids and tsetse flies that house endosymbiotic bacteria such as species of *Buchnera*

bacteriome: organ-like assemblage of endosymbiont-containing bacteriocytes that is found in some insects

chromatophore: the alphacyanobacterial endosymbiont or nascent photosynthetic organelle of the cercozoan amoeba *Paulinella chromatophora*

endosymbiotic gene transfer (EGT): horizontal transfer of genes from an endogenous endosymbiont to the nuclear genome of its host

infection thread: invagination in the membrane of a root hair in leguminous plants through which Rhizobia are taken up and sequestered into nodule cells

horizontal transmission: repeated uptake of symbionts by a host organism from the environment or from another host

kleptoplast: temporarily retained functional plastids obtained from ingested algal prey

root nodule: a morphological structure on roots of leguminous plants that contains rhizobial endosymbionts that fix atmospheric nitrogen

trophosome: specialized organ in the coelomic cavity of siboglinid tubeworms that houses endosymbiotic bacteria that oxidize sulfur compounds

12.2 Diversity, evolution, and stability of endosymbiotic relationships

12.2.1 Diversity of endosymbionts and their physiological functions

Compared to eukaryotes, prokaryotes have evolved an astonishing diversity of metabolic capabilities. This genetic and metabolic diversity makes prokaryotes attractive symbiotic partners, and indeed the greatest endosymbiont diversity occurs among prokaryotes. Host dietary restrictions seem to be an important selective force that drives the evolution of (endo)symbioses. Examples of biochemical abilities provided by prokaryotic endosymbionts are (i) the synthesis of essential amino acids and cofactors (e.g. nutritional endosymbionts of insects with restricted or specialized diets such as aphids and psyllids that feed on plant sap (▶ Fig. 12.1 a) or tsetse flies that feed on vertebrate blood (▶ Fig. 12.1 l)), (ii) the production of fixed carbon through photosynthesis (e.g. alphacyanobacterial endosymbionts called chromatophores in the amoeba *Paulinella chromatophora* (▶ Fig. 12.1 f)), (iii) chemoautotrophic processes (e.g. sulfur-oxidizing gammaproteobacteria in marine invertebrates (▶ Fig. 12.1 c and [6])), and (iv) methanotrophy (e.g. methane-oxidizing bacteria in deep-sea mytilid mussels (▶ Fig. 12.1 n and [7, 8])), (v) N_2 fixation (e.g. Rhizobia in legumes (▶ Fig. 12.1 b)), and (vi) the production of defense compounds that act as deterrents to pathogens or parasites (e.g. [9–11]). Physiological traits conferred to similar host species by different endosymbionts can contribute to the success of the partnership in various habitats [12, 13].

Probably the largest number of endosymbiotic associations has been described for multicellular hosts, in particular for insects. Among vertebrates, endosymbiotic associations are virtually unknown. One exception is a green algal endosymbiont that lives intracellularly in embryos of the salamander *Ambystoma maculatum* [14, 15]. However, there are also a large number of unicellular algae, amoebae, and ciliates in natural populations that house intracellular bacteria [16–20]. Endosymbiotic bacteria have been shown to have a broad phylogenetic distribution, although known bacterial endosymbionts are particularly frequent among the *Alpha-* and *Gammaproteobacteria* [21]. In a number of cases, multi-species endosymbiotic communities provide nutrients to their hosts (e.g. [22–25]). However, for many bacterial-protist associations little more than a morphological description is available, precluding conclusions about the physiological roles of the bacteria in the association and whether the interactions should be classified as endosymbiosis, parasitism, or predator-prey events.

While eukaryotes evolved many times to live in intracellular environments, as exemplified by protist pathogens within trypanosomatids, apicomplexans, green algae (*Prototheca*), heterokonts (*Phytophtora*), or fungi (microsporidia), mutualistic eukaryotic endosymbionts appear primarily restricted to photosynthetic species. A counterexample has been noted for insects, which can harbor yeast-like symbionts that synthesize vitamins and digestive enzymes, or degrade toxic compounds in the host diet [26]. However, most of the non-photosynthetic eukaryotic associations identified appear to be ectosymbiotic (not endosymbiotic).

In some relatively unusual associations that range somewhere between a predator-prey and host-symbiont association, only the photosynthetic organelles (plastids) of a prey organism are maintained within a host. For example, some sacoglossan mollusks or sea slugs ingest different species of green, red, or heterokont algae and maintain the alga's plastids within gland cells of finely branching diverticula of the digestive tract. These plastids provide the mollusk with fixed carbon (▶ Fig. 12.1 e and E). In the sea slug *Elysia chlorotica* this association can be maintained for ~10 months before the plastid is degraded [27–29]. The use of so-called kleptoplasts (i.e. temporarily retained functional plastids obtained from ingested algal prey) is not exclusive to sea slugs and is frequently observed in unicellular hosts that include ciliates [30, 31] and foraminifera [32].

12.2.2 Evolutionary routes to establish and maintain endosymbiosis

An obvious question about the evolution of endosymbiosis concerns the mechanisms by which endosymbionts invade host cells. Phylogenetic maps of host-association traits suggest that bacterial endosymbionts evolved (i) from environmental bacteria that acquired symbiosis island genes by horizontal gene transfer (HGT; e.g. in Rhizobia), (ii) from environmental bacteria captured as prey by phagocytic uptake, (iii) or from intracellular pathogens [37]. Unless the genetic repertoire required to establish a symbiotic association is obtained by HGT of symbiotic plasmids, it has to evolve in single steps. For this to happen, a long term association between the future endosymbiont and host would have to be sustained, potentially through continuous predator-prey or host-pathogen interactions. In the latter case, loss of virulence with prolonged infection times would drive the system towards endosymbiosis. In accord with this concept, it is sometimes unclear at just what point an intracellular organism is considered a parasite or endosymbiont. For example, the association of the alphaproteobacterium *Wolbachia* with several arthropod hosts protects the host from viral and bacterial infections, while also inducing a number of reproductive alterations in the host including feminization, male-killing, parthenogenesis, and cytoplasmic incompatibility [38].

Maintenance of endosymbiosis between host generations requires a mechanism for symbiont transmission. This transmission can be achieved in remarkably diverse ways (reviewed in [39]), with perhaps the most prominent mechanism involving horizontal transmission, i.e. repeated uptake of endosymbionts from an environmental pool. This mode of transmission is used by legumes (▶ Fig. 12.1 B and [40]), tubeworms (▶ Fig. 12.1 C and [41, 42]), and many corals (▶ Fig. 12.1 D and [43]), among others. For horizontally transmitted endosymbionts, the symbiotic life-style is facultative, while for the host it is often obligate (e.g. tubeworms, corals). Thus, successful recapture of the symbiont becomes vital for host survival. Therefore, sophisticated molecular machineries have evolved to attract endosymbionts and mediate inter-partner recognition [39]. A symbiont recapture process that has been characterized in detail is represented by nodule formation in legumes [40]: flavonoids synthesized and excreted by legumes induce production of nod factors by Rhizobia. Nod factors (a group of lipo-chito-oligosaccharides with strain-specific chemical decorations) are perceived by specific host receptors and elicit various host responses, such as root hair curling, root hair invasion, and the formation of a nodule meristem. Bacteria enter the plant through root hairs in a process that resembles endocytosis. The plant forms a structure called an infection thread, the bacteria traverse this structure and ultimately thousands of bacteria, each encased in a peribacteroid membrane, are deposited in the cytoplasm of root nodule cells (▶ Fig. 12.1 B). An interesting problem for hosts that associate with multiple genotypes of symbionts -such as legumes- is the occurrence of 'cheaters', i.e. individual symbionts that cooperate to a lesser extent, or do not cooperate at all, but are able to indirectly benefit from the cooperation of other symbionts [44]. Since these cheaters can limit their own costs (e.g. for N_2 fixation) by 'free-riding' on other symbionts, their selection and proliferation would be favored if there were no interventions by the host. It is not always possible to establish an efficacious partnership based on partner selection through biochemical signaling since signaling can be mimicked by potential cheaters and a cooperating symbiont can be transformed into a cheater as a consequence of a single mutation. Thus, to stabilize cooperation between two organisms, monitoring of symbiont performance by the host and sanctions that discriminate among partners based on actual symbiotic performance, are necessary. In soybean, experimentally decreasing the nodule's N_2-fixing performance resulted in reduced rhizobial reproduction within nodules along with a potential host-imposed reduction in the supply of O_2 to the nodule [45].

Since horizontal transmission of the endosymbiont between generations depends on their successful recapture, there is some possibility of failure. A more robust strategy involves the strict vertical transmission of the endosymbionts to the next generation of hosts. For protists, vertical transmission can be achieved by sorting a defined number of symbionts to the daughter cell upon cell division, as in the photosynthetic amoeba *P. chromatophora* ([46] and ▶ Fig. 12.1 F) or symbiont-harboring trypanosomatids ([35] and ▶ Fig. 12.1 G). In multicellular hosts, mechanisms of vertical transmission range from transfer of endosymbionts within the body of the host to germ line cells (e.g. the endosymbiont *Buchnera aphidicola* in aphids; see ▶ Fig. 12.1 A), to behavioral mechanisms that ensure delivery of endosymbionts to offspring (reviewed in [39]). An interesting system that exploits both vertical and horizontal transmission involves the katablepharid flagellate *Hatena arenicola* [36, 47]. *H. arenicola* harbors the green algal symbiont *Nephroselmis* sp. Upon cell division, only one of the daughter cells inherits the symbiont, resulting in an autotrophic symbiont-bearing flagellate and a symbiont-lacking (colorless) cell that develops a feeding apparatus *de novo* and ingests prey cells. However, capture of a suitable *Nephroselmis* cell can reestablish the symbiotic lifestyle (▶ Fig. 12.1 H). This phenomenon illustrates the need for synchronized division of host and endosymbiont and sorting of a defined number of symbionts to each daughter cell in order to sustain a continuous symbiotic lifestyle.

At this stage we have a significant understanding of the way in which an endosymbiont evolved into a plastid or mitochondrion, the complex metabolic interactions between organelle and host cytoplasm, and some of the mechanisms and host- and organelle-encoded factors that coordinate metabolism, gene expression, and organelle division [48, 49]. However, for endosymbionts, there is still little known about how the biological processes of the two organisms are integrated and coordinated, including those processes that synchronize host and endosymbiont DNA replication and cell division.

12.2.3 Stability and the age of endosymbioses

Strictly vertical transmission of the endosymbiont to the next generation of hosts leads to long term co-evolution of the host and its endosymbiont. This is reflected by congruent phylogenies of host and endosymbiont marker genes; such congruent phylogenies are observed in numerous symbiotic associations (e.g. clams [50], aphids [51], carpenter ants [52], psyllids [53], symbiont-harboring trypanosomatids [54]). If fossil records are available for the host, knowledge of co-speciation can be used to determine the age of the symbioses, which is otherwise notoriously difficult to deduce. This approach has enabled the dating of some insect-bacterial endosymbioses to up to 260 Ma old [55–58]. Because endosymbiotic interaction might have originated before the common ancestor for each of the clades in which the partner organisms are found, age determined in this way might be an underestimation. Incongruence of host and symbiont phylogenies is an indicator of horizontal transmission of symbionts [59, 60].

12.3 Genome evolution in endosymbiotic bacteria

12.3.1 Reductive genome evolution in endosymbionts

Horizontal transfer of specific genes into the progenitor of the endosymbiont might facilitate the establishment of symbiotic associations. Comparative genomics of various sulfur-oxidizing gammaproteobacterial endosymbionts of marine invertebrates suggest that HGT is involved in tailoring the metabolism of endosymbionts to specific ecological niches [61]. Furthermore, ample evidence supports the notion that HGT of plasmids/gene islands encoding key symbiotic functions has played a crucial role in Rhizobia evolution. However, experimental work has revealed that acquisition of the plasmid alone might not be enough to turn a bacterium into an endosymbiont; the phytopathogenic betaproteobacterium *Ralstonia solanacearum* transfected with a symbiotic rhizobial plasmid was only able to induce nodulation and populate nodules after mutations occurred that inhibited aspects of R. solanacearum virulence [62]. Hence, natural selection of bacterial adaptive changes in the host environment may be a major driver in the evolution of endosymbiotic associations.

Once a bacterium becomes an endosymbiont, adaptive evolution can result in genome reduction and deletion of costly functions that may be supplied by the host, as well as functions dispensable in a stable intracellular environment. This loss of genetic potential establishes a metabolic dependence of the endosymbiont on its host. Comparisons of facultative and obligate symbionts, such as Richelia intracellularis and Calothrix rhizosoleniae, two N_2-fixing cyanobacteria that associate with diatoms, illustrate this genetic 'streamlining'. R. intracellularis is a filamentous, heterocyst-forming Cyanobacterium that is an obligate symbiont present in the periplasm (between siliceous cell wall and plasma membrane) of the diatom Hemiaulus sp. (▶ Fig. 12.1 m). In contrast, the facultative symbiont C. rhizosoleniae (also fixes N_2 and develops heterocysts) is found free-living in the open ocean, but also attaches to the cell wall of the diatom Chaetoceros sp. While the R. intracellularis genome is reduced to 2.2 Mb (making it the N_2-fixing, heterocyst-forming cyanobacterium with the smallest genome) and lacks specific N-metabolism genes, which promote N_2 fixation and metabolic cooperation with its host, the genome of the closely related C. rhizosoleniae is 6.0 Mb and is more similar to genomes of other free-living heterocyst-forming cyanobacteria [63].

Different aspects of reductive genome evolution in endosymbiotic bacteria are exhaustively covered in a number of excellent recent reviews [64–69] and therefore will only be briefly mentioned here. Typical patterns of genome reduction involve the loss of genes for proteins integral to metabolic pathways that are redundant in the host, pathways for the synthesis of nutrients that are abundant in the host diet, outer membrane proteins, proteins involved in the biosynthesis of cell wall components, transporters and regulatory elements, DNA repair components, and hypothetical proteins of unknown function. Many of the proteins of unknown function may be involved in the responses of the organism to changing environmental conditions. Mechanistically, genome reduction seems to be driven by the combination of reduced selection pressure for maintaining numerous functions, small effective population sizes that cause rapid genetic drift, and the separation of endosymbionts from environmental organisms, which prevents HGT and the gain or repair of lost functions. The loss of genes involved in DNA repair during genome reduction increases the mutation rate, which likely accelerates sequence evolution. These factors lead to an accumulation of pseudogenes and subsequent pseudogene loss and genome compaction resulting from deletional bias [70]. Thus, in the establishment of endosymbiosis, genome reduction proceeds from the initial large, gene-dense environmental genome to an intermediate sized, gene-poor genome, and finally to a small, compact genome in a more mature symbiotic association. Furthermore, the lack of DNA repair functions causes a greater tendency of the genome to experience the universal GC to AT mutational bias [71], which likely causes the conspicuous predominance of highly AT-rich genomes among endosymbiotic bacteria.

12.3.2 Evolution toward an organelle and beyond

Endosymbiont genomes may evolve to organellar genome size. The genome of *Sulcia muelleri*, the bacteroidetes endosymbiont associated with the glassy-winged sharpshooter, is 245 kbp and encodes 228 proteins [72]. The genome of *Carsonella ruddii*, the gammaproteobacterial endosymbiont of psyllids, is 160 kbp and encodes 182 proteins [73]. And the genome of *Hodgkinia cicadicola*, the alphaproteobacterial symbiont of the singing cicadas, is 144 kbp and encodes 189 genes [74]. Survival of the endosymbionts, even though their genomes lack many essential genes, is explained by the cooperation between the host and endosymbiont (or sometimes interaction between the host and a second symbiont that co-occurs in the same host, see ▶ Tab. 12.1). Cooperation between the various partners includes the import of metabolites from the host cytoplasm. However, in some cases genome reduction is accompanied by the loss of genes involved in DNA replication, transcription, and translation, functions that occur in endosymbionts but that are not readily compensated for at the metabolite level [24, 65, 72, 75, 76]. For example, *H. cicadicola* encodes only two genes involved in DNA replication, both *H. cicadicola* and *S. muelleri* (co-occurring in cicada) encode only 17 of 20 amino acyl-tRNA synthetases, and *H. cicadicola* seems to contain an incomplete set of tRNAs. Thus, sequence information suggests that proteins and tRNAs encoded on the genome of the host cell, either of host origin or acquired through HGT from the endosymbiont (referred to as endosymbiotic gene transfer or EGT) or other organisms, must be imported into the endosymbiont, although presently there is no experimental evidence to support this hypothesis.

The dependence of the endosymbiont on expression and import of nuclear-encoded proteins and tRNAs is regarded as a hallmark of an endosymbiont-derived organelle (i.e. mitochondria and plastids). Organelle evolution has not only been accompanied by the loss of many genes from the endosymbiont genome, but also by the transfer of a subset of these genes into the host nuclear genome. The host genome therefore has become a mosaic of host and endosymbiont genes. In the reference plant *Arabidopsis thaliana* ~18% of the nuclear genes originated from the endosymbiont that evolved into the plastid [77]. As a consequence, complex import machineries evolved in plastids and mitochondria, the TIC/TOC and TIM/TOM complex, respectively, that enable the organelles to import the majority of their proteins from those synthesized on the 80S cytoplasmic ribosomes [78, 79]. This reductive evolution precludes the organelle from sustaining itself as an independent organism.

Recently, the import of proteins with photosynthetic function into the chromatophore, a nascent photosynthetic organelle, has been demonstrated in the amoeba *Paulinella chromatophora* (Cercozoa, Rhizaria) [80]. Genes encoding the photosystem I (PSI) subunits, PsaE and two copies of PsaK, were transferred from the chromatophore genome into the host nuclear genome. The corresponding proteins are synthesized in the cytoplasm of the amoeba and traffic (likely through the Golgi) into the chromatophore, where they assemble with chromatophore-encoded PSI subunits into PSI

Table 12.1. Summary of the symbiotic associations discussed. The table includes the most obvious physiological function of the symbiont (although symbionts might perform additional, less obvious functions). In most cases one species is given as an example for a whole group.

Host Scientific name	Common name	Group	Symbiont Name	Group	Main Function	Localization	Type[a]	Fig.
Acyrthosiphon pisum	aphids	animals	*Buchnera aphidicola*	Gammaproteobacteria	production of amino acids and cofactors	bacteriome	endo	1 a
Ambystoma maculatum	salamanders	animals	*Oophila amblystomatis*	green algae	photosynthesis	in developing embryos	endo	–
Anemonia viridis	sea anemones	animals	*Symbiodinium* sp.	dinoflagellates	photosynthesis	endoderm	endo	–
Angomonas deanei	–	trypanosomatids	Ca. Kinetoplastibacterium crithidii	Betaproteobacteria	production of amino acids and cofactors	cytoplasm	endo	1 g
Bathymodiolus spp.	deep-sea mussels	animals	type I methanotrophs	Gammaproteobacteria	methanotrophy	gill bacteriocytes	endo	1 n
Diceroprocta semicincta	singing cicadas	animals	coresident *Sulcia muelleri* and *Hodgkinia cicadicola*	Bacteroidetes and Alphaproteobacteria	together: production of amino acids and cofactors	bacteriome	endo	–
Elysia chlorotica	sea slugs	animals	kleptoplasts from *Vaucheria litorea*	heterokontophytes	photosynthesis	gland cells of gut diverticula	klepto	1 e
Euprymna scolopes	squids	animals	*Vibrio fischeri*	Gammaproteobacteria	chemoluminescence	light organ in the mantle cavity	ecto	1 j
Glossina morsitans	tsetse flies	animals	*Wigglesworthia glossinidia*	Gammaproteobacteria	production of amino acids and cofactors	bacteriome	endo	1 l
Hatena arenicola	–	katablepharids	*Nephroselmis* sp.	green algae	photosynthesis	cytoplasm	endo	1 h
Hemiaulus sp.	–	diatoms	*Richelia intracellularis*	filamentous, heterocyst-forming Cyanobacteria	N_2-fixation	periplasm	endo[b]	1 m
Homalodisca vitripennis	sharpshooters	animals	coresident *Sulcia muelleri* and *Baumannia cicadellinicola*	Bacteroidetes and Gammaproteobacteria	together: production of amino acids and cofactors	bacteriome	endo	–
Medicago truncatula	legumes	land plants	Rhizobia	diverse soil bacteria of the Alpha- and Betaproteobacteria	N_2-fixation	root nodules	endo	1 b
Pachypsylla venusta	psyllids	animals	*Carsonella ruddii*	Gammaproteobacteria	production of amino acids and cofactors	bacteriome	endo	–
Paulinella chromatophora	–	cercozoa	chromatophores	alphacyanobacteria	photosynthesis	cytoplasm	endo	1 f
Planococcus citri	mealybugs	animals	nested symbionts Ca. Moranella endobia and Ca. Tremblaya princeps	Gamma- and Betaproteobacteria	together: production of amino acids and cofactors	bacteriome	endo	–
Philanthus triangulum	beewolves	animals	Ca. *Streptomyces philanthi*	Actinobacteria	production of antibiotics	antennal gland reservoirs	ecto	1 i
Riftia pachyptila	tubeworms	animals	Ca. *Endoriftia persephone*	Gammaproteobacteria	thioautotrophy	trophosome	endo	1 c
Sitophilus oryzae	weevils	animals	Sitophilus primary endosymbiont (SPE)	Gammaproteobacteria	production of amino acids and cofactors	bacteriome	endo	–
various species	corals	animals	*Symbiodinium* sp.	dinoflagellates	photosynthesis	endoderm	endo	1 d
various species	coelomate animals	animals	microbiota	diverse microorganisms	degradation of complex nutrients	intestines	ecto	1 k

[a] Abbreviations: ecto, ectosymbiosis; endo, endosymbiosis; klepto, kleptoplasty.
[b] The symbiont localizes between cell wall and plasma membrane, so its characterization as endo- or ectosymbiont is debatable.

complexes. Although the exact mechanism of protein import without a TIC/TOC complex is yet to be elucidated, these results contribute to the blurring of boundaries that distinguish an endosymbiont from an organelle.

It is important to note that plastids and mitochondria do not represent an end point of evolution. However, in most habitats there is strong selection pressure to maintain photosynthetic and respiratory functions. In fact, the ability to photosynthetically fix carbon is so advantageous in most photic habitats that following the establishment of primary plastids from a cyanobacterial endosymbiont (that led to the three lineages within the plantae: the *Viridiplantae* [green algae and land plants], *Rhodophyta* [red algae], and glaucophyta [freshwater microscopic algae]), photosynthetic ability spread to other eukaryotic lineages through secondary endosymbioses. In secondary endosymbioses a red or green alga served as an endosymbiont, leading to the establishment of secondary plastids in heterokontophytes, cryptophytes, and haptophytes (in the chromists), dinoflagellates and apicomplexa (in the alveolates), chlorarachniophytes (in the rhizaria), and euglenophytes (in the excavata) [81, 82]. In most cases the secondary eukaryotic symbiont is only detected through traces of one or two extra membranes around the plastids; however, in the cryptophytes and chlorarachniophytes, which evolved from a red alga and a green alga, respectively, a residual eukaryotic nucleus (called a nucleomorph) with a highly reduced genome is still present in the periplastidial compartment [83]. When selective constraints to maintain photosynthetic function are relaxed through the establishment of a parasitic or saprotrophic lifestyle, the plastids can lose their ability to perform photosynthetic CO_2 fixation. Reductive genome evolution with loss of photosynthesis-related genes has been documented in plastids from achlorophyllic lineages among land plants [84, 85], green algae [86–88], euglenoids [89], and apicomplexa [90, 91]. In the case of the parasitic land plant *Rafflesia lagascae* and the non-photosynthetic green alga *Polytomella* spp. an apparent complete loss of the plastid genomes has been reported [92, 93]. Physiological functions of non-photosynthetic plastids are best documented in apicoplasts, the reduced non-photosynthetic plastids of apicomlexan parasites such as *Plasmodium falciparum* (the malaria parasite) or *Toxoplasma gondii*. Apicoplasts have a genome of 35 kb and have an essential role in the synthesis of isoprenoid precursors, type II fatty acids, and heme [94]. Similarly, in anaerobic ecological niches, selective constraints to maintain respiration are relaxed, and mitochondria evolved into various specialized organelles called hydrogenosomes, mitosomes, or mitochondria-like organelles [95]. This divergent evolution led to a reduction [96–98] or complete loss of the mitochondrial genome [95].

As a consequence of endosymbiosis, the acquisition of both new biosynthetic abilities and new compartments have increased the complexity of host cells. The evolutionary loss of organelles seems to be very rare. As described above, even when the primary function of an organelle (e.g. photosynthesis or respiration) is lost and its genome becomes highly reduced or completely lost, the symbiont-derived compartment is usually retained. This suggests that compartmentalization achieved through endosymbiosis is advantageous, and that it might provide a specialized environment that can be exploited by the cell. Compartmentalization could: (i) enable the use of distinctly different chemical environments within the same cell when the optimum operating environments for different enzymes are different; (ii) allow targeting of specific nuclear-encoded proteins into different compartments to enable the same enzyme to fulfill different physiological functions within distinct cellular environments; (iii) prevent undesired cross-talk between different metabolic pathways; or (iv) provide membrane barriers for energetic processes that require them (e.g. sequestration of metabolites, metal ions, establishment of proton gradients).

12.4 Evolution of the host genome as shaped by endosymbiosis

So far, symbiont genomes have received more attention than the genomes of their hosts, partially because they are much smaller than those of the host. However, endosymbiont evolution can only be understood in the context of host physiology and genetic potential. As deep sequencing is becoming less expensive and assembly technologies advancing, information from large, more complex host genomes are becoming accessible, making it possible to unveil molecular mechanisms used by host cells to control and manipulate their endosymbionts. Interesting features of host genomes that are starting to emerge include: (i) the complementary metabolic abilities of the host and endosymbiont; (ii) the evolution of transport mechanisms for exchanging metabolites, reductant, and energetic compounds between the host and endosymbiont; (iii) the contribution of HGT, gene duplications, and *de novo* evolution of genes in the evolution of the genetic repertoire that mediates host-endosymbiont interactions; and (iv) the control of endosymbiosis by host immune functions that have been recruited and re-functionalized.

12.4.1 Complementarity of host and endosymbiont metabolic abilities

Comparisons of cellular functions encoded on host and endosymbiont genomes reveals that there is often little overlap in retained genes involved in the production of specific important metabolites. Either complete pathways for the synthesis of specific nutrients are exclusively encoded on the genome of one symbiotic partner or the pathway genes are represented by a patchwork of genes present on the host and endosymbiont genomes [99–102]. This complementarity is most striking in multi-species symbioses in which the pathways are partitioned between different symbionts, such as in the tripartite symbiosis between the mealybug *Planococcus citri* and two nested bacterial endosymbionts, the gammaproteobacterium '*Ca.* Moranella endobia' that lives inside the betaproteobacterium '*Ca.* Tremblaya princeps' [24, 75, 103].

Thus, genomic data suggest an intricate metabolic interaction between endosymbiont and host, which would require a large repertoire of specific transport systems to enable control of metabolite fluxes between the two symbiotic partners. However, transport capacities encoded on the genomes of endosymbiotic bacteria are often extremely limited, suggesting that significant transport capacity has to be encoded on the host genome and that these cytoplasmically-synthesized transporters are recruited, in a way that has not yet been defined, to the symbiont-delimiting membranes.

12.4.2 Acquisition of symbiotic potential

For the establishment of an efficient symbiotic relationship, new genetic capabilities would be needed that underlie the potential for symbiotic interaction (here referred to as 'symbiotic potential'). Three important mechanisms to establish new cellular functions are HGT, gene duplication and diversification, and *de novo* evolution of genes (establishment and evolution of so called orphan genes).

12.4.2.1 Acquisition of symbiotic potential through HGT

Intriguingly, in many endosymbiotic associations, the host's contribution to metabolic pathways that are partitioned between host and endosymbiont genomes appears to involve a combination of native host-derived genes and bacterial genes acquired by HGT. For example, in trypanosomatid flagellates harboring a betaproteobacterial endosymbiont multiple amino acid biosynthetic pathways are chimeras of host- and endosymbiont-encoded enzymes. Particularly in parts of the pathways that are missing from many other eukaryotes such as methionine/cysteine and arginine/ornithine biosynthesis, several host-encoded genes were obtained by HGT [99]. These bacterial genes were preferentially obtained from *Firmicutes*, *Bacteroidetes*, and *Gammaproteobacteria*. Only a single nuclear-encoded gene encoding ornithine cyclodeaminase might have been obtained by EGT from the betaproteobacterial endosymbiont. Similarly, sap-feeding insects have acquired multiple genes of bacterial origin [100, 104–107]: psyllids, aphids, and mealybugs acquired at least 10, 12, and 22 bacterial genes, respectively. Interestingly, most of these genes were shown to be preferentially expressed in the bacteriome, suggesting important roles in mediating the symbiosis [100, 104–107]. Phylogenetic analyses of the acquired bacterial genes identified only sporadically the endosymbiont (or a bacterium closely related to the endosymbiont) as the donor; instead, the genes appear to have been acquired from a broad array of different bacterial phyla [100, 105–107]. This is a surprising finding, particularly for multicellular organisms, since in these organisms stable gene transfer depends on access of the bacterial DNA to the germ line. Access to the germ line is given for the transovarially transmitted endosymbionts, and although a wide variety of bacteria are associated with insects (gut microbiota), it is unclear how the bacterial DNA would be delivered to host germline cells. In all of these cases it should be emphasized that frequent HGT of metabolic genes among bacteria and strong sequence divergence in the endosymbiont genomes can impede or completely preclude identification of the original gene donor.

In none of the systems described above is there evidence for massive movement of genes from the endosymbiont into the host nuclear genome; this contrasts with the evolutionary history of plastids and mitochondria and also with initial studies of *P. chromatophora* in which a minimum of 32 genes (conservative estimate) were transferred from the chromatophore into the host nuclear genome [108]. Most of these genes encode short polypeptides with functions related to photosynthesis and light-acclimation. Additionally, in *P. chromatophora* 10 expressed genes of cyanobacterial, but not clearly alphacyanobacterial origin, were identified; they likely represent HGT from prey organisms and may have been acquired before the photoautotrophic lifestyle was established. These genes encode more diverse functions than those of the EGT genes.

12.4.2.2 Acquisition of symbiotic potential through gene duplications

The merging of two organisms and integration of their physiological processes could benefit from gene duplications accompanied by modifications of gene function. For example, the evolution of a new repertoire of transport functions within a host genome might help coordinate metabolite exchange between the host and endosymbiont. In accord with this concept, aphid and citrus mealybug genomes have undergone massive expansions of amino acid transporter genes and some paralogs became highly expressed in bacteriocytes; this latter finding, suggests a function in mediating interactions with the endosymbiont [109, 110].

Another primary challenge in establishing optimal metabolic integration between host and endosymbiont would involve the capacity of the host organism to exploit the outputs of photosynthesis in the endosymbiont and coordinate the production of those outputs with both external factors and the ability of the host to access and utilize the fixed carbon generated in the endosymbiont. An inability of the host to use photosynthetic outputs under conditions when the endosymbiont is harvesting light energy (absorption by the phycobilisome) would cause hyper-reduction of the photosynthetic electron transport system and the generation of potentially damaging, reactive oxygen species. In *P. chromatophora*, 13 of the 32 nuclear genes identified as being derived from the chromatophore genome, and likely arising from gene duplications, encode a class of proteins designated Hlips, or high-light inducible proteins. These small proteins are proposed to be the progenitors of the light-harvesting chlorophyll a, b binding proteins in green algae and vascular plants and have been shown to be critical for the acclimation of cyanobacteria to elevated light levels [111, 112].

Also in the sea anemone *Anemonia viridis* (cnidaria), which houses the endosymbiotic dinoflagellate *Symbiodinium* in endoderm cells, several cnidarian-specific gene duplications occurred for genes preferentially expressed in the symbiotic anemone (relative to aposymbiotic anemones) and in the endodermal tissue (relative to the ectodermal tissue) [113]. The functions of the proteins encoded by these duplicated genes include cell adhesion, metabolite transport, and signaling; they have been proposed to be involved in mediating symbiotic interactions.

12.4.2.3 Acquisition of symbiotic potential through *de novo* gene evolution

Novel genes can evolve *de novo* from proto-genes generated by basic translational activity in non-genic regions. These genes, which are called orphan genes, are usually short, display no significant sequence similarity with genes identified in other sequenced organisms, and have been associated with the evolution of lineage specific traits [114]. A draft assembly of the pea aphid genome identified approximately 20% of the total number of genes as orphan genes [115]. While this large number of orphan genes may at least partially reflect high rates of false positive gene predictions, a later study found that many of the most highly expressed genes in the aphid bacteriome are orphan genes. Among the 30 genes with the highest transcript levels in bacteriocytes, 10 represent orphan genes and encode short proteins with N-terminal signal peptides [116]. Expression of these genes occurs at a developmental stage that coincides with strict incorporation of endosymbionts in host cells that become part of the bacteriome, and bacteriocyte-specific expression is maintained throughout the life of the aphid. These findings suggest that *de novo* evolved genes could contribute to the establishment and maintenance of endosymbiotic associations.

12.4.3 Redefinition of immune functions

Components of the host immune system, a system specialized for recognition and control of pathogenic bacteria, appear to be readily modified to assume key regulatory functions in controlling symbiont-host interactions. In symbiotic associations in which symbionts are repeatedly captured from the environment, components of the immune system can be recruited to control symbiont specificity. In legumes, a Toll-interleukin receptor/nucleotide-binding site/leucine-rich repeat (TIR-NBS-LRR) class of plant resistance (R) proteins was found to be involved in controlling genotype-specific infection and nodulation [117]. While incompatible rhizobial strains can induce root hair curling, the infection threads fail to grow, probably because of defense responses triggered by recognition of yet unidentified rhizobial effectors by the host R proteins. In accord with this finding, many rhizobial strains possess a type III secretion system that enables delivery of effector proteins, which are known to modulate host range, into the host cells. Some rhizobial effector proteins are homologous to effector proteins of pathogenic bacteria, suggesting a similar recognition mechanism associated with symbiotic and pathogenic host–bacteria interactions [118]. The type III effector genes used by plant pathogens to counter host defense strategies display genetic patterns indicative of rapid evolution that likely reflect an antagonistic arms race between the pathogen and host. In contrast, rhizobial type III effector genes are highly conserved, potentially reflecting selective pressures to maintain symbiotic interactions with their host plants [119].

One strategy to control the growth and division of beneficial symbionts, while protecting them from fierce immune reactions that have evolved to eliminate pathogenic bacteria within the host body, is the localization of endosymbionts to specific 'protected' tissues i.e. bacteriome, trophosome, nodules, and others. A symbiosis-stabilizing environment is created by expression of 'adjusted' immune system-derived genes, while at the same time suppressing expression of those immune system genes that would adversely impact the association. Numerous studies report differential expression of immune system-related genes within bacteriocytes of various insects (e.g. [104, 120–122]). Furthermore, in the hydrothermal vent tubeworm *Ridgeia piscesae*, specific pattern recognition receptors (including peptidoglycan recognition proteins) and a Toll-like receptor, both classes of proteins previously associated with innate immune functions, were reported to show distinct accumulation in the trophosome compared to plume tissue [123]. The functional significance of these and similar findings remain to be clarified in most systems.

Two systems in which regulation of endosymbiosis through immune-system derived host genes has been characterized in some detail are the weevil *Sitophilus* spp., which hosts gammaproteobacterial endosymbionts in its bacteriome, and the legume *Medicago truncatula*, which hosts rhizobial endosymbionts in root nodules. In the weevil, the insect gene coleoptericin A (*colA*), which encodes an antimicrobial peptide, is over-expressed in bacteriocytes [121]. Immunohistochemistry using an antibody against ColA confirmed not only *colA* expression in all endosymbiont-bearing tissues, but also localized the ColA peptide to the endosymbiont cytoplasm [124]. Interestingly, ColA has a bacteriostatic effect on gram-negative bacteria (such as the endosymbiont) and causes bacterial gigantism in *Escherichia coli*, which resembles the elongated morphology of the endosymbiont in the bacteriome. Knock down of *colA* transcripts in weevil larvae resulted in resumption of endosymbiont cytokinesis and loss of the elongated cell morphology, suggesting that ColA is a component that potentially regulates the growth of endosymbionts, their escape from bacteriocytes, and their spread in larvae. Since ColA was found to interact with bacterial outer membrane proteins (Omps), it was hypothesized that ColA enters the symbiont through Omps. Similarly, the legume *M. truncatula* targets short secreted peptides that evolved from effectors of the plant's innate immune system into rhizobial endosymbionts harbored in host-derived membrane vesicles in nodule cells [125]. These immune system-derived peptides have a major impact on the fate and morphological and physiological characteristics of the endosymbiont.

12.5 Conclusions and future directions

Through establishment of an endosymbiotic association a eukaryotic host can rapidly co-opt new metabolic capabilities, thereby extending its ecological versatility. Genome-based studies of endosymbiotic associations (in particular with insects) are generating an expanding reservoir of knowledge concerning patterns of bacterial endosymbiont reductive genome evolution. However, for gaining a deeper understanding of factors that drive endosymbiont genome evolution, it will be important to define the actual intracellular environment of the symbiont, which is largely defined by the host. Determining transport capacities for metabolites and proteins across the membranes that delineate the symbiont and identifying transporters that control these fluxes will provide new insights into the exact mechanisms of exchange and how this exchange might be key for integrating the physiologies of the partner organisms.

Only recently has additional information become available on ways in which endosymbiosis shapes functionalities encoded in host genomes. The picture that is emerging is one that includes significant asymmetry of functions. As soon as the endosymbiosis becomes obligate for the symbiont, the symbiont loses numerous functions, transferring most control to the host organism. In the process of refining host-endosymbiont interactions, the host might gain novel functions through HGT, EGT, gene duplications and divergence, and possibly also *de novo* gene evolution.

However, how host-encoded functions compensate for gene losses in the endosymbiont, except in a few instances (e.g. import of nuclear-encoded photosystem I subunits in *P. chromatophora*), is unknown. In situations where metabolic pathways are partitioned between host and endosymbiont compartments, cooperation could occur (i) either through multiple transport processes that traffic pathway intermediates between the compartments, (ii) or through transport of enzymes of a given pathway into a single compartment (e.g. through the development of protein transport mechanisms). Differentiating between these scenarios is challenging since both elucidating metabolic fluxes between the host and endosymbiont and proteomic characterizations of endosymbiotic associations are difficult. In a shotgun proteomic analysis of *Buchnera* endosymbionts isolated from hundreds of bacteriomes of aphid larvae, no imported host-encoded proteins could be identified with confidence [126]. However, identification of a specific protein in such experiments depends on the abundance and resolution of that protein in a highly complex protein mixture. A more thorough molecular understanding of endosymbiotic interactions will ultimately require sophisticated proteomics and the establishment of gene transfer systems that will allow disruption and knock down of endogenous genes as well as the introduction of heterologous genes.

Finally, the number of cases in which a protist has been described to harbor an endosymbiont probably represents a small fraction of the true diversity of such relationships in nature. There are estimates that 5% of all eukaryotic algae and amoeba carry intracellular bacteria [17, 20]. This makes it important for future researchers to systematically examine more protists for the occurrence of endosymbionts, determine their phylogenetic origins, physiological functions, and the ways in which the partner organisms have integrated their metabolic machinery.

Acknowledgments

This work was supported by National Science Foundation Grant MCB-1157627.

References

[1] M. Kaltenpoth, T. Engl. Defensive microbial symbionts in Hymenoptera. Funct Ecol 2014, 28, 315–327.
[2] M. Kaltenpoth, W. Göttler, G. Herzner, E. Strohm. Symbiotic bacteria protect wasp larvae from fungal infestation. Curr Biol 2005, 15, 475–479.
[3] S. Koehler, J. Doubský, M. Kaltenpoth. Dynamics of symbiont-mediated antibiotic production reveal efficient long-term protection for beewolf offspring. Front Zool 2013, 10, 3.
[4] B. W. Jones, M. K. Nishiguchi. Counterillumination in the Hawaiian bobtail squid, *Euprymna scolopes* Berry (Mollusca: Cephalopoda). Marine Biol 2004, 144, 1151–1155.
[5] S. Devkota, E. B. Chang. Nutrition, microbiomes, and intestinal inflammation. Curr Opin Gastroenterology 2013, 29, 603–607.
[6] N. Dubilier, C. Bergin, C. Lott. Symbiotic diversity in marine animals: the art of harnessing chemosynthesis. Nat Rev Microbiol 2008, 6, 725–740.
[7] E. G. DeChaine, C. M. Cavanaugh. Symbioses of methanotrophs and deep-sea mussels (Mytilidae: Bathymodiolinae). Prog Mol Subcell Biol 2006, 41, 227–249.
[8] J. M. Petersen, N. Dubilier. Methanotrophic symbioses in marine invertebrates. Environ Microbiol Rep 2009, 1, 319–335.
[9] K. M. Oliver, J. A. Russell, N. A. Moran, M. S. Hunter. Facultative bacterial symbionts in aphids confer resistance to parasitic wasps. Proc Natl Acad Sci USA 2003, 100, 1803–1807.
[10] C. L. Scarborough, J. Ferrari, H. C. J. Godfray. Aphid protected from pathogen by endosymbiont. Science 2005, 310, 1781.
[11] A. Nakabachi, R. Ueoka, K. Oshima, et al. Defensive bacteriome symbiont with a drastically reduced genome. Curr Biol 2013, 23, 1478–1484.
[12] Y. Fujiwara, K. Takai, K. Uematsu, S. Tsuchida, J. C. Hunt, J. Hashimoto. Phylogenetic characterization of endosymbionts in three hydrothermal vent mussels: Influence on host distributions. Marine Ecol Prog Ser 2000, 208, 147–155.
[13] R. A. Beinart, J. G. Sanders, B. Faure, et al. Evidence for the role of endosymbionts in regional-scale habitat partitioning by hydrothermal vent symbioses. Proc Natl Acad Sci USA 2012, 109, E3241-E50.
[14] E. R. Graham, S. A. Fay, A. Davey, R. W. Sanders. Intracapsular algae provide fixed carbon to developing embryos of the salamander *Ambystoma maculatum*. J Exp Biol 2013, 216, 452–459.
[15] R. Kerney, E. Kim, R. P. Hangarter, A. A. Heiss, C. D. Bishop, B. K. Hall. Intracellular invasion of green algae in a salamander host. Proc Natl Acad Sci USA 2011, 108, 6497–6502.
[16] S. I. Fokin. Bacterial endocytobionts of ciliophora and their interactions with the host cell. In: International Review of Cytology – a Survey of Cell Biology; 2004:181–249.
[17] M. Horn. Chlamydiae as symbionts in eukaryotes. Annu Rev Microbiol 2008, 62, 113–131.
[18] E. C. M. Nowack, M. Melkonian. Endosymbiotic associations within protists. Phil Trans R Soc Lond B-Biol Sci 2010, 365, 699–712.
[19] S. Schmitz-Esser, E. R. Toenshoff, S. Haider, et al. Diversity of bacterial endosymbionts of environmental *Acanthamoeba* isolates. Appl Environ Microbiol 2008, 74, 5822–5831.
[20] B. Surek, M. Melkonian. Intracellular bacteria in the Euglenophyceae: Prolonged axenic culture of an alga – bacterial system. In: H. E. A. Schenk, W. Schwemmler, eds. Endocytobiology. Berlin; New York: de Gruyter; 1983:475–486.
[21] C. Toft, S. G. E. Andersson. Evolutionary microbial genomics: Insights into bacterial host adaptation. Nat Rev Genet 2010, 11, 465–475.
[22] M. J. Gosalbes, A. Lamelas, A. Moya, A. Latorre. The striking case of tryptophan provision in the cedar aphid *Cinara cedri*. J Bacteriol 2008, 190, 6026–6029.

[23] Wu D, S. C. Daugherty, S. E. Van Aken, et al. Metabolic complementarity and genomics of the dual bacterial symbiosis of sharpshooters. PLoS Biol 2006, 4, 1079–1092.
[24] J. P. McCutcheon, C. D. Von Dohlen. An interdependent metabolic patchwork in the nested symbiosis of mealybugs. Curr Biol 2011, 21, 1366–1372.
[25] S. Duperron, T. Nadalig, J. C. Caprais, et al. Dual symbiosis in a *Bathymodiolus* sp. mussel from a methane seep on the Gabon Continental Margin (Southeast Atlantic): 16S rRNA phylogeny and distribution of the symbionts in gills. Appl Environ Microbiol 2005, 71, 1694–1700.
[26] C. M. Gibson, M. S. Hunter. Extraordinarily widespread and fantastically complex: Comparative biology of endosymbiotic bacterial and fungal mutualists of insects. Ecology Letters 2010, 13, 223–234.
[27] J. de Vries, G. Christa, S. B. Gould. Plastid survival in the cytosol of animal cells. Trends Plant Sci 2014, 19, 347–350.
[28] K. Händeler, H. Wägele, U. Wahrmund, M. Rüdinger, V. Knoop. Slugs' last meals: Molecular identification of sequestered chloroplasts from different algal origins in Sacoglossa (Opisthobranchia, Gastropoda). Mol Ecol Resour 2010, 10, 968–978.
[29] M. E. Rumpho, F. P. Dastoor, J. R. Manhart, J. Lee. The Kleptoplast. In: K. WRRHJ, ed. The Structure and Function of Plastids. Dordrecht: Springer; 2006:451–473.
[30] G. F. Esteban, B. J. Finlay, K. J. Clarke. Sequestered organelles sustain aerobic microbial life in anoxic environments. Environ Microbiol 2009, 11, 544–550.
[31] D. K. Stoecker, A. E. Michaels, L. H. Davis. Large proportion of marine planktonic ciliates found to contain functional chloroplasts. Nature 1987, 326, 790–792.
[32] S. L. Richardson. Endosymbiont change as a key innovation in the adaptive radiation of Soritida (Foraminifera). Paleobiology 2001, 27, 262–289.
[33] R. Koga, X. Y. Meng, T. Tsuchida, T. Fukatsu. Cellular mechanism for selective vertical transmission of an obligate insect symbiont at the bacteriocyte-embryo interface. Proc Natl Acad Sci USA 2012, 109, E1230-E7.
[34] D. A. Stahl, M. Hullar, S. Davidson. The structure and function of microbial communities. In: M. Dworkin, S. Falkow, E. Rosenberg, K. H. Schleifer, E. Stackebrandt, eds. The Prokaryotes: Symbiotic associations, Biotechnology, Applied Microbiology: Springer Verlag; 2006.
[35] M. C. M. Motta, C. M. C. Catta-Preta, S. Schenkman, et al. The bacterium endosymbiont of *Crithidia deanei* undergoes coordinated division with the host cell nucleus. PLoS ONE 2010, 5, e12415.
[36] N. Okamoto, I. Inouye. *Hatena arenicola* gen. et sp. nov., a katablepharid undergoing probable plastid acquisition. Protist 2006, 157, 401–419.
[37] J. L. Sachs, R. G. Skophammer, J. U. Regus. Evolutionary transitions in bacterial symbiosis. Proc Natl Acad Sci USA 2011, 108, 10800–10807.
[38] A. Saridaki, K. Bourtzis. Wolbachia: more than just a bug in insects genitals. Curr Opin Microbiol 2010, 13, 67–72.
[39] M. Bright, S. Bulgheresi. A complex journey: Transmission of microbial symbionts. Nat Rev Microbiol 2010, 8, 218–230.
[40] K. E. Gibson, H. Kobayashi, G. C. Walker. Molecular determinants of a symbiotic chronic infection. Annu Rev Genet 2008, 42, 413–441.
[41] A. D. Nussbaumer, C. R. Fisher, M. Bright. Horizontal endosymbiont transmission in hydrothermal vent tubeworms. Nature 2006, 441, 345–348.
[42] T. L. Harmer, R. D. Rotjan, A. D. Nussbaumer, et al. Free-living tube worm endosymbionts found at deep-sea vents. Appl Environ Microbiol 2008, 74, 3895–3898.
[43] A. H. Baird, J. R. Guest, B. L. Willis. Systematic and biogeographical patterns in the reproductive biology of scleractinian corals. In: Annu Rev Ecol Evol Syst; 2009:551–571.

[44] E. T. Kiers, R. F. Denison. Sanctions, cooperation, and the stability of plant-rhizosphere mutualisms. In: Annu Rev Ecol Evol Syst; 2008:215–236.
[45] E. T. Kiers, R. A. Rousseau, S. A. West, R. F. Denlson. Host sanctions and the legume-rhizobium mutualism. Nature 2003, 425, 78–81.
[46] H. R. Hoogenraad. Zur Kenntnis der Fortpflanzung von *Paulinella chromatophora* Lauterb. Zool Anz 1927, 72, 140–150.
[47] N. Okamoto, I. Inouye. A secondary symbiosis in progress? Science 2005, 310, 287.
[48] S. Miyagishima. Origin and evolution of the chloroplast division machinery. J Plant Res 2005, 118, 295–306.
[49] K. W. Osteryoung, J. Nunnari. The division of endosymbiotic organelles. Science 2003, 302, 1698–1704.
[50] A. S. Peek, R. A. Feldman, R. A. Lutz, R. C. Vrijenhoek. Cospeciation of chemoautotrophic bacteria and deep sea clams. Proc Natl Acad Sci USA 1998, 95, 9962–9966.
[51] M. A. Clark, N. A. Moran, P. Baumann, J. J. Wernegreen. Cospeciation between bacterial endosymbionts (*Buchnera*) and a recent radiation of aphids (*Uroleucon*) and pitfalls of testing for phylogenetic congruence. Evolution 2000, 54, 517–525.
[52] C. Sauer, E. Stackebrandt, J. Gadau, B. Holldobler, R. Gross. Systematic relationships and cospeciation of bacterial endosymbionts and their carpenter ant host species: Proposal of the new taxon *Candidatus* Blochmannia gen. nov. Int J Syst Evol Microbiol 2000, 50, 1877–1886.
[53] M. L. Thao, N. A. Moran, P. Abbot, E. B. Brennan, D. H. Burckhardt, P. Baumann. Cospeciation of psyllids and their primary prokaryotic endosymbionts. Appl Environ Microbiol 2000, 66, 2898–2905.
[54] M. M. G. Teixeira, T. C. Borghesan, R. C. Ferreira, et al. Phylogenetic validation of the genera *Angomonas* and *Strigomonas* of Trypanosomatids harboring bacterial endosymbionts with the description of new species of trypanosomatids and of proteobacterial symbionts. Protist 2011, 162, 503–524.
[55] A. W. Spaulding, C. D. Von Dohlen. Phylogenetic characterization and molecular evolution of bacterial endosymbionts in psyllids (Hemiptera: Sternorrhyncha). Mol Biol Evol 1998, 15, 1506–1513.
[56] N. A. Moran, P. Tran, N. M. Gerardo. Symbiosis and insect diversification: An ancient symbiont of sap-feeding insects from the bacterial phylum Bacteroidetes. Appl Environ Microbiol 2005, 71, 8802–8810.
[57] M. A. Munson, P. Baumann, M. A. Clark, et al. Evidence for the establishment of aphid-eubacterium endosymbiosis in an ancestor of four aphid families. J Bacteriol 1991, 173, 6321–6324.
[58] N. A. Moran, M. A. Munson, P. Baumann, H. Ishikawa. A molecular clock in endosymbiotic bacteria is calibrated using the insect hosts. Proc R Soc Lond Ser B-Biol Sci 1993, 253, 167–171.
[59] C. Decker, K. Olu, S. Arnaud-Haond, S. Duperron. Physical proximity may promote lateral acquisition of bacterial symbionts in vesicomyid clams. PLoS ONE 2013, 8.
[60] F. J. Stewart, C. M. Cavanaugh. Pyrosequencing analysis of endosymbiont population structure: Co-occurrence of divergent symbiont lineages in a single vesicomyid host clam. Environ Microbiol 2009, 11, 2136–2147.
[61] M. Kleiner, J. M. Petersen, N. Dubilier. Convergent and divergent evolution of metabolism in sulfur-oxidizing symbionts and the role of horizontal gene transfer. Curr Opin Microbiol 2012, 15, 621–631.
[62] M. Marchetti, D. Capela, M. Glew, et al. Experimental evolution of a plant pathogen into a legume symbiont. PLoS Biol 2010, 8, e1000280.

[63] J. A. Hilton, R. A. Foster, H. James Tripp, B. J. Carter, J. P. Zehr, T. A. Villareal. Genomic deletions disrupt nitrogen metabolism pathways of a cyanobacterial diatom symbiont. Nature Commun 2013, 4, 1767.
[64] P. Baumann. Biology of bacteriocyte-associated endosymbionts of plant sap-sucking insects. Annu Rev Microbiol 2005, 59, 155–189.
[65] J. P. McCutcheon. The bacterial essence of tiny symbiont genomes. Curr Opin Microbiol 2010, 13, 73–78.
[66] J. P. McCutcheon, N. A. Moran. Extreme genome reduction in symbiotic bacteria. Nat Rev Microbiol 2012, 10, 13–26.
[67] N. A. Moran, J. P. McCutcheon, A. Nakabachi. Genomics and evolution of heritable bacterial symbionts. Annu Rev Genet 2008, 42, 165–190.
[68] A. Moya, J. Peretó, R. Gil, A. Latorre. Learning how to live together: genomic insights into prokaryote-animal symbioses. Nat Rev Genet 2008, 9, 218–229.
[69] E. Zientz, T. Dandekar, R. Gross. Metabolic interdependence of obligate intracellular bacteria and their insect hosts. Microbiol Mol Biol Rev 2004, 68, 745–770.
[70] A. Mira, H. Ochman, N. A. Moran. Deletional bias and the evolution of bacterial genomes. Trends Genet 2001, 17, 589–596.
[71] R. Hershberg, D. A. Petrov. Evidence that mutation is universally biased towards AT in bacteria. PLoS Genet 2010, 6, e1001115.
[72] J. P. McCutcheon, N. A. Moran. Parallel genomic evolution and metabolic interdependence in an ancient symbiosis. Proc Natl Acad Sci USA 2007, 104, 19392–19397.
[73] A. Nakabachi, A. Yamashita, H. Toh, et al. The 160-kilobase genome of the bacterial endosymbiont *Carsonella*. Science 2006, 314, 267.
[74] J. P. McCutcheon, B. R. McDonald, N. A. Moran. Origin of an alternative genetic code in the extremely small and GC-rich genome of a bacterial symbiont. PLoS Genet 2009, 5, e1000565.
[75] J. P. McCutcheon, B. R. McDonald, N. A. Moran. Convergent evolution of metabolic roles in bacterial co-symbionts of insects. Proc Natl Acad Sci USA 2009, 106, 15394–15399.
[76] J. Tamames, R. Gil, A. Latorre, J. Peretó, F. J. Silva, A. Moya. The frontier between cell and organelle: genome analysis of *Candidatus* Carsonella ruddii. BMC Evol Biol 2007, 7, 7.
[77] W. Martin, T. Rujan, E. Richly, et al. Evolutionary analysis of *Arabidopsis*, cyanobacterial, and chloroplast genomes reveals plastid phylogeny and thousands of cyanobacterial genes in the nucleus. Proc Natl Acad Sci USA 2002, 99, 12246–12251.
[78] J. Gross, D. Bhattacharya. Mitochondrial and plastid evolution in eukaryotes: An outsiders' perspective. Nat Rev Genet 2009, 10, 495–505.
[79] E. Schleiff, T. Becker. Common ground for protein translocation: Access control for mitochondria and chloroplasts. Nat Rev Mol Cell Biol 2011, 12, 48–59.
[80] E. C. M. Nowack, A. R. Grossman. Trafficking of protein into the recently established photosynthetic organelles of *Paulinella chromatophora*. Proc Natl Acad Sci USA 2012, 109, 5340–5345.
[81] J. M. Archibald. The puzzle of plastid evolution. Curr Biol 2009, 19, R81-R8.
[82] S. B. Gould, R. R. Waller, G. I. McFadden. Plastid evolution. Annu Rev Plant Biol 2008, 59, 491–517.
[83] B. A. Curtis, G. Tanifuji, F. Burki, et al. Algal genomes reveal evolutionary mosaicism and the fate of nucleomorphs. Nature 2012, 492, 59–65.
[84] K. H. Wolfe, C. W. Morden, J. D. Palmer. Function and evolution of a minimal plastid genome from a nonphotosynthetic parasitic plant. Proc Natl Acad Sci USA 1992, 89, 10648–10652.
[85] S. Wicke, K. F. Müller, C. W. de Pamphilis, et al. Mechanisms of functional and physical genome reduction in photosynthetic and nonphotosynthetic parasitic plants of the broomrape family. Plant Cell 2013, 25, 3711–3725.

[86] T. Borza, C. E. Popescu, R. W. Lee. Multiple metabolic roles for the nonphotosynthetic plastid of the green alga *Prototheca wickerhamii*. Eukaryot Cell 2005, 4, 253–261.
[87] A. P. de Koning, P. J. Keeling. The complete plastid genome sequence of the parasitic green alga *Helicosporidium* sp. is highly reduced and structured. BMC Biol 2006, 4, 12.
[88] A. Tartar, D. G. Boucias. The non-photosynthetic, pathogenic green alga *Helicosporidium* sp. has retained a modified, functional plastid genome. FEMS Microbiol Lett 2004, 233, 153–157.
[89] G. Gockel, W. Hachtel. Complete gene map of the plastid genome of the nonphotosynthetic Euglenoid flagellate *Astasia longa*. Protist 2000, 151, 347–351.
[90] S. Köhler, C. F. Delwiche, P. W. Denny, et al. A plastid of probable green algal origin in Apicomplexan parasites. Science 1997, 275, 1485–1489.
[91] R. J. M. I. Wilson, P. W. Denny, P. R. Preiser, et al. Complete gene map of the plastid-like DNA of the malaria parasite *Plasmodium falciparum*. J Mol Biol 1996, 261, 155–172.
[92] J. Molina, K. M. Hazzouri, D. Nickrent, et al. Possible loss of the chloroplast genome in the parasitic flowering plant *Rafflesia lagascae* (Rafflesiaceae). Mol Biol Evol 2014, 31, 793–803.
[93] D. R. Smith, R. W. Lee. A plastid without a genome: Evidence from the nonphotosynthetic green algal genus *Polytomella*. Plant Physiol 2014, 164, 1812–1819.
[94] L. Sheiner, A. B. Vaidya, G. I. McFadden. The metabolic roles of the endosymbiotic organelles of *Toxoplasma* and *Plasmodium* spp. Curr Opin Microbiol 2013, 16, 452–458.
[95] A. M. Shiflett, P. J. Johnson. Mitochondrion-related organelles in eukaryotic protists. In: Annu Rev Microbiol; 2010:409–429.
[96] R. M. De Graaf, G. Ricard, T. A. Van Alen, et al. The organellar genome and metabolic potential of the hydrogen-producing mitochondrion of *Nyctotherus ovalis*. Mol Biol Evol 2011, 28, 2379–2391.
[97] V. Pérez-Brocal, C. G. Clark. Analysis of two genomes from the mitochondrion-like organelle of the intestinal parasite *Blastocystis*: Complete sequences, gene content, and genome organization. Mol Biol Evol 2008, 25, 2475–2482.
[98] I. Wawrzyniak, M. Roussel, M. Diogon, et al. Complete circular DNA in the mitochondria-like organelles of *Blastocystis hominis*. Int J Parasitol 2008, 38, 1377–1382.
[99] J. M. P. Alves, C. C. Klein, F. M. Da Silva, et al. Endosymbiosis in trypanosomatids: The genomic cooperation between bacterium and host in the synthesis of essential amino acids is heavily influenced by multiple horizontal gene transfers. BMC Evol Biol 2013, 13, 190.
[100] F. Husnik, N. Nikoh, R. Koga, et al. Horizontal gene transfer from diverse bacteria to an insect genome enables a tripartite nested mealybug symbiosis. Cell 2013, 153, 1567–1578.
[101] S. Nygaard, G. Zhang, M. Schiøtt, et al. The genome of the leaf-cutting ant *Acromyrmex echinatior* suggests key adaptations to advanced social life and fungus farming. Genome Res 2011, 21, 1339–1348.
[102] A. C. C. Wilson, P. D. Ashton, F. Calevro, et al. Genomic insight into the amino acid relations of the pea aphid, *Acyrthosiphon pisum*, with its symbiotic bacterium *Buchnera aphidicola*. Insect Mol Biol 2010, 19, 249–258.
[103] C. D. Von Dohlen, S. Kohler, S. T. Alsop, W. R. McManus. Mealybug beta-proteobacterial endosymbionts contain gamma-proteobacterial symbionts. Nature 2001, 412, 433–436.
[104] A. Nakabachi, S. Shigenobu, N. Sakazume, et al. Transcriptome analysis of the aphid bacteriocyte, the symbiotic host cell that harbors an endocellular mutualistic bacterium, *Buchnera*. Proc Natl Acad Sci USA 2005, 102, 5477–5482.
[105] N. Nikoh, J. P. McCutcheon, T. Kudo, S. Y. Miyagishima, N. A. Moran, A. Nakabachi. Bacterial genes in the aphid genome: Absence of functional gene transfer from *Buchnera* to its host. PLoS Genet 2010, 6, e1000827.
[106] N. Nikoh, A. Nakabachi. Aphids acquired symbiotic genes via lateral gene transfer. BMC Biol 2009, 7, 12.

[107] D. B. Sloan, A. Nakabachi, S. Richards, et al. Parallel histories of horizontal gene transfer facilitated extreme reduction of endosymbiont genomes in sap-feeding insects. Mol Biol Evol 2014, 31, 857–871.
[108] E. C. M. Nowack, H. Vogel, M. Groth, A. R. Grossman, M. Melkonian, G. Glöckner. Endosymbiotic gene transfer and transcriptional regulation of transferred genes in *Paulinella chromatophora*. Mol Biol Evol 2011, 28, 407–422.
[109] R. P. Duncan, F. Husnik, J. T. Van Leuven, et al. Dynamic recruitment of amino acid transporters to the insect/symbiont interface. Mol Ecol 2014, 23, 1608–1623.
[110] D. R. G. Price, R. P. Duncan, S. Shigenobu, A. C. C. Wilson. Genome expansion and differential expression of amino acid transporters at the aphid/*Buchnera* symbiotic interface. Mol Biol Evol 2011, 28, 3113–3126.
[111] M. Havaux, G. Guedeney, Q. He, A. R. Grossman. Elimination of high-light-inducible polypeptides related to eukaryotic chlorophyll a/b-binding proteins results in aberrant photoacclimation in *Synechocystis* PCC6803. Biochim Biophys Acta – Bioenerg 2003, 1557, 21–33.
[112] Q. F. He, N. Dolganov, O. Björkman, A. R. Grossman. The high light-inducible polypeptides in *Synechocystis* PCC6803 – Expression and function in high light. J Biol Chem 2001, 276, 306–314.
[113] P. Ganot, A. Moya, V. Magnone, D. Allemand, P. Furla, C. Sabourault. Adaptations to endosymbiosis in a Cnidarian-Dinoflagellate association: Differential gene expression and specific gene duplications. PLoS Genet 2011, 7, e1002187.
[114] A. R. Carvunis, T. Rolland, I. Wapinski, et al. Proto-genes and *de novo* gene birth. Nature 2012, 487, 370–374.
[115] IAGC. Genome sequence of the pea aphid *Acyrthosiphon pisum*. PLoS Biol 2010, 8, e1000313.
[116] S. Shigenobu, D. L. Stern. Aphids evolved novel secreted proteins for symbiosis with bacterial endosymbiont. Proc R Soc Lond Ser B-Biol Sci 2013, 280, 20121952.
[117] S. Yang, F. Tang, M. Gao, H. B. Krishnan, H. Zhu. R gene-controlled host specificity in the legume-rhizobia symbiosis. Proc Natl Acad Sci USA 2010, 107, 18735–18740.
[118] K. Kambara, S. Ardissone, H. Kobayashi, et al. Rhizobia utilize pathogen-like effector proteins during symbiosis. Mol Microbiol 2009, 71, 92–106.
[119] J. A. Kimbrel, W. J. Thomas, Y. Jiang, et al. Mutualistic co-evolution of type III effector genes in *Sinorhizobium fredii* and *Bradyrhizobium japonicum*. PLoS Pathogens 2013, 9, e1003204.
[120] A. Vigneron, D. Charif, C. Vincent-Monégat, et al. Host gene response to endosymbiont and pathogen in the cereal weevil *Sitophilus oryzae*. BMC Microbiol 2013, 12, S14.
[121] C. Anselme, V. Pérez-Brocal, A. Vallier, et al. Identification of the Weevil immune genes and their expression in the bacteriome tissue. BMC Biol 2008, 6, 43.
[122] C. Ratzka, R. Gross, H. Feldhaar. Gene expression analysis of the endosymbiont-bearing midgut tissue during ontogeny of the carpenter ant *Camponotus floridanus*. J Insect Physiol 2013, 59, 611–623.
[123] S. V. Nyholm, P. Song, J. Dang, C. Bunce, P. R. Girguis. Expression and putative function of innate immunity genes under in situ conditions in the symbiotic hydrothermal vent tubeworm *Ridgeia piscesae*. PLoS ONE 2012, 7, e38267.
[124] F. H. Login, S. Balmand, A. Vallier, et al. Antimicrobial peptides keep insect endosymbionts under control. Science 2011, 334, 362–365.
[125] W. van de Velde, G. Zehirov, A. Szatmari, et al. Plant peptides govern terminal differentiation of bacteria in symbiosis. Science 2010, 327, 1122–1126.
[126] A. Poliakov, C. W. Russell, L. Ponnala, et al. Large-scale label-free quantitative proteomics of the pea aphid-*Buchnera* symbiosis. Mol Cell Proteomics 2011, 10, M110.007039.

Fabia U. Battistuzzi and Anais Brown
13 Rates of evolution under extreme and mesophilic conditions

13.1 Overview

Prokaryotes are among the most versatile species on Earth in terms of their adaptability to a wide array of environmental conditions. The distribution of species in space and time has varied following the, sometimes drastic, planetary changes that Earth has gone through. The most evident example is the shift in atmospheric composition from anoxic to oxic, approximately 2.5 billion years ago, that drove lineages thriving in common anoxic environments to become niche-specialists (e.g., methanogens) [1, 2]. The hallmark of the longevity and persistence of prokaryotes is in the plasticity of their genomes, which are shaped by generally high evolutionary rates, frequent gene exchange events (horizontal gene transfer), and loss/gain of genes (genome reduction and duplication events) [3–8]. All of these phenomena alter the genetic make-up of an organism leading to the evolution or acquisition of potentially new functions. The rate at which genetic changes are accumulated is, therefore, a proxy for the rate of functional innovations. For simplicity, we can divide these genetic changes into two types: changes caused by alterations of the nucleotides/amino acids, estimated via mutation or substitution rates, and those caused by gene acquisitions or losses (duplication/loss rates, rate of horizontal gene transfers). Both of these categories are regulated by selective forces and genetic drift, which vary in different environments. Among mesophilic and extreme environments, the latter ones are usually regarded as more stressful, implying that stronger selective pressures might be acting on organisms that occupy these niches [9–12]. Here, we review current knowledge on evolutionary rates in prokaryotes focusing on the mesophile-extremophile dichotomy.

Prokaryotes are known for their high evolutionary rates that, coupled with their haploid state, allow them to quickly react to environmental changes. Similarly, gaining of new functions in prokaryotes is mediated by frequent gene acquisition events often associated with the colonization of new habitats [4, 13–15]. However, despite these general trends, very little is known about the rate (i.e., the number of events per time unit) at which these changes happen. Evolutionary rates can describe either changes within or between species that correspond, respectively, to the accumulation of polymorphisms (i.e., mutations that are not fixed in a population) or substitutions (i.e., mutations that have gone to fixation and are part of the genetic make-up of the species). Known mutation rates compiled from the literature have an average of 8.4×10^{-9} mutations per site per year (range: 6.7×10^{-10}–2.7×10^{-8}) while the average of substitution rates is 9.2×10^{-10} substitutions per site per year (range: 7×10^{-11}–3.4×10^{-8}), see ▶ Tab. 13.1 [16–20].

Table 13.1. Compilation of evolutionary rates in Bacteria and Archaea. Substitution rates (number of substitutions per site per year) represent comparisons between species. Comparisons between different strains represent mutation rates (number of mutations per site per year).

Species	Between-species Substitution rate (#subst/site/yr)		Within-species Mutation rate (#mut/site/yr)
	This study[a]	Other studies	Other studies
BACTERIA, average		9.23E-10	8.45E – 09
Aeromonas (hydrophila/salmonicida)	–	1.03E-08 [b] [18]	–
Anabaena variabilis[e]	1.63E-10	–	1.08E-08[b] [18]
Anaeromyxobacter dehalogenans[e]	1.45E-10	–	4.54E-09[b] [18]
Aquifex aeolicus VF5	1.13E-10	–	–
Aster yellows witches-broom phytoplasma	2.61E-10	–	–
Bacillus (anthracis/cereus)[e]	9.53E-11	–	1.99E-09[b] [18]
Bacteroides fragilis NCTC 9343	1.95E-10	–	–
Bartonella (henselae/quintana)[e]	2.52E-10	–	1.07E-08[b] [18]
Bdellovibrio bacteriovorus HD100	2.39E-10	–	–
Bifidobacterium longum NCC2705	2.24E-10	–	–
Bordetella bronchiseptica RB50	1.53E-10	–	–
Borrelia (turicatae /hermsii)	–	–	4.88E-09[b] [18]
Borrelia burgdorferi B31	2.43E-10	–	–
Bradyrhizobium japonicum USDA110	1.71E-10	–	–
Brucella ovis /Ochrobactrum anthropi	–	1.24E-08 [b] [18]	–
Brucella abortus	1.02E-10	–	–
Buchnera aphidicola str. APS	4.19E-10	–	2.20E-07[d] [19]
Burkholderia (ambifaria /pseudomallei)	–	–	1.64E-08[b] [18]
Campylobacter jejuni (subsp. doylei 269.97 /RM1221)	1.71E-10	–	1.98E-09[b] [18] 2.80E-05[d] [19]
"*Candidatus* Phytoplasma asteris (AY /OY)"[f]		–	6.19E-09
"*Candidatus* Blochmannia florida"	2.30E-10	–	–
Carboxydothermus hydrogenoform	1.03E-10	–	–
Caulobacter crescentus CB15	1.72E-10	–	–
Chlamydophila abortus S26/3	2.51E-10	–	–
Chlorobium chlorochromatii CaD	1.96E-10	–	–
Citrobacter koseri Enterobacter sp.	–	3.43E-08 [b] [18]	–
Clavibacter michiganensis (str. ATCC 33113[f] /subsp. michiganensis NCPPB 382)	–	–	3.29E-09[b] [18]
Clostridium botulinum (E3 str. Alaska E43 /B str. Eklund 17B)	–	–	1.99E-09[b] [18]
Clostridium acetobutylicum ATCC 824	1.55E-10	–	–
Colwellia psychrerythraea 34H	1.87E-10	–	–
Corynebacterium diphtheriae NCTC 13129	3.67E-10	–	–
Coxiella burnetii RSA 493	1.86E-10	–	–
Dechloromonas aromatica RCB	1.24E-10	–	–
Dehalococcoides ethenogenes str. CBDB1	2.00E-10	–	1.19E-08[b] [18]
Deinococcus radiodurans R1	2.03E-10	–	–

Table 13.1 (cont.). Compilation of evolutionary rates in Bacteria and Archaea. Substitution rates (number of substitutions per site per year) represent comparisons between species. Comparisons between different strains represent mutation rates (number of mutations per site per year).

Species	Between-species Substitution rate (#subst/site/yr)		Within-species Mutation rate (#mut/site/yr)
	This study[a]	Other studies	Other studies
Desulfitobacterium hafniense Y51	1.28E-10	–	–
Desulfovibrio desulfuricans G20	1.75E-10	–	–
Escherichia coli /Salmonella enterica	–	6-8E-09 [c] [16]	–
		4.50E-09 [c] [17]	
		2.803E-08 [b,c] [18]	
Ehrlichia canis str. Jake	2.66E-10	–	–
Erythrobacter litoralis HTCC2594	1.63E-10	–	–
Francisella tularensis subsp. holarctica	2.21E-10	–	–
Frankia sp. CcI3	1.39E-10	–	–
Fusobacterium nucleatum	1.51E-10	–	–
Geobacter metallireducens GS-15	1.15E-10	–	–
Gloeobacter violaceus PCC 7421	1.83E-10	–	–
Gluconobacter oxydans 621H	1.63E-10	–	–
Haemophilus ducreyi 35000HP	1.29E-10	–	–
Hahella chejuensis KCTC 2396	1.23E-10	–	–
Helicobacter (acinonychis/pylori)	–	–	1.84E-08[b] [18]
			< 2E-05 [d] [19]
Herminiimonas arsenicoxydans/Janthinobacterium	–	1.24E-08 [b] [18]	–
Idiomarina loihiensis L2TR	1.25E-10	–	–
Klebsiella pneumoniae (342 /MGH 78578)	–	–	5.22E-09[b] [18]
Lactobacillus acidophilus NCFM	1.87E-10	–	–
Lactococcus lactis (subsp. cremoris MG1363 /subsp. lactis IL1403)	–	–	6.46E-09[b] [18]
Legionella pneumophila	1.70E-10	–	–
Leifsonia xyli subsp. xyli str. CTCB07	1.12E-10	–	–
Leptospira interrogans serovar Copenhageni	1.99E-10	–	–
Listeria monocytogenes (EGD-e 4b /serotype 4b str. F2365)	–	–	1.38E-09[b] [18]
Listeria innocua Clip11262	1.09E-10	–	–
Mesoplasma florum L1	2.31E-10	–	–
Mesorhizobium loti MAFF303099	1.43E-10	–	–
Methylobacterium (extorquens /populi)	–	–	7.44E-09[b] [18]
Methylococcus capsulatus	1.25E-10	–	–
Moorella thermoacetica ATCC 39073	1.04E-10	–	–
Mycobacterium avium	2.36E-10	–	–
Mycoplasma (mycoides /capricolum)[e]	2.81E-10	–	4.27E-09[b] [18]
Mycoplasma genitalium G37	3.20E-10	–	–
Neisseria (gonorrhoeae /meningitidis)	–	–	1.30E-08[b] [18]

Table 13.1 (cont.). Compilation of evolutionary rates in Bacteria and Archaea. Substitution rates (number of substitutions per site per year) represent comparisons between species. Comparisons between different strains represent mutation rates (number of mutations per site per year).

Species	Between-species Substitution rate (#subst/site/yr)		Within-species Mutation rate (#mut/site/yr)
	This study[a]	Other studies	Other studies
Neisseria gonorrhoeae FA 1090	1.24E-10	–	–
Nitrosococcus oceani ATCC 19707	1.21E-10	–	–
Nitrosomonas europaea ATCC 19718	1.95E-10	–	–
Nocardia farcinica IFM 10152	1.68E-10	–	–
Pelobacter carbinolicus DSM 2380	1.15E-10	–	–
Porphyromonas gingivalis W83	1.87E-10	–	–
Prochlorococcus marinus (AS9601 /MIT 9312)	–	–	4.10E-09[b] [18]
Propionibacterium acnes KPA171202	1.46E-10	–	–
Pseudoalteromonas haloplanktis	1.40E-10	–	–
Pseudomonas aeruginosa (PA7 /PAO1)	1.13E-10	–	6.68E-10[b] [18]
Pseudomonas putida (F1 /GB-1)	–	–	1.31E-09[b] [18]
Pseudomonas syringae (str. 1448A[f] /pv. syringae B728a)	–	–	1.98E-09[b] [18]
Psychrobacter (cryohalolentis /arcticus)[e]	2.27E-10	–	1.01E-08[b] [18]
Ralstonia eutropha[e]	1.38E-10	–	1.30E-08[b] [18]
Rhizobium etli (CFN 42 /CIAT 652)	1.74E-10	–	3.98E-09[b] [18]
Rhodobacter sphaeroides (ATCC 17029 /ATCC 17025)	–	–	1.51E-08[b] [18]
Rhodoferax ferrireducens DSM 15236	1.96E-10	–	–
Rhodopirellula baltica SH 1	2.15E-10	–	–
Rhodopseudomonas palustris (HaA2 /BisB5)	–	–	2.73E-09[b] [18]
Rhodospirillum rubrum ATCC 11170	1.19E-10	–	–
Rickettsia (canadensis /akari)	–	–	2.45E-08[b] [18]
Rickettsia conorii str. Malish 7	2.14E-10	–	–
Salinibacter ruber DSM 13855	2.11E-10	–	–
Salinispora (tropica /arenicola)	–	–	4.61E-09[b] [18]
Shewanella oneidensis[e]	1.22E-10	–	2.68E-08[b] [18]
Silicibacter pomeroyi DSS-3	1.55E-10	–	–
Sinorhizobium (medicae /meliloti)	–	–	2.71E-09[b] [18]
Solibacter usitatus Ellin6076	1.76E-10	–	–
Staphylococcus aureus (122 /RF122)	1.37E-10	–	–
Stenotrophomonas maltophilia (R551-3 /K279a)	–	–	6.50E-09[b] [18]
Streptococcus agalactiae 2603VR	1.62E-10	–	–
Streptomyces avermitilis MA-4680	1.24E-10	–	–
Symbiobacterium thermophilum IAM14863	1.39E-10	–	–
Synechococcus elongatus PCC 6301	1.90E-10	–	–
Synechococcus sp. JA-2-3B'a(2-13)	1.71E-10	–	–

Table 13.1 (cont.). Compilation of evolutionary rates in Bacteria and Archaea. Substitution rates (number of substitutions per site per year) represent comparisons between species. Comparisons between different strains represent mutation rates (number of mutations per site per year).

Species	Between-species Substitution rate (#subst/site/yr)		Within-species Mutation rate (#mut/site/yr)
	This study[a]	Other studies	Other studies
Synechocystis sp. PCC 6803	2.38E-10	–	–
Thermoanaerobacter tengcongensis	1.07E-10	–	–
Thermobifida fusca YX	1.64E-10	–	–
Thermosynechococcus elongatus	1.69E-10	–	–
Thermotoga (maritima /petrophila)[e]	6.99E-11	–	7.06E-09[b] [18]
Thiobacillus denitrificans ATC[f]	1.12E-10	–	–
Thiomicrospira crunogena XCL-2	1.49E-10	–	–
Thiomicrospira denitrificans A[f]	2.23E-10	–	–
Tropheryma whipplei TW08/27	3.35E-10	–	–
Vibrio (harveyi /parahaemolyticus)	–	–	5.44E-09[b] [18]
Vibrio cholerae O1 biovar El Tor	1.08E-10	–	6.70E-05[d] [19]
Wolbachia pipientis (wMel /wPip)[f]	–	–	2.70E-08[b] [18]
Xanthomonas (axonopodis /campestris)[e]	1.45E-10	–	1.29E-09[b] [18]
Yersinia (enterocolitica /pestis)	–	–	2.28E-08[b] [18]
ARCHAEA, average	1.55E-10		4.60E – 09
Archaeoglobus fulgidus DSM 4304	1.34E-10	–	–
Haloarcula marismortui ATCC 43049	2.15E-10	–	–
Methanocaldococcus jannaschii	8.21E-11	–	–
Methanococcus maripaludis S2	1.57E-10	–	–
Methanopyrus kandleri AV19	1.44E-10	–	–
Methanosarcina barkeri str. Fusario	1.42E-10	–	–
Methanosphaera stadtmanae DSM[f]	1.48E-10	–	–
Methanospirillum hungatei JF-1	1.64E-10	–	–
Nanoarchaeum equitans Kin4-M	1.75E-10	–	–
Picrophilus torridus DSM 9790	2.12E-10	–	–
Pyrococcus abyssi GE5	9.13E-11	–	–
Sulfolobus solfataricus P2	1.43E-10	–	–
Sulfolobus islandicus	–	–	4.60E-09 [20]
Thermoplasma acidophilum DSM 1728	2.04E-10	–	–

[a] substitution rates are calculated as the averaged ratio of estimated root-to-tip genetic distance and divergence times from 25 genes used in Battistuzzi et al. [67]; [b] rate based on 16S rRNA; [c] synonymous substitution rate; [d] synonymous mutation rate; [e] different strains; [f] currently absent from NCBI.

Simulation and *in vivo* studies have started shedding light on the short-term causes of mutation rates, which include mutations in genes key to the correct functioning of repair systems (e.g., *mutS, mutL*) and response to variable environments [21–23]. Hypervariable strains of known lineages are often identified in association with stressful environments, such as hosts treated with antibiotics. Lessons learned from these hypervariable strains have highlighted the relative importance of deleterious, advantageous, and neutral mutations for the long-term stability of a lineage in a predictable or unpredictable environment. Despite being more likely to develop advantageous new functions because of higher mutation rates, high mutator strains are unstable in the long term, even in highly stressful environments, suggesting that the negative load of deleterious mutations quickly out-weighs the benefits of the rare positive adaptations.

These studies have focused on the short-term accumulation of genetic variability in prokaryotes providing much needed insight into the mechanisms at the basis of different mutation rates in populations. More elusive are the evolutionary patterns at long timescales, millions to hundreds of millions of years. The known substitution rates vary from 7×10^{-11} to 3.4×10^{-8} substitutions per site per year (▶ Tab. 13.1), with many of them not being independent estimates, but rather rates inferred with a comparative approach from estimates of other species. A fundamental difference between mutation and substitution rates is that they measure two different events: mutation rates mark the appearance of genetic changes in the sequence of one or few individuals of a population, while substitution rates represent the fixation of mutations in the vast majority of individuals of the species. Because of their nature, these two rates act on dramatically different timescales, short (in some cases even the span of a few generations) for mutation rates and long (millions to thousands of millions of years) for substitution rates. Therefore, the methods to estimate these two types of rates are different. On the one hand, mutation rates can be estimated from experimental set-ups that collect time series data for a few generations [7, 19]; on the other hand, substitution rates require timescales over long geologic periods, which preclude an ad hoc experimental set-up [24]. Instead, information regarding long timescales can be obtained from two primary sources, the geologic record (fossils) or the molecular record (molecular clocks).

The classic approach of estimating substitution rates is to correlate timelines inferred from the fossil record with genetic distances obtained from the pairwise comparisons of multiple DNA or protein orthologous sequences. This approach is routinely applied to eukaryotes such as mammals, rodents, and primates, where it has led to the identification of speed-ups and slow-downs in evolutionary rates for groups of lineages [25–28]. Unfortunately, the prokaryote fossil record, as well as other geological evidence such as biomarkers, is extremely sparse [29–31]. Timelines of prokaryotes must, therefore, rely on molecular data.

A seminal paper by Ochman et al. [16] was the first to use a molecular clock to estimate substitution rates of two ancient microbial lineages (*Salmonella typhimurium* and *Escherichia coli*, estimated to have diverged approximately 160 Ma ago) and many studies still use this estimated rate as a reference for substitution rates in prokaryotes (e.g., [19, 32]). The basis for this estimation was the approximately linear correlation of neutral substitutions with time in different genomes, the primary tenet of the molecular clock hypothesis [33, 34]. Briefly, according to the original formulation of the molecular clock, divergence times of speciation events can be estimated from known genetic distances and evolutionary rates:

$$T = \frac{D}{r}$$

where T is divergence time, D is the genetic distance, and r is the evolutionary rate.

It therefore follows that evolutionary rates can be obtained applying a known divergence time to the genetic divergence of pairwise comparisons:

$$r = \frac{D}{2T}$$

Unfortunately, the interdependency of genetic distances, evolutionary rates, and divergence times poses a circular reasoning issue, such that rates are necessary to obtain times but times are necessary to obtain rates. This circularity is solved with the use of calibration points, which are nodes in a phylogeny with known time information obtained from a non-molecular record (e.g., fossil record). While first generation molecular clocks (global clocks) applied a single constant rate of evolution obtained using a calibration to produce timelines, second and third generation methods (local and relaxed clocks) allow group- or branch-specific evolutionary rates to accommodate the well-known phenomenon of rate variation among lineages [35]. This flexibility in rate estimation is achieved with complex iterative algorithms that simultaneously optimize phylogeny, branch lengths, and divergence times or any combination of these variables [36–39] (for a review see [40]). In principle, the sophistication of modern molecular clock methods allows the use of their outputs, timetrees, to obtain lineage-specific evolutionary rates. In practice, some key divergences in prokaryote evolution have highly variable estimates of divergence time differing, in some cases, by more than 500 Ma (e.g, *Gammaproteobacteria* vs. *Chlorobium* [41, 42]). This leads to variable evolutionary rates for these lineages and, therefore, uncertainties in the pace of evolutionary innovations.

13.2 How do we estimate rates of genetic change?

In this genomic era much attention is given to automated processes that can quickly analyze large-scale patterns for hundreds to thousands of species. The estimation of evolutionary rates is not an exception. While lineage-specific experiments are extremely valuable to gain insights into the mechanisms triggering or suppressing mutation rates, large-scale phylogenetic methods are required to identify trends of rate speed-ups and slow-downs shared by groups of lineages. Based on phylogenies, methods to estimate substitution rates fall into two general categories: those that estimate relative rates and those producing absolute estimates (usually expressed in number of substitutions per site per year). Both these categories have strengths and weaknesses and can be applicable based on the type of information available.

13.2.1 Relative rate estimation

One of the simplest methods used to estimate relative substitution rates is the comparison of branch lengths in a phylogeny. Irrespective of the substitution model used, branch lengths are proportional to the number of substitutions in pairwise comparisons and are, therefore, representative of the frequency of genetic changes through time. Because two sister lineages share the same common ancestor and have evolved for the same amount of time (assuming all lineages are extant), longer branches denote faster rates of substitution while shorter branches represent slower rates. This is the basis for relative rate tests that compare branches of sister lineages to an outgroup lineage (▶ Fig. 13.1 (a)) [43, 44]. Any statistically significant difference between the two sequences is recorded as a rate shift. Methods such as this one have been widely used to estimate rates in mammals and, in particular, rate changes between the rodent and the primate lineage (e.g., [27, 43]). These methods are also used to test the molecular clock hypothesis, and were fundamental when the assumption of rate constancy was crucial to the time estimation process (global clocks) [45].

A variation of this approach that allows a quick evaluation of all branches in a large phylogeny entails calculating root-to-tip distances for each lineage and comparing each of these estimates to their average or median. This approach can show

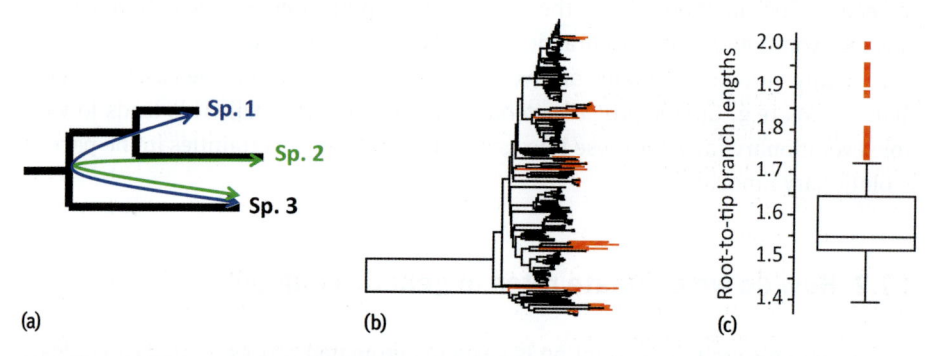

Fig. 13.1. Rate variation among lineages. (a). Schematic of a relative rate test that uses pairwise comparisons of the branch lengths from leaves (Sp. 1 and 2) to outgroup (Sp. 3) to estimate variation in evolutionary rates. (b). Phylogenetic tree from 218 prokaryote species [67] showing rate outliers in orange. (c). Box plot of root-to-tip branch lengths for the tree in panel B showing outliers in orange.

branches that are outliers compared to the overall trend in the tree. While these approaches do not produce any absolute estimate for the rate of changes, they do inform on lineages or groups of lineages that have undergone a shift in substitution rate (▶ Fig. 13.1 (b) and (c)). However, disadvantages include (i) low discriminatory power in case of small rate changes, and (ii) a strong dependency on accurate estimations of branch lengths, which is known to be affected by multiple factors, including substitution models and the shape of the tree [24, 46].

13.2.2 Absolute rate estimation

Information gained from relative rate methods can be transformed into absolute values if a reference timeline is available. Because branch lengths of a phylogeny are scaled to pairwise genetic distances, the age of the ancestral node of two lineages can be used to frame such distance within time units, therefore producing the familiar units of substitution rates expressed in changes per site per time unit (e.g. years). When accurate, estimations obtained with this approach provide information on the number of substitutions accumulated in a given amount of time, information that can be correlated with diversification and adaptive events (e.g. [47, 48]). However, the dependency of this method on a timeline of evolution poses a conundrum in those cases in which node ages are estimated with high uncertainty. This is the case for microbial organisms for which the only source of extensive chronological information is molecular timetrees, which are particularly controversial given the unique evolutionary characteristics of microbial genomes (e.g., common horizontal gene transfer events). The two major issues in estimating accurate divergence times for prokaryotes affect two fundamental parameters in the molecular clock formulation: genetic distances and calibration points.

13.2.2.1 The genetic distances problem

As mentioned above, an evolutionary rate is a measure of genetic changes (substitutions) in a given amount of time, with the number of genetic changes being calculated in a comparative framework as a pairwise genetic distance (see above). Based on the phylogenetic group being analyzed, genetic distances (D) range from a small to a large number of substitutions relative to the total length of the sequence alignment. While the methods to estimate rates based on genetic distances and times are the same irrespective of the overall value of D, branch lengths calculated based on very small or very large D values are more likely to be inaccurate because of the limited amount of information available. For example, many genomes at the sub-species (strain or serovar) level are now available for different prokaryotic species. These lineages are highly similar with only a few substitutions to distinguish them (these estimates often do not include variability in plasmids that are mobile and, therefore, can be more divergent among sub-species) [17, 49, 50]. In cases such as these, small errors in the estimation of substitutions can have a large effect on the evolutionary rate obtained. On the other end of the spectrum, lineages that are very distantly related are likely to show varying degrees of saturation in their sequences leading to an underestimation of genetic distances and, therefore, observed evolutionary rates slower than expected. This issue is in part mitigated by the use of amino acid changes (i.e., non-synonymous substitutions) instead of DNA changes (synonymous and non-synonymous) that, because of their larger number of states (20 instead of 4) and their overall slower accumulation, are likely to be less saturated. Unfortunately, studies on the correlation between divergence times and genetic saturation of sequences are mostly lacking [51–53].

13.2.2.2 The calibration points problem

The second biasing factor in the estimation of divergence times for prokaryotes is related to the availability of calibration points. As mentioned above, these are nodes in the phylogeny that allow estimating absolute divergence times (e.g., in millions of years) based on chronological information from non-molecular records. While lineages in the Phanerozoic have an abundant fossil record, direct fossil evidence of prokaryotes reaches, at most, 2.1 billion years ago with the fossilized spores of cyanobacteria (akinetes) [54]. Indirect evidence, such as biomarkers or fossilized microbial structures (e.g., stromatolites), can be found deeper in time but their biogenicity is controversial, they are not lineage specific and, therefore, not applicable to any specific node of a phylogenetic tree (e.g. [55–60]). This means that the timetree of prokaryotes relies on few calibration points, requiring strong extrapolations over large genetic distances to obtain node ages for the whole phylogeny. It has been shown in simulation studies that this can lead to inaccurate estimates, a condition reflected in the usually large credibility intervals associated with deep node times [52, 61–63].

In order to address the issues posed by both strategies, those producing relative and absolute evolutionary rates, it would be advisable to increase the number of calibrations in the prokaryotic tree and decrease the phylogenetic distance between lineages in order to decrease the length between ancestor-descendant nodes. Unfortunately, this is not feasible in many cases. An alternative is to use a recently developed method, RelTime, which, thanks to a simple approach to estimating relative times and rates, can mitigate some of the issues encountered by other methods. For example, by using a tip-to-root approach in the analysis of the phylogenetic tree, it uses the fact that extant sequences have an implicit calibration point of 0 to estimate branch-specific relative rates based on a large number of intrinsic calibration points. By expanding this approach to progressively more inclusive clades, it calculates evolutionary rates on each external and internal branch that can then be used to evaluate patterns of rate shift in subtrees [39].

13.3 How do we model evolutionary rates?

Despite our poor understanding of their variation patterns, evolutionary rates are routinely incorporated in molecular clock analyses during which they are modeled according to two different patterns: the autocorrelated and the uncorrelated model. The fundamental difference between the two models is in the amount of evolutionary rate change allowed between ancestral and descendant lineages with the autocorrelated model constraining it and the uncorrelated model leaving it unconstrained [64]. Both strategies are based on biological properties of organisms but the focus changes from short-term to long-term evolutionary patterns. Evolutionary rates at the short-term timescale are expected to be driven mostly by the genetic similarity of closely related sequences, which therefore will also share similar (but not identical) evolutionary rates. This correlation between ancestral-descendant rates breaks down at larger timescales that include distantly-related lineages [64]. Therefore, autocorrelated rates could be expected to be more likely in closely-related lineages and uncorrelated rates in highly divergent ones. Unfortunately, most datasets include both close and distant lineages, leading to a potential mismatch of assumed and empirical rate models. Simulations have shown this to be a biasing factor in the accuracy of molecular clock trees [65] and, therefore, for the estimation of empirical evolutionary rates.

13.4 Environments and evolutionary rates

One of the basic principles of evolution is that genetic variability is favored in the presence of stressful environmental conditions (e.g. [3, 7]). This is because higher mutation rates and recombination rates are more likely to produce favorable genetic innovations that will allow adaptations to new physicochemical and biological parameters. Sexually reproducing organisms are an excellent example of this as they are favored in harsh environments compared to asexual lineages [66]. However, other strategies are implemented by asexual organisms, such as prokaryotes, to adapt to stressful environments. *In vivo* and *in vitro* studies have illustrated the dynamics of populations of bacteria, such as *E. coli* and *Pseudomonas*, under changing environmental conditions focusing on the rise and spread of "mutator" strains. These are organisms with mutations (often in the *mutS* and *mutL* genes) that release constraints on the amount of changes that can accumulate in their genomes. Whether due to lower efficiency of DNA repair mechanisms or to intrinsically higher mutation rates, the net effect is that these organisms accumulate mutations at a faster rate, therefore raising the probability of evolving new advantageous variants and increasing the genetic polymorphism of the population in general [21–23]. Once adapted to the new conditions, the mutator strains decrease in frequency because of the high cost of frequent deleterious mutations and the population reverts to evolutionary rates comparable to those before the environmental shift.

The dynamic exemplified in these experiments is at the short-term population level but it has implications for more generalized evolutionary processes. Irrespective of the niche in which a prokaryote lives (e.g., extreme or mesophilic), shifts in evolutionary rates are most likely caused by changes of selective pressures caused by changing environmental conditions, rather than by the conditions themselves. It is therefore expected that whether extreme or not, rate shifts would not be common in lineages living in predictable environments. Rather, the expectation would be to find heterogeneous rates in lineages associated with unpredictable environments, irrespective of the physicochemical parameters they may have. Analyses of a 218 prokaryote dataset from Battistuzzi et al. [67] follow these predictions. In a box plot of the estimated root-to-tip distances in prokaryotic species occupying a variety of habitats and having different lifestyles (mesophiles, extremophiles, free-living, pathogenic), the majority of the outliers (15/25 or 60%) were pathogens and parasites that all showed faster evolutionary rates than the median (▶ Fig. 13.1 (c)). Similarly, when a relative rate test (the two cluster test) was used to identify faster or slower-evolving lineages, 13 out of 25 lineages with biased evolutionary rates were found to be pathogens or parasites. In contrast, fewer than 5 of these outliers were extremophiles (for temperature and pH).

While these results do not show a correlation between extremophiles and biased evolutionary rates, comparative studies carried out among microenvironments colonized by the cyanobacterium *Nostoc linckia* have shown a faster rate of evolution in those populations living in environments that are more extreme [68–71]. However, in this case, the extreme nature of the environment coincided with strong shifts in temperature and aridity during night-day cycles. Therefore, this result also supports the observation that evolutionary rates are altered during environmental shifts and not necessarily in extreme environments *per se*. An exception to this general trend could be organisms living in environments exposed to high UV radiation as these have been linked to DNA damage and, therefore, mutations. Organisms such as *Deinococcus radiodurans* have evolved numerous mechanisms to shield their DNA from harmful radiation and repair fast-occurring mutations but there is currently no clear evidence of what effect such an environment has on its evolutionary rates [72–76]. None of our analyses shows *D. radiodurans* as an outlier for evolutionary rates but additional evidence from other lineages of radiation tolerant microbes are needed.

13.4.1 Evolutionary rates of pathogens

The majority of the lineages found to have biased evolutionary rates in ours and other's analyses are prokaryotic symbionts of eukaryotic organisms. Irrespective of their relation to the host (e.g. pathogen or mutualist), these species have consistently shown fast evolutionary rates that lead, in many cases, to dramatic genome reductions [77, 78]. The mechanisms at the basis of these rate changes in intracellular prokaryotes are dependent on two primary factors: (i) the relation with the host species and (ii) the population size of the symbiont. The first factor determines the response of the intracellular organism based on either a predictable or unpredictable host environment (e.g. adaptations to frequent changes in host immune response) while the second affects the relative strengths of selection and drift and how new mutations spread into the symbionts' populations. While it will be difficult to isolate the effects of these two parameters, it is possible to speculate through the use of comparative approaches of closely-related pairs of intracellular and free-living lineages and quantify the nature of rate change based on the predictability of the environment. Such an approach would provide a valuable bioinformatics tool to investigate symbiont-host relationships but will require access to genome information for multiple strains of free-living and intracellular prokaryotes.

13.5 Large-scale genomic changes: duplications/loss and horizontal gene acquisition

Evolutionary rates are a way of measuring genetic innovations and how these might correlate with environmental factors. While many algorithms used to investigate evolutionary rates focus on number of substitutions per site per time unit, genetic innovation is also achieved by acquisition of new genes and, potentially, new functions. In prokaryotes, gene acquisition is primarily mediated by two mechanisms: gene duplications and horizontal gene transfers (HGT). While gene duplications are a within-genome mechanism that increases the number of copies of one or more genes, horizontal gene transfers are between-species events that copy DNA from a donor to a receiver species. Although widely different in terms of mechanism, once the new gene is present in the genome the evolutionary pressures to which it is exposed are similar: if the gene provides an advantage to the organism, its likelihood of retention will be higher than if negative effects on fitness are caused by the acquired gene. In light of the long evolutionary history of microbial species, identifiable acquired genes are those that have survived this "fate-determination" phase and are now a stable complement of the genome [79]. Acquired genes that are retained in the genome can be further analyzed to identify point mutations among copies that are at the basis of the new function gained by the organism, therefore providing a connection between rates of gene acquisition and substitution rates. Given their generally neutral or positive effect on fitness, acquired genes are of extreme interest in understanding how genomes respond to changing and stressful environmental conditions.

13.5.1 Rates of gene duplication and loss

Genome size in prokaryotes is extremely variable ranging from less than 0.5 Mb to more than 14 Mb. Size is generally positively correlated with genome plasticity that provides the necessary genomic background to survive in variable environments [5, 8]. One of the factors contributing to larger genome sizes is the number of paralogs found in each genome potentially linking duplication rates with genome plasticity [14, 80, 81].

Multiple studies have shown how duplicated genes are involved in adaptive strategies to overcome environmental challenges, whether by providing new functions or by augmenting current ones [14, 82, 83]. Some of the challenges addressed by duplications include temperature stress, high concentrations of heavy metals, presence of antibiotics, and, in general, changes in chemical compositions in the environment (e.g., increase in oxygen concentration). Support in favor of gene duplication as a strategy to achieve environmental adaptation comes from higher duplication rates for functional categories associated with metabolism, energy production and conversion, and defense mechanisms [14, 81, 84]. Additionally, a long timescale analysis of the expansion and contraction of gene families has suggested spikes in rates of gene acquisition and loss at different times during the history of Earth in correlation with environmental changes at a planetary scale [15].

Despite these general trends, no clear correlation has emerged between rates of gene duplications and extremophilic conditions. We conducted a survey of > 2,400 prokaryote genomes to investigate the relation between extremophilic lifestyles and the degree of duplication within each genome (defined as the ratio of the number of genes with at least one paralog and the total number of genes). On average the degree of duplication is 47.1% (range: 27.1–65.1%) with the outliers belonging predominantly to intracellular symbionts that are not considered extremophiles in the classic sense (▶ Fig. 13.2).

On the other end of the spectrum are very small genomes that have limited duplications within their genomes. Although counterintuitive at first, gene loss can also be adaptive as it increases efficiency and speed of replication and alters the expression of other genes [77, 85, 86]. Most of these genomes belong to symbionts (pathogens or mutualists) that are not considered extremophiles but rather highly specialized organisms with very restricted and stable niches. However, a recent case of genome reduction has been identified in a fast-evolving free-living prokaryote, *Prochlorococcus*, which shows different adaptations to low-light and high-light environments [32]. Although not considered an extremophile in the classic sense (e.g. extreme pH, salinity, or temperature), this discovery expands the role of gene loss as a strategy to adapt not only to host conditions but also to environments with variable physicochemical properties. Interestingly, lineages found to be outliers by the relative rate approaches (relative rate test and box plot) do not have a corresponding trend in duplication rates. In fact, even pathogens with small genomes such as the onion yellows phytoplasma that has a < 1 Mb genome size can have duplication rates comparable to genomes of free-living species.

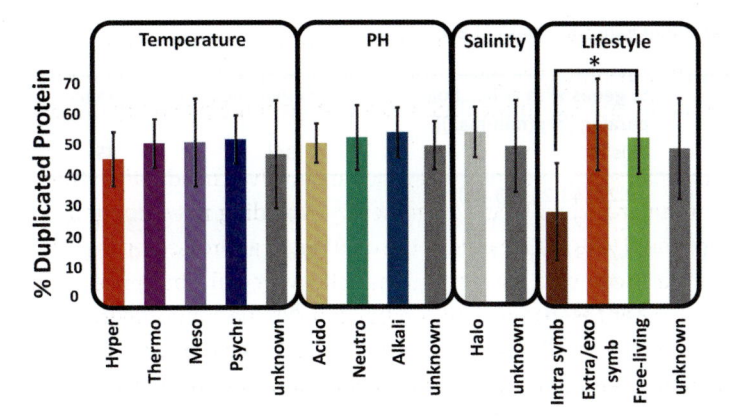

Fig. 13.2. Degree of duplication per extremophile classification and lifestyle. The degree of duplication was calculated using a within-genome BLAST approach and the identification of orthologous groups. For each category we are showing the average and standard deviation (vertical bars). The temperature classification includes the categories hyperthermophile (Hyper), thermophile (Thermo), mesophile (Meso), psychrophile (Psychr), and unknown. The pH classification includes the categories acidophile (Acido), neutrophile (Neutro), and alkaliphile (Alkali), and unknown. The salinity classification includes the categories halophile (Halo) and unknown. The lifestyle classification includes the categories intracellular symbiont (Intra symb), extracellular and exosymbiont (Extra/Exo symb), free-living, and unkown. All classifications were taken from the Genome Online database for a total of 2,439 Bacteria and Archaea.

13.5.2 Highways of horizontal gene transfers

Movement of genes through a non-vertical route is a common force in the evolution of prokaryotic genomes (▶ Tab. 13.2 and 13.3) [87, 88]. Despite large differences in genome architecture and composition of different species, these events do not seem to be constrained by phylogenetic relationships but rather can happen between closely or distantly related species and even between different domains of life [89]. The common evolutionary pressure that favors these transfers, irrespective of the identity of the species involved, is the acquisition of new functions that can improve the fitness of a species within a given environment. For this reason, many HGTs happen between species that co-inhabit the same niche: the extremophiles *Aquifex* and *Thermotoga*, for example, have exchanged genes with *Epsilonproteobacteria* and many archaea that occupy the same high-temperature environments. Similarly, mesophiles within *Cyanobacteria* and *Alpha-*, *Beta-*, and *Gamma-proteobacteria* also show a high degree of HGT due to their presence in soil, marine, or host habitats [89, 91].

These patterns show that HGT behaves in a similar way as gene duplication/loss or even point mutations: the rate at which these events happen does not appear to be related to a specific environmental condition, such as high temperature, but rather seems to be driven by selective forces imposed by an unstable environment (whether extreme or mesophilic).

Table 13.2. Averages of acquired genes per class of Bacteria.

Class (# species)	% genes with at least one paralog [avg(min-max)] This study	% of genome from HGT events [84]	[85]
Acidobacteriia (5)	49.21 (43.21–53.35)	–	–
Actinobacteria (256)	45.76 (7.27–69.19)	3.0	2.38–6.74
Alphaproteobacteria (228)	40.74 (1.48–65.12)	3.6–9.7	0–2.31
Anaerolineae (1)	48.47	–	–
Aquificae (11)	35.49 (31.24–37.48)	8.7	5.39
Bacilli (389)	46.71 (32.63–94.69)	2.4–28.8	0.7–3.2
Bacteroidia (23)	43.12 (28.45–59.31)	–	–
Betaproteobacteria (141)	45.40 (0–67.91)	0.5–4.4	1.39
Caldilineae (1)	56.97	–	–
Caldisericia (1)	37.66	–	–
Chlamydiia (78)	21.65 (18.67–38.55)	2.8–3.1	0–0.3
Chlorobia (11)	38.86 (34.84–42.91)	–	–
Chloroflexi (6)	53.54 (50.72–55.07)	–	–
Chrysiogenetes (1)	41.12	–	–
Clostridia (122)	49.32 (31.15–65.89)	–	–
Cytophagia (11)	49.41 (39.23–59.30)	–	–
Deferribacteres (4)	42.71 (41.18–45.58)	–	–
Dehalococcoidia (8)	36.66 (33.87–40.10)	–	–
Deinococci (19)	41.98 (33.58–52.06)	3	0.88
Deltaproteobacteria (55)	48.35 (29.19–61.66)	–	–
Dictyoglomia (2)	44.11 (43.97–44.25)	–	–
Elusimicrobia (1)	32.42	–	–
Epsilonproteobacteria (90)	29.60 (23.67–45.69)	0.5	0.35–2.72
Erysipelotrichia (1)	36.59	–	–
Fibrobacteria (1)	44.03	–	–
Flavobacteriia (42)	30.44 (9.14–52.53)	–	–
Fusobacteriia (7)	43.99 (37.24–57.85)	0–13.1	0.43–3.57
Gammaproteobacteria (520)	43.99 (37.24–57.85)	–	–
Gemmatimonadetes (1)	44.47	–	–
Gloeobacteria (1)	53.62	–	–
Ignavibacteria (2)	47.67 (46.99–48.35)	–	–
Mollicutes (73)	35.76 (19.03–80.24)	0–2.1	1.4–4.02
Negativicutes (5)	43.19 (36.92–49.40)	–	–
Nitrospira (4)	42.09 (38.13–48.76)	–	–
Opitutae (2)	46.26 (42.42–50.11)	–	–
Phycisphaerae (1)	34.9	–	–
Planctomycetia (5)	46.57 (40.08–57.43)	–	–
Solibacteres (1)	65.27	–	–
Sphingobacteriia (8)	54.05 (39.37–60.71)	–	–
Spirochaetia (47)	38.84 (17.29–78.62)	1.1–32.6	0–1.39
Synergistia (4)	40.34 (35.42–42.36)	–	–
Thermodesulfobacteria (2)	36.67 (32.28–41.06)	–	–
Thermomicrobia (2)	50.78 (46.21–55.35)	–	–
Thermotogae (15)	42.31 (37.40–47.48)	–	–
Verrucomicrobiae (1)	35.3	–	–

Table 13.3. Averages of acquired genes per class of Archaea.

Class (# species)	% genes with at least one paralog [avg(min-max)] This study	% of genome from HGT events [84]	% of genome from HGT events [85]
Archaeoglobi (5)	41.83 (33.49–47.18)		
Halobacteria (24)	50.04 (40.76–62.08)	15.6	1.33
Methanobacteria (9)	42.32 (38.22–51.93)	8.6	2.5
Methanococci (15)	36.14 (30.98–41.32)	4.8	3.37
Methanomicrobia (25)	49.41 (40.43–60.70)	–	–
Methanopyri (1)	36.98	–	–
Thermococci (15)	43.43 (39.39–47.43)	4.6–6.3	0.35–0.38
Thermoplasmata (4)	36.09 (33.16–40.01)	11.5	0.0
Thermoprotei (47)	38.30 (23.40–54.05)	4.3–14	2.04–4.73

13.6 Conclusions

While the categorization of prokaryotes into extremophiles and mesophiles is useful from a classification and phylogenetic point of view, its application to the understanding of evolutionary rate variation remains uncertain. Classically, extremophiles are identified based on "extreme" (from an anthropocentric point of view) physicochemical preferences in the habitats they live in. Temperature, pH, and salinity are among the most common properties considered to define an extreme environment. However, many other types of "extremes" could be considered as grounds for classification of an organism as an extremophile. An exclusive association with a host, for example, causes unique genetic adaptations driven by very strong selective pressures which can be akin to the unique set of adaptations required to survive in high temperature environments. A highly variable environment could also be considered extreme as it requires organisms to be able to survive in a suite of different conditions.

If we consider extremophiles in the classic sense, no clear correlation between extreme environments and rate variations is present. Organisms living in extreme (but stable) environments have evolutionary rates (whether fast or slow) that resemble those of their mesophilic relatives. However, this observation changes if we use an expanded view of extremophiles with symbionts, for example, showing slower duplication rates than free-living prokaryotes and pathogens showing higher mutation rates than benign organisms.

From these considerations, it seems that the long-standing question of what drives changes in evolutionary rates among lineages is unlikely to be answered anytime soon. However, hope resides in increasingly large datasets that are allowing us to investigate the extent of rate variation in a comparative framework for closely related lineages and to identify general trends that will likely affect our current definition of mesophiles and extremophiles.

References

[1]   H. D. Holland. Volcanic gases, black smokers, and the Great Oxidation Event, Geochim Cosmochim Ac, 2002, 21, 3811–3826.
[2]   D. E. Canfield, M. T. Rosing, C. Bjerrum, Early anaerobic metabolisms, Philos Trans R Soc Lond B Biol Sci, 2006, 361, 1819–1836.
[3]   D. Romero and R. Palacios, Gene Amplification and Genomic Plasticity in Prokaryotes, Annu Review Genet, 1997, 31, 91–111.
[4]   T. J. Treangen and E. P. C. Rocha, Horizontal Transfer, Not Duplication, Drives the Expansion of Protein Families in Prokaryotes, PLoS Genet, 2011, 7, e1001284.
[5]   B. Ryall, G. Eydallin, and T. Ferenci, Culture History and Population Heterogeneity as Determinants of Bacterial Adaptation: the Adaptomics of a Single Environmental Transition, Microbiol Mol Biol Rev., 2012, 76, 597–625.
[6]   K. T. Elliott, L. E. Cuff, and E. L. Neidle, Copy number change: evolving views on gene amplification, Future Microbiol, 2013, 8, 887–899.
[7]   Z. Bao, P. V. Stodghill, C. R. Myers, H. Lam, H.-L. Wei, S. Chakravarthy, B. H. Kvitko, A. Collmer, S. W. Cartinhour, P. Schweitzer, and B. Swingle, Genomic Plasticity Enables Phenotypic Variation of Pseudomonas syringae pv. tomato DC3000, PLoS ONE, 2014, 9, e86628.
[8]   Y. Zhang and S. M. Sievert, Pan-genome analyses identify lineage- and niche-specific markers of evolution and adaptation in Epsilonproteobacteria, Front Microbiol, 2014, 5.
[9]   P. A. Parsons, Habitats, stress, and evolutionary rates, Journal of Evolutionary Biology, 1994, 7, 387–397.
[10] R. Cavicchioli, R. Amils, D. Wagner, and T. McGenity, Life and applications of extremophiles, Environ Microbiol, 2011, 13, 1903–1907.
[11] M. Grover, S. Z. Ali, V. Sandhya, A. Rasul, and B. Venkateswarlu, Role of microorganisms in adaptation of agriculture crops to abiotic stresses, World J Microbiol Biotechnol, 2011, 27, 1231–1240.
[12] G. Storz and R. Hengge, Bacterial stress responses. Washington D.C., USA, ASM Press, 2000.
[13] M. H. Serres, A. R. Kerr, T. J. McCormack, and M. Riley, Evolution by leaps: gene duplication in bacteria, Biol Direct, 2009, 4, 46.
[14] M. S. Bratlie, J. Johansen, B. T. Sherman, D. W. Huang, R. A. Lempicki, and F. Drabløs, Gene duplications in prokaryotes can be associated with environmental adaptation, BMC Genomics, 2010, 11, 588.
[15] L. A. David and E. J. Alm, Rapid evolutionary innovation during an Archaean genetic expansion, Nature, 2011, 469, 93–96.
[16] H. Ochman and A. C. Wilson, Evolution in bacteria: evidence for a universal substitution rate in cellular genomes, J Mol Evol, 1987, 26, 74–86.
[17] H. Ochman and I. B. Jones, Evolutionary dynamics of full genome content in Escherichia coli, The EMBO Journal, 2000, 19, 6637–6643.
[18] C.-H. Kuo and H. Ochman, Inferring clocks when lacking rocks: the variable rates of molecular evolution in bacteria, Biol Direct, 2009, 4, 35.
[19] G. Morelli, X. Didelot, B. Kusecek, S. Schwarz, C. Bahlawane, D. Falush, S. Suerbaum, and M. Achtman, Microevolution of Helicobacter pylori during Prolonged Infection of Single Hosts and within Families, PLoS Genet, 2010, 6, e1001036.
[20] M. L. Reno, N. L. Held, C. J. Fields, P. V. Burke, and R. J. Whitaker, Biogeography of the Sulfolobus islandicus pan-genome, Proc Natl Acad Sci USA, 2009, 106, 8605–8610.
[21] J. M. J. Travis and E. R. Travis, Mutator dynamics in fluctuating environments, Proc R Soc Lond B, 2002, 269, 591–597.

[22] M. M. Tanaka, C. T. Bergstrom, and B. R. Levin, The Evolution of Mutator Genes in Bacterial Populations: The Roles of Environmental Change and Timing, Genetics, 2003, 164, 843–854.
[23] E. Denamur and I. Matic, Evolution of mutation rates in bacteria, Mol Microbiol, 2006, 60, 820–827.
[24] R. Lanfear, J. J. Welch, and L. Bromham, Watching the clock: Studying variation in rates of molecular evolution between species, Trends Ecol Evol, 2010, 25, 495–503.
[25] Li WH, Wu CI, and C. C. Luo, A new method for estimating synonymous and nonsynonymous rates of nucleotide substitution considering the relative likelihood of nucleotide and codon changes, Mol Biol Evol, 1985, 2, 150–174.
[26] R. K. Blackman and M. Meselson, Interspecifc nucleotide sequence comparisons used to identify regulatory and structural features of the Drosophila hsp82 gene, J Mol Biol, 1986, 188, 499–515.
[27] Li W-H, D. L. Ellsworth, J. Krushkal, B. H.-J. Chang, and D. Hewett-Emmett, Rates of Nucleotide Substitution in Primates and Rodents and the Generation–Time Effect Hypothesis, Mol Phyl Evol, 1996, 5, 182–187.
[28] Yi S, D. L. Ellsworth, and Li W-H, Slow Molecular Clocks in Old World Monkeys, Apes, and Humans, Mol Biol Evol, 2002, 19, 2191–2198.
[29] J. Kazmierczak and W. Altermann, Neoarchean biomineralization by benthic cyanobacteria, Science, 2002, 298, 2351–2351.
[30] W. Altermann and J. Kazmierczak, Archean microfossils: a reappraisal of early life on Earth, Res Microbiol, 2003, 154, 611–617.
[31] J. J. Brocks and A. Pearson, Building the biomarker tree of life, Rev MineralGeochem, 2005, 59, 233–258.
[32] A. Dufresne, L. Garczarek, and F. Partensky, Accelerated evolution associated with genome reduction in a free-living prokaryote, Genome Biol, 2005, 6, R14.
[33] E. Zuckerkandl and L. Pauling, Molecular disease, evolution, and genetic heterogeneity, in Horizons in Biochemistry, eds. M. Marsha. and B. Pullman. New York City, NY, USA, Academic Press, 1962, 189–225.
[34] E. Zuckerkandl and L. Pauling, Evolutionary divergence and convergence in proteins, in Evolving genes and proteins, eds. V. Bryson and H. J. Vogel, New York City, NY, USA, Academic Press, 1965, 97–166.
[35] S. Kumar, Molecular clocks: four decades of evolution, Nature Rev Genet, 2005, 6, 654–662.
[36] N. Lartillot, T. Lepage, and S. Blanquart, PhyloBayes 3: a Bayesian software package for phylogenetic reconstruction and molecular dating, Bioinf, 2009, 25, 2286–2288.
[37] M. D. Reis and Z. Yang, Approximate Likelihood Calculation on a Phylogeny for Bayesian Estimation of Divergence Times, Mol Biol Evol, 2011, 28, 2161–2172.
[38] A. J. Drummond, M. A. Suchard, D. Xie, and A. Rambaut, Bayesian Phylogenetics with BEAUti and the BEAST 1.7, Mol Biol Evol, 2012, 29, 1969–1973
[39] K. Tamura, F. U. Battistuzzi, P. Billing-Ross, O. Murillo, A. Filipski, and S. Kumar, Estimating Divergence times in large molecular phylogenies, Proc Natl Acad Sci USA, 2012, 109, 19333–19338
[40] T. Lepage, D. Bryant, H. Philippe, and N. Lartillot, A General Comparison of Relaxed Molecular Clock Models, Mol Biol Evol, 2007, 24, 2669–2680.
[41] P. P. Sheridan, K. H. Freeman, and J. E. Brenchley, Estimated Minimal Divergence Times of the Major Bacterial and Archaeal Phyla, Geomicrobiol J, 2003, 20, 1–14.
[42] F. U. Battistuzzi, A. Feijao, and S. B. Hedges, A genomic timescale of prokaryote evolution: insights into the origin of methanogenesis, phototrophy, and the colonization of land, BMC Evol Biol, 2004, 4, 44.

[43] Wu CI and Li WH, Evidence for higher rates of nucleotide substitution in rodents than in man, PNAS, 1985, 82, 1741–1745.
[44] Li W-H and M. Tanimura, The molecular clock runs more slowly in man than in apes and monkeys, Nature, 1987, 326, 93–96.
[45] N. Takezaki, A. Rzhetsky, and M. Nei, Phylogenetic Test of the Molecular Clock and Linearized Trees, Mol Biol Evol, 1995, 12, 823–833.
[46] A. F. Hugall and M. S. Y. Lee, The Likelihood Node Density Effect and Consequences for Evolutionary Studies of Molecular Rates, Evol, 2007, 61, 2293–2307.
[47] Ho SYW, R. Lanfear, L. Bromham, M. J. Phillips, J. Soubrier, A. G. Rodrigo, and A. Cooper, Time-dependent rates of molecular evolution, Mol Ecol, 2011, 20, 3087–3101.
[48] B. Shapiro and Ho SYW, Ancient hyaenas highlight the old problem of estimating evolutionary rates, Mol Ecol, 2014, 23, 499–501.
[49] U. Dobrindt, F. Agerer, K. Michaelis, A. Janka, C. Buchrieser, M. Samuelson, C. Svanborg, G. Gottschalk, H. Karch, and J. Hacker, Analysis of Genome Plasticity in Pathogenic and Commensal Escherichia coli Isolates by Use of DNA Arrays, J Bacteriol, 2003, 185, 1831–1840.
[50] R. A. Welch, V. Burland, G. Plunkett, P. Redford, P. Roesch, D. Rasko, E. L. Buckles, S.-R. Liou, A. Boutin, J. Hackett, D. Stroud, G. F. Mayhew, D. J. Rose, S. Zhou, D. C. Schwartz, N. T. Perna, H. L. T. Mobley, M. S. Donnenberg, and F. R. Blattner, Extensive mosaic structure revealed by the complete genome sequence of uropathogenic Escherichia coli, Proc Nat Acad Sci USA, 2002, 99, 17020–17024.
[51] Ho SYW, M. J. Phillips, A. Cooper, and A. J. Drummond, Time Dependency of Molecular Rate Estimates and Systematic Overestimation of Recent Divergence Times, Mol Biol Evol, 2005, 22, 1561–1568.
[52] A. J. Roger and L. A. Hug, The origin and diversification of eukaryotes: problems with molecular phylogenetics and molecular clock estimation, Philos Trans R Soc Lond B Biol Sci, 2006, 361, 1039–1054.
[53] Z. H. Yang and B. Rannala, Bayesian estimation of species divergence times under a molecular clock using multiple fossil calibrations with soft bounds, Mol Biol Evol, 2006, 23, 212–226.
[54] A. Tomitani, A. H. Knoll, C. M. Cavanaugh, and T. Ohno, The evolutionary diversification of cyanobacteria: molecular-phylogenetic and paleontological perspectives, Proc Natl Acad Sci USA, 2006, 103, 5442–5447.
[55] M. D. Brasier, O. R. Green, A. P. Jephcoat, A. K. Kleppe, M. J. Van Kranendonk, J. F. Lindsay, A. Steele, and N. V. Grassineau, Questioning the evidence for Earth's oldest fossils, Nature, 2002, 416, 76–81.
[56] J. W. Schopf, A. B. Kurdryavtsev, D. G. Agresti, T. J. Wdowiak, and A. D. Czaja, Laser-Raman imagery of Earth's earliest fossils, Nature, 2002, 416, 73–76.
[57] M. Brasier, O. Green, J. Lindsay, and A. Steele, Earth's oldest (similar to 3.5 Ga) fossils and the 'Early Eden hypothesis': Questioning the evidence, Origins Life Evolution B, 2004, 34, 257–269.
[58] M. Brasier, N. McLoughlin, O. Green, and D. Wacey, A fresh look at the fossil evidence for early Archaean cellular life, Philos Trans R Soc Lond B Biol Sci, 2006, 361, 887–902.
[59] J. W. Schopf, Fossil evidence of Archaean life, Philos Trans R Soc Lond B Biol Sci, 2006, 361, 869–885.
[60] J. W. Schopf, A. B. Kudryavtsev, A. D. Czaja, and A. B. Tripathi, Evidence of archean life: Stromatolites and microfossils, Precambrian Res, 2007, 158, 141–155.
[61] D. J. Cutler, Estimating divergence times in the presence of an overdispersed molecular clock, Mol Biol Evol, 2000, 17, 1647–1660.
[62] S. B. Hedges and S. Kumar, Precision of molecular time estimates, Trends Genet, 2004, 20, 242–247.

[63] L. A. Hug and A. J. Roger, The impact of fossils and taxon sampling on ancient molecular dating analyses, Mol Biol Evol, 2007, 24, 1889–1897.
[64] A. J. Drummond, Ho SY, M. J. Phillips, and A. Rambaut, Relaxed phylogenetics and dating with confidence, PLoS Biol, 2006, 4, e88.
[65] F. U. Battistuzzi, A. Filipski, S. B. Hedges, and S. Kumar, Performance of relaxed-clock methods in estimating evolutionary divergence times and their credibility intervals, Mol Biol Evol, 2010, 27, 1289–1300.
[66] J. Heitman, S. Sun, and T. Y. James, Evolution of fungal sexual reproduction, Mycologia, 2013, 105, 1–27.
[67] F. U. Battistuzzi and S. B. Hedges, A Major Clade of Prokaryotes with Ancient Adaptations to Life on Land, Mol Biol Evol, 2009, 26, 335–343.
[68] E. Nevo, Evolution of genome–phenome diversity under environmental stress, Proc Nat Acad Sci USA, 2001, 98, 6233–6240.
[69] V. Dvornyk, O. Vinogradova, and E. Nevo, Long-term microclimatic stress causes rapid adaptive radiation of kaiABC clock gene family in a cyanobacterium, Nostoc linckia, from 'Evolution Canyons' I and II, Israel, Proc Nat Acad Sci USA, 2002, 99, 2082–2087.
[70] E. Nevo, Evolution Under Environmental Stress at Macro- and Microscales, Genome Biol Evol, 2011, 3, 1039–1052.
[71] T. M. Hoehler and B. B. Jørgensen, Microbial life under extreme energy limitation, Nat Rev Microbiol, 2013, 11, 83–94.
[72] V. Mattimore and J. R. Battista, Radioresistance of Deinococcus radiodurans: Functions necessary to survive ionizing radiation are also necessary to survive prolonged desiccation, J Bacteriol, 1996, 178, 633–637.
[73] L. Albuquerque, C. Simoes, M. F. Nobre, N. M. Pino, J. R. Battista, M. T. Silva, F. A. Rainey, and M. S. da Costa, Truepera radiovictrix gen. nov., sp. nov., a new radiation resistant species and the proposal of Trueperaceae fam. nov, FEMS Microbiol Lett, 2005, 247, 161–169.
[74] M. M. Cox and J. R. Battista, Deinococcus radiodurans – the consummate survivor, Nat Rev Microbiol, 2005, 3, 882–892.
[75] M. V. Omelchenko, Y. I. Wolf, E. K. Gaidamakova, V. Y. Matrosova, A. Vasilenko, M. Zhai, M. J. Daly, E. V. Koonin, and K. S. Makarova, Comparative genomics of Thermus thermophilus and Deinococcus radiodurans: divergent routes of adaptation to thermophily and radiation resistance, BMC Evol Biol, 2005, 5, 57.
[76] M. Shukla, R. Chaturvedi, D. Tamhane, P. Vyas, G. Archana, S. Apte, J. Bandekar, and A. Desai, Multiple-stress tolerance of ionizing radiation-resistant bacterial isolates obtained from various habitats: correlation between stresses, Curr Microbiol, 2007, 54, 142–148.
[77] H. Ochman, Genomes on the shrink, Proc Nat Acad Sci USA, 2005, 102, 11959–11960.
[78] J. P. McCutcheon and N. A. Moran, Extreme genome reduction in symbiotic bacteria, Nat Rev Microbiol, 2012, 10, 13–26.
[79] H. Innan and F. Kondrashov, The evolution of gene duplications: classifying and distinguishing between models, Nat Rev Genet, 2010, 11, 97–108.
[80] I. K. Jordan, K. S. Makarova, J. L. Spouge, Y. I. Wolf, and E. V. Koonin, Lineage-Specific Gene Expansions in Bacterial and Archaeal Genomes, Genome Res, 2001, 11, 555–565.
[81] R. Pushker, A. Mira, and F. Rodriguez-Valera, Comparative genomics of gene-family size in closely related bacteria, Genome Biol, 2004, 5, R27.
[82] K. Yamanaka, L. Fang, and M. Inouye, The CspA family in Escherichia coli : multiple gene duplication for stress adaptation, Mol Microbiol, 1998, 27, 247–255.
[83] S. Hooper and O. Berg, On the nature of gene innovation: Duplication patterns in microbial genomes, Mol Biol Evol, 2003, 20, 945–954.

[84] D. Gevers, K. Vandepoele, C. Simillion, and Y. Van de Peer, Gene duplication and biased functional retention of paralogs in bacterial genomes, Trends Microbiol, 2004, 12, 148–154.

[85] N. Nakata, T. Tobe, I. Fukuda, T. Suzuki, K. Komatsu, M. Yoshikawa, and C. Sasakawa, The absence of a surface protease, OmpT, determines the intercellular spreading ability of Shigella: the relationship between the ompT and kcpA loci, Mol Microbiol, 1993, 9, 459–468.

[86] H. Ochman and N. A. Moran, Genes Lost and Genes Found: Evolution of Bacterial Pathogenesis and Symbiosis, Science, 2001, 292, 1096–1099.

[87] E. V. Koonin, K. S. Makarova, and L. Aravind, Horizontal gene transfer in prokaryotes: Quantification and classification, Annu Rev Microbiol, 2001, 55, 709–742.

[88] Ge F, L. S. Wang, and J. Kim, The cobweb of life revealed by genome-scale estimates of horizontal gene transfer, Plos Biol, 2005, 3, 1709–1718.

[89] R. G. Beiko, T. J. Harlow, and M. A. Ragan, Highways of gene sharing in prokaryotes, Proc Nat Acad Sci USA, 2005, 102, 14332–14337.

[90] B. Boussau, L. Guéguen, and M. Gouy, Accounting for horizontal gene transfers explains conflicting hypotheses regarding the position of aquificales in the phylogeny of Bacteria, BMC Evol Biol, 2008, 8, 272.

[91] O. Zhaxybayeva, K. S. Swithers, P. Lapierre, G. P. Fournier, D. M. Bickhart, R. T. DeBoy, K. E. Nelson, C. L. Nesbø, W. F. Doolittle, J. P. Gogarten, and K. M. Noll, On the chimeric nature, thermophilic origin, and phylogenetic placement of the Thermotogales, Proc Nat Acad Sci USA, 2009, 106, 5865–5870.

Index

16S rRNA 58, 70

A

ABC-type transporters 79, 129
abiotic stress 7, 196
abiotic synthesis 59
absolute rate 255
Ace Lake, Antarctica 46
Acetobacterium woodii 176
acetogen 172–179
acid mine drainage 1, 19, 25
Acidianus 67, 210, 215
acidic amino acids 8, 97, 99, 104, 106
acidic hot springs 10, 64, 66, 77, 158, 209
acidic proteome 99, 100, 104–106
Acidilobus 66
Acidithiobacillus 19, 21, 23, 24
– *A. ferrivorans* 135
acidocalcisomes 136
acidophile 4, 9, 19–27, 68, 130, 133, 135, 153, 261
adaptive radiation 189
adaptive trait 8
Afipia felis 129
algae 20, 33, 38, 41, 70, 158, 160, 227, 234, 237, 240
Alkalilimnicola halodurans 135
alkaline 2, 60, 65, 68–71, 76, 79, 130, 133, 137, 153, 156, 174, 175
alkaline phosphatase 127, 136–138
alkaliphile 4, 97, 105, 130, 131, 153, 261
allopatric speciation 26
allopatry 70
Alteromonadales 40
amino acid composition 37
aminoacyl-tRNA synthetases 80, 232
anaerobic 66, 75, 97, 104, 105, 132, 133, 135, 138, 157, 172, 179, 234
anhydrobiosis 161
Antarctic 9, 32, 33, 38, 40, 43, 102, 106, 187, 189, 213–215
Antarctica
– Dry Valleys 134, 156
Anthropocene 140
antifragility 188

antifreeze proteins 35, 37, 38
Aquifex 2, 65, 261
– *A. aeolicus* 77, 215
Aquificae 65, 70, 132
Aquificales 9, 64, 79
Archaeoglobus 66, 75
Archean 58, 119
Arctic 10, 32, 38, 40, 42, 189, 192, 213, 214
arid soil 120, 134, 258
aromatic hydrocarbon 191
– assimilation 192, 196
aromatic interactions 73
asexual 158, 188, 189, 257
Aspergillus 188, 191, 192, 195
astrobiology 156
AT mutational bias 231
Atacama Desert 156
ATPase 174, 176
Aureobasidium
– *A. melanogenum* 186, 193
– *A. namibiae* 186
– *A. pullulans* 186, 192, 193, 196, 198
– *A. subglaciale* 186
autocorrelated model 257
autotroph 19, 65, 67, 68, 129–132, 172, 174, 179, 227, 229, 237
auxotroph 78
Azores, Portugal 68, 120

B

Bacillus 135
– *B. alkalidiazotrophicus* 135
– *B. anthracis* 4
– *B. pumilis* 113
– *B. subtilis* 117
– TA2.A1 130
Beaufort Sea 43, 214
bedrock 59, 61, 63
biofilm 3, 21–25, 37, 76, 190–192, 196–198
biogenicity 256
bioleaching 20, 24
biomarkers 252, 256
biomining 20
biomolecules 1, 31, 77, 136
black yeast 99, 185, 186, 191, 193, 195

Borrelia burgdorferi 137
bottleneck 6–8, 35, 187
branch lengths 253–255
Buchnera 225, 226, 240
– *B. aphidicola* 4–6, 229

C

Caldisphaera 67
Caldivirga 66, 67
calibration points 253–256
Calothrix rhizosoleniae 231
Candida 195
– *C. albicans* 190
– *C. guilliermondii* 194
– *C. parapsilosis* 192, 194
carbon dioxide-concentrating mechanism 131
carbon fixation 129–133, 138, 172
– acetyl-CoA pathway 132, 133, 172, 175, 178, 179
carbonic anhydrase 127, 131, 138
Carsonella ruddii 232
Caudovirales 210
Cedars, California 131
Celerinatantimonas yamalensis 134
cell fusion 102
cheaters 138, 229
chemiosmotic 173–175
chemolithoautotroph 65, 67, 129, 131
chemosynthesis 69
chemotaxis 77
chimera 236
Chlorobium 253
chlorophyll 70, 105, 137, 234, 237
Chromohalobacter salexigens 100, 101
Chroococcidiopsis 120, 156
circumneutral 65, 68–71, 79, 137
Cladosporium sphaerospermum 189, 192
Clostridium
– *C. alkalicellulosum* 135
– *C. fervidus* 130
cobalt 138
co-evolution 230
cold environments 31, 46, 212–214, 260, 261
cold shock proteins 37
Colwellia psychrerythraea 2, 37, 40, 43, 46, 134
compatible solutes 10, 37, 40, 42, 46, 74, 75, 99, 189
competitive 72, 82, 197, 212
– exclusion 26

conductivity 61, 63
convergent evolution 8, 35, 99–101, 103
copiotrophic 35, 40
core genome 6, 21
co-speciation 230
Crenarchaeota 19, 20, 40, 64, 66, 72, 133
CRISPR 19, 23, 37, 79, 209, 212, 215
Crocosphaera watsonii 135, 138
Cryomyces antarcticus 187
cryosphere 31–47
cryptic genetic variation 187, 189
cryptobiosis 161, 187
cryptoendolith 158
Cyanidiophyceae 158
Cyanidioschyzon merolae 159
cyanobacteria 41, 66, 70, 76, 103, 114, 131, 134–136, 138, 156, 157, 159, 160, 212, 227, 231, 234, 237, 256, 261
cyanophycin 134
cyclic-2,3-bisphosphoglycerate 75
cyclopentyl 72
cytochromes 173, 176, 178, 179
cytoplasmic bridges 102

D

de novo gene creation 4, 5, 235, 236, 238, 240
Dead Sea 100
Deep Lake 9, 102, 103, 106
deep sea 2, 37, 105, 120, 131, 134, 209, 212, 215, 227
deep subsurface 2, 11, 60, 158, 173
dehydration 120, 154, 155, 161, 162
Deinococcus 120
– *D. radiodurans* 114, 115, 127, 130, 154, 258
deleterious mutations 252, 257
Deltaproteobacteria 66, 97
denaturation 1, 75, 99
desiccation 10, 103, 120, 155–158, 160, 161, 185, 187, 190, 192, 197
Desulfurella 66
diatoms 41, 215, 231
diazotrophy 128, 134, 135
DIC 130–132
dipole 73
dishwasher 1, 191–198
– fungi 195
dispersal 32–35, 38, 69
– aerial 43

– efficiency 187
– limitation 70
dissolved inorganic carbon 130–132
disulfide bridges 73
divergence 3, 5, 8, 25, 41, 70, 78, 80, 133, 174, 176, 189, 212, 234, 236, 240, 253, 256, 257
– time 253, 255
DNA repair 10, 20, 22, 119, 154, 161, 231
– mechanisms 257
double-strand breaks 10, 112, 120, 154
dsrAB 66
Dunaliella salina 2, 99, 106

E
early life 58, 171–179
ecotype 5, 78, 103, 134, 135
ectoine 100, 101, 105
ectosymbiosis 223
elemental sulfur 64–68, 77
emergence 58, 81, 174, 178
empirical evolutionary rates 257
Enceladus 31, 34
endemism 38, 189
endospores 117
endosymbiont 5, 9, 10, 132, 159, 223–240, 261
endosymbiosis 2, 223–240, 261
– secondary 234
energy stress 210
enzyme efficiency 37
epiphyte 40
Epsilonproteobacteria 261
Escherichia coli 5, 6, 77, 106, 113, 171, 239, 253, 257
ester 72, 130, 171
Europa 31, 34
Europan ice shell 34
Euryarchaeota 8, 20, 40, 64, 66, 72, 97, 114
evolutionary history 3, 4, 7, 8, 22, 25, 35, 237, 259
exaptation 9, 118, 195
Exophiala dermatitidis 191, 192, 195
extracellular DNA 36, 43, 102
extracellular polymeric substances (EPS) 37, 40–42, 46, 76, 196, 198
extremotolerant 2, 187, 189, 191

F
facilitative 82
fate-determination phase 259

fatty acids 46, 73, 129, 234
[FeFe]-hydrogenase 71
fermentation 64, 71, 97, 104, 135, 157, 174
Ferroplasma 19, 22, 24
filamentous 76, 135, 215, 231
Firmicutes 66, 97, 99, 103, 133, 236
Fischer-Tropsch type synthesis 176
Flexistipes sinusarabici 105
food web 78
fossil record 230, 252, 253, 256
free DNA 1, 9, 21, 42, 102
frost flowers 41, 43
frozen environments 31–47, 193
fumaroles 59–61
functional innovations 247
Fuselloviridae 215

G
Galdieria sulphuraria 20, 158
gas vacuole 41
GDGT 72
gene duplication 4, 5, 21, 80, 81, 235, 237, 240, 259–261
gene fusions 57, 80
gene island 9, 230
gene loss 5, 8, 10, 57, 66, 79–81, 162, 231, 240, 260
gene transfer agents 5
generalist 36, 186, 187, 189, 195, 196
genetic distance 252–256
genetic drift 5, 6, 8, 11, 35, 70, 105, 187, 231, 247
genetic redundancy 188
genome plasticity 25, 247, 259, 260
genome reduction 10, 231, 232, 247, 258, 260
genome restructuring 188, 198
genome, streamlined 136
genomic islands 22, 24, 26, 46, 136
genomics, comparative 20, 22, 25, 78, 210, 212, 230
Geoalkalibacter ferrihydritucus 135
geographic boundaries 3
geographic data 68
geographic isolation 6, 8, 69, 70
geologic record 46, 171, 252
geological history 3
geological timescale 57, 252
geysers 59, 61, 76
glacial ice 33, 35, 185, 192, 194, 196

Glaciecola 40
glacier 31–34, 185, 189
glycine betaine 46, 100, 101, 104, 105
GMGT 72
Greenland 33, 36
Guarapari, Brazil 115
Guerrero Negro 100

H
Halanaerobiales 97, 99, 104–106
Halanaerobium
– *H. hydrogeniformans* 104
– *H. praevalens* 104
haloalkaliphile 135, 211
haloarchaea 8, 103, 211
Haloarcula hispanica 211
Halobacillus halophilus 106
Halobacteriaceae 97, 101, 106
Halobacterium
– *H. halobium* 130, 211
– *H. salinarum* 101, 157
Halobacteroides halobius 104
Haloferax 102, 103, 106
– *H. volcanii* 102, 130
Halohasta 103
Halomonas 137
– *H. elongata* 101
halophilic 8, 10, 40, 66, 97–107, 157, 190, 211, 261
– fungus 187
Haloquadratum walsbyi 102
Halorhodospira 105
– *H. halophila* 99
Halorubrum 102, 106, 211
– *H. lacusprofundi* 103
Halothermothrix orenii 104
halotolerant 40, 99
– fungus 186
heat shock 10, 161
Helicobacter pylori 4
helix 73
heterogeneous rates 258
high mutator strains 252
high temperature 2, 10, 57–83, 119, 121, 129, 134, 158, 195, 197, 211, 261
Hodgkinia cicadicola 232
homologous recombination 9, 154

horizontal gene transfer 4, 8, 9, 20, 34, 36, 40, 42, 46, 57, 79, 101–103, 106, 133, 136, 212–215, 228, 230, 232, 236, 247, 259, 261
– *endosymbiotic, EGT* 236
horizontal transmission 226, 228–230
Hortaea werneckii 99, 186
hot springs 3, 10, 23, 57–83, 129, 134, 135, 139, 209–213
– acidic 10, 64, 66, 77, 158, 209
– alkaline 68, 76
– circumneutral 65, 68–71, 79, 137
hydA 71
hydraulic fracturing 192
hydrogen bonding 73
hydrogen generation (Ni-S based) 175
hydrogen sulfide 59, 61, 129, 162
Hydrogenobacter 65, 79
Hydrogenobaculum 65, 79
hydrolysis 21, 71, 104, 113
hydrophobic 73, 99, 106, 176, 191
hydrothermal system 3, 57–83, 134, 175
– continental 57–83
hydrothermal vent 10, 57–83, 119, 120, 130–132, 134, 153, 174–177, 212–214, 239
hydroxyectoine 100
hydroxyl radicals 112
hypersaline 3, 10, 99–107, 134, 137, 153, 156, 157, 185–187, 189, 192, 194, 198, 209–212
hypersalinity gene island 9, 103
hyperthermophile 3, 9, 20, 57–83, 129, 130, 132, 134, 139, 210, 215, 261
hypervariable 23, 104, 252

I
ice sheet 31–42
ice-binding proteins 37, 38, 42, 46, 215
indoor habitat 186, 189–192, 194, 195, 197, 198
insertion sequence 24, 36, 101
integrative conjugative elements 4, 21
integron 4
intracellular pathogens 5, 228
ionic 61, 73, 74, 100, 105, 106, 155
ionizing radiation 2, 10, 111–121, 154–157, 189, 190, 192
ionizing radiation resistance 10, 111–121, 154–157
ion-pair 74
iron 19, 59, 64, 71, 75, 134, 135, 137, 138, 190, 194, 196

iron reduction 59
iso/anteiso 73
isoelectric point 100
isoprenoid 72, 73, 234

K
Kamchatka 60, 66, 67

L
Lactobacillus 5
Lake Matano, Indonesia 137
Leptospirillum 19, 21–25, 65
– *L. ferrodiazotrophum* 135
lichens 156, 159–161, 197
lipids 37, 72, 112, 136
loci 23, 68, 188
Lost City 175
low temperatures 1, 3, 10, 31–42, 120, 133, 134, 161, 175, 186, 196, 213–215
low water activity 37, 106, 185, 189, 190, 192
LUCA 58, 66, 172, 176

M
magmatic 57–83
mannosylglycerate 75
Mars 31, 34, 117, 156
Mars-like conditions 156, 158, 160, 187, 189
Martian permafrost 34
melanin 189, 190, 196
membrane 1, 20, 37, 72, 99, 129, 136, 161, 173, 174, 176, 177, 179, 229, 231, 234, 235, 239
– fluidity 37
– saturation 37
mesophilic 20, 58, 74, 77, 81, 130, 135, 247–263
metabolome 104
metagenomic 3, 6, 7, 19, 25, 44, 68–70, 100, 102, 103, 106, 129, 190, 211, 214
metal resistance 8, 9, 20, 22
metalloenzyme 137–139
Metallosphaera 67
metatranscriptomic 19, 25, 44
methane 59, 60, 130, 136, 139, 173, 175, 178, 227
Methanobacter thermoautotrophicus 74
Methanobacterium marburgensis 178
Methanococcoides burtonii 2, 46, 215
Methanococcus
– *M. jannaschii* 74

– *M. thermolithotrophicus* 134
– *M. voltae* 212
methanogen 8, 65, 66, 75, 81, 97, 134, 172–179, 247
methanogenesis 8, 59, 171, 179
Methanopyrus kandleri 75, 132
methanotrophy 227
Methylacidiphilum fumariolicum 135, 139
Methylobacterium radiotolerans 139
Meyerozyma guilliermondii 194, 196
microaerophilic 135
microbial species
– definition 6, 11, 34, 190
Miller/Urey 97
minerals 32, 59, 60, 63, 127
mobilome 21
model systems 1–12, 43
molecular clock 46, 133, 252–257
molecular record 252, 253, 256
Mono Lake, California 135, 137
mudpots 19, 59, 61
multilocus sequence analysis 26, 102, 186
mutation 35, 209, 247
mutation rate 8–10, 20, 25, 103, 247–263
– increased 188, 197, 198, 231
mutational bias 5, 231
mutator 252, 257
mycosporine 189, 196

N
Na^+-coupled secondary transporters 130
'Nanohaloarchaea' 97
Natranaerobiales 97, 105
Natranaerobius thermophilus 10, 105
Natronobacterium magadii 211
Natronomonas pharaonis 10
naturally competent 36, 42, 44
negative selection 11, 35
Neisseria meningitides 4
Nesterenkonia sp. AN1 10
New Zealand 66
next generation sequencing 35, 41, 43, 82
niche 3, 21, 23, 57, 65, 78, 80, 119, 135, 139, 154, 158, 173, 189–191, 194, 224, 230, 234, 247, 258, 260, 261
nickel 138, 158, 178
nifH 65, 71, 79, 105

nitrogen 71, 127, 130, 133–136, 160, 173, 178, 226
– $^{15}N_2$ 65
– cycle 64
– fixation 9, 81, 133, 159, 226, 227, 229, 231
– limitation 71
nitrogenase 71, 74, 79, 105, 133–135
– Fe 71, 74, 134
– Mo 74, 81
Nitrosopumilus maritimus 133
Northern Atlantic 43, 136
Nostoc linckia 258
nutrient limitation 127–140
nutrient partitioning 78

O

Octadecabacter 40, 46
Oklo, Gabon 116
oligomerization 73
oligotroph 2, 11, 35, 40, 41, 135–139, 153, 190, 194, 196
open reading frames 82
opportunistic pathogen 190–198
ORFans 5
organelle 136, 226, 228, 230, 232–235
organic osmotic solutes 99, 100, 104–106
Orientia tsutsugamuchi 4
orphan genes 5, 236, 238
osmolyte 74–76, 100, 187
oxidation 20, 24, 64–66, 129
– ammonia 64, 65, 68
– hydrogen 58, 64–68, 71, 81
– iron 19
– methane 139
– sulfide 61, 64, 65
– sulfur 19, 65, 67
oxidative stress 31, 37, 131, 189, 191, 196

P

pairwise comparisons 252–255
pan genome 6, 102, 104
panspermia 116–118
paralog 5, 81, 237, 260–263
pathogenicity 5, 194, 195
pathogens 4, 5, 37, 137, 188–198, 223, 227, 228, 238, 258, 260, 263
Paulinella chromatophora 225–227, 232
Pelagibacter ubique 2, 40, 135
pelagic 3, 40

permafrost 2, 31–36, 134
– active layer 32
permeability 59, 72, 73
Persephonella 65
phage 5, 10, 19–23, 26, 36–43, 102, 209–217
– transducing 40–42, 44
Phanerozoic 256
phenotypic plasticity 187, 189, 194, 197, 198
photosynthesis 41, 59, 66, 69, 70, 76, 82, 129, 135, 138, 159, 212, 227, 232, 234, 237
phylogenetic barriers 80
phylogenetic ecology 57, 68, 71, 82
phytoplankton 38, 41, 130
phytoplasma 260
Pichia guilliermondii 194
Picrophilus torridus 20
piezophile 8, 9, 11, 153
piezotolerance 10
pigments 70, 154, 160, 190, 192, 196
plasmid 19, 21–24, 36, 42–44, 101, 130, 210, 211, 215, 228, 230, 231, 255
Polaribacter 40, 46
polyextremophile 2, 10, 97, 153–162
polyextremotolerant 2, 10, 185–198
– fungi 185–198
polyphosphate 130, 136
polyploid 103, 154
polysulfide 64
polythionite 64
positive selection 5, 35, 78
pre-adaptation 9, 118, 192, 195, 196
precipitation 38, 60, 73
predictable environments 258
Prochlorococcus 5, 9, 131, 135, 136, 138, 139, 212, 260
prophage 10, 22, 212, 214
protein flexibility 37
protein-protein 74
proteomics 19, 24, 25, 27, 36, 44, 78, 134, 187, 240
proteorhodopsin 46
Proterozoic 130
Pseudoalteromonas 43
pseudogene 5, 231
Pseudomonas 257
– *P. fluorescens* 6
Psychroflexus torquis 2, 46
Psychromonas 40

psychrophile 3, 9, 10, 35–46, 130, 132, 134, 139, 153, 210, 213, 215, 261
pyrite synthesis 175
Pyrococcus 77, 78, 80, 81
– *P. abyssi* 78, 215
– *P. furiosus* 74, 129

R

radiation 10, 31, 57, 70, 103, 111–121, 127, 153, 162, 185, 188, 189, 192, 258
radiation resistant 103, 111–121, 139, 153–162, 258
radiolytic hydrolysis 71
radionuclides 112, 115, 116
Ralstonia solanacearum 230
rare earth elements 128, 139
rate shift 254, 256, 258
rate variation 253, 254, 263
reactive oxygen species 119, 237
recharge 59, 60, 70
recombination 4–9, 19, 21–26, 34, 36, 154, 188, 209, 210, 257
recombination rates 257
red algae 20, 158, 159, 234
redox couple 174, 178
relative rate tests 253–260
relaxed clock 157, 253
remote sensing 33
residence time 33, 61
respiratory chains 174
Rhizobia 9, 225–231, 238, 239
Rhodotorula mucilaginosa 194, 195
ribosome 8, 174, 232
Richelia intracellularis 225, 231
Roseiflexus 76
Roseobacter 40
Rubrobacter 120

S

Salinibacter 99, 100, 103–106
– *S. ruber* 8, 97, 104
Salinibacter ruber 97, 104
salinity 3, 7, 31, 32, 39, 75, 99–102, 106, 156, 185, 191, 192, 196, 211, 260, 261, 263
– sea ice 39
salinixanthin 103
Salisaeta 211

Salmonella
– *S. enterica* 113
– *S. typhimurium* 252
salterns 3, 97, 100, 102, 104, 185, 189
'salt-in' strategy 99, 103–105
SAM 81
SAR11 (*Pelagibacter*) 40, 41, 135, 136
Sargasso Sea 136, 137
saturation 37, 73, 97, 99, 100, 256
scavenging 127, 136, 139
sea ice 3, 31–47, 185, 213, 214
seasonal 3, 31–35, 38, 41, 70, 73
secondary endosymbiosis 234
selective pressure 19, 25, 27, 33, 41, 65, 75, 80, 111, 113, 117–119, 133, 209, 238, 247, 258, 263
selective sweep 5, 11, 25, 35, 43
serpentinization 71, 175
Serpentinomonas 132
siderophore 137, 138, 196
simulation 156, 160, 252, 256, 257
snow 31–33, 35, 39, 41, 185
soda lake 3, 10, 105, 135
solfataric field 65, 68
Sonoran desert 120
Southern Ocean 38, 137
Staphylococcus aureus 4
STIV 19, 211, 212
storage 127, 128, 130, 136, 138, 139, 155
Streptococcus agalactiae 6
Streptomyces 134, 223, 225
stress metabolites 160
stress tolerance 186, 193, 195, 198
stress-induced mutations 9
stromatolites 256
structural determinants 74
subseafloor 212, 214
substitution of compounds 137, 191
substitution of elements 127, 136, 139
substitutions 4, 35, 36, 247, 254–256
– neutral 253
– non-synonymous 256
– rates 9, 25, 26, 162, 247–256, 259
– synonymous 256
succession 25, 76
Sulcia muelleri 232
sulfate reducers 66, 173, 179
sulfide 61, 64, 65, 70
Sulfolobales 66–68, 75

Sulfolobus 19, 20, 23, 67, 69, 211, 215
– *S. solfataricus* 20, 130
sulfur metabolism 59
sulfur reduction 179
Sulfurisphaera 67
Sulfurococcus 67
sulfur-oxidizing 227, 230
superoxide dismutase 138
superoxide metabolism 134
surface attachment 43, 76, 77
symbiont 4, 223–240, 258–261, 263
Synechococcus 76, 78, 131, 134, 136–139, 212
– WH8102 131

T
tardigrades 161, 162
taxonomic 31, 40, 59–82, 120, 186
temporal 7, 8, 23, 32, 57, 70, 130
Thaumarchaeota 40, 41, 133
Thermatoga maritima 215
thermoacidophile 68, 130, 135, 139
Thermococcus 75, 114, 212
Thermocrinis 65, 79
thermophile 10, 57–83, 104, 129–134, 137, 172, 261
Thermoplasma 20, 66, 74
Thermoplasmata 8, 19
thermostable 73–76, 132, 211
thermotaxis 77
Thermotoga 261
– *T. maritima* 74, 130
Thermotogales 66, 75, 79
Thialkalispira microaerophila 135
Thiomicrospira crunogena 131
Thiomonas 21
thiosulfate 64
timetrees 253, 255, 256
Titan 31, 34
toxic metal ions 158
toxin-antitoxin 210
transcription 5, 78, 80, 134, 209, 232
transformation 5, 21, 36, 40, 42, 43, 77, 102, 213
– frequency 36, 42
transition metals 128, 157, 172, 174, 175, 178

transposon 22–25, 36, 188, 197
trehalose 75, 105, 129, 130, 161, 189
Trichodesmium 135, 136, 138, 139
triclosan 191
trophic level 78
Trupera 120
tungsten 139, 178
two cluster test 258
type IV secretion system 5

U
ultraviolet radiation (UV) 119, 154, 156, 157, 160, 175, 189, 192, 258
uncorrelated model 257
unpredictable environments 252, 258, 259

V
vapor 60, 61, 64
vertical descent 4, 57, 78
vertical transmission 229, 230
Vibrio 42
– *V. cholera* 4
– *V. fischeri* 223, 225
viral defense systems 10
viral production 40
virion 210, 211, 216
virophage 214
viruses 5, 10, 19, 23, 27, 31, 34, 36, 39, 40, 42, 43, 209–217
Vulcanisaeta 67

W
Wallemia ichthyophaga 99, 187
weathering 61
Wolbachia 228

X
Xanthorhodopsin 103

Y
Yellowstone National Park 23, 60–76, 134, 137, 213

Z
zinc 19, 23, 64, 138